SEX SOUNDS

Vectors of Difference in Electronic Music

DANIELLE SHLOMIT SOFER

The MIT Press
Cambridge, Massachusetts
London, England

The MIT Press would like to thank the anonymous peer reviewers who provided comments on drafts of this book. The generous work of academic experts is essential for establishing the authority and quality of our publications. We acknowledge with gratitude the contributions of these otherwise uncredited readers.

This book was set in Adobe Garamond Pro and Berthold Akzidenz Grotesk by Westchester Publishing Services. Printed and bound in the United States of America.

Portions of chapter 7 were previously published in Danielle Sofer, "Breaking Silence, Breaching Censorship: 'Ongoing Interculturality' in Alice Shields's Electronic Opera Apocalypse," *American Music* 36, no. 2 (2018): 135–162.

Portions of Chapter 9 were previously published in Danielle Sofer, "The Macropolitics of Microsound: Gender and Sexual Identities in Barry Truax's Song of Songs," *Organised Sound* 23, no. 1 (April 2018): 80–90.

Library of Congress Cataloging-in-Publication Data

Names: Sofer, Danielle, author.
Title: Sex sounds : vectors of difference in electronic music / Danielle Shlomit Sofer.
Description: Cambridge, Massachusetts : MIT Press, 2022. | Includes bibliographical references and index.
Identifiers: LCCN 2021033957 | ISBN 9780262045193 (paperback)
Subjects: LCSH: Electronic music—History and criticism. | Sex in music.
Classification: LCC ML1092 .S64 2022 | DDC 786.7—dc23
LC record available at https://lccn.loc.gov/2021033957

10   9   8   7   6   5   4   3   2   1

SEX SOUNDS

Contents

Acknowledgments

Thank you to all the haters.

I want to thank all those institutions and individuals that have so firmly enforced boundaries. The musical, social, categorical, political, geographical borders, the static and staid canons up against which I have been continually thrust through the writing process and ever since growing up in this world as a Jewish, mixed-race, nonbinary person. Colonizers, white supremacists, Lacanians, proponents of "Great work of Great men," science-allied music theorists, pinkwashing homonormative gays, the TERFiest separatist feminists have all had a hand in this book. This book would have been much worse without their persistent bigotry, intolerance, and indifference. The injustices of this world came to new heights while I was writing this book. I watched in horror at the quick dissolution of Jeffrey Epstein's trial. I toiled over his influence on MIT, the very institution whose name underwrites the contents of this book, and whose press, at that time, had already generously contracted my book. Denying the words of this book to the world would not have been useful to those who suffered intentional harm or by way of indifference at such violently upheld boundaries. Instead, I vow to reallocate financial proceeds accrued personally from this publication by donating a portion of the accumulated royalties to resources supporting trans youth of color.

Though only a small token compared to what they've given me, I reserve my sincerest gratitude to those who actively contributed to this project. First and foremost, thank you to the artists for the sounds, and to those who also offered time and insight into their respective workings, Juliana Hodkinson, Alice Shields, Annie Sprinkle, Barry Truax, Lucy Neale, Niels Rønsholdt, Robert Normandeau, and Stas THEE Boss. I thank those who suggested repertoire for this book, Elizabeth Hoffman, John Young, Eric Lyon, and Amy Williamson. I am grateful to great friends, Glenda Bates who introduced me to Janelle Monáe and Esperanza Spalding before they collaborated on record; Anna Benedikt, Visda Goudarzi, David Pirro commiserated, members of the Graz Stammtisch community for the good times; and the humorous deflagrations of music Twitter. I am indebted to readers of the manuscript at various stages of preparedness, including many, many anonymous reviewers; my dissertation readers Gascia Ouzounian and Gerhard Eckel; individuals who gave time to the project, Susan McClary, Robert Fink, Margaret Schedel, Piero Guimaraes; members of the TAXIS reading group Magnus Schaefer, Ezra Teboul, Eamonn Bell, and Joe Pfender; my UW-Madison cohort—especially Kelly Hiser and Ilana Schroeder; past and present members of the Society for Music Theory's (SMT) Queer Resource Group; and to the SMT Committee on the Status of Women's Proposal Mentoring Program and Virtual Research Group Program, where I was met with gracious support from Sara Bakker, Antares Boyle, Elizabeth Monzingo, and Carissa Reddick. Amy Williamson's generous editing aroused in me an incomparable feeling of combined relief and confidence in my writing and ideas. Many kudos to Josh Rutner for his absolutely critical precision.

This project would not exist without the continued communication and support from Noah J. Springer and the entire editorial team at the MIT Press. Nor would the project have made very many waves without the generous financial support of the Förderprogramm Forschung 2013+ and the Graduate School and Institute for Music Aesthetics at the University of Music and Performing Arts Graz, which funded my dissertation and PhD studies, as well as subventions from the Society for Music Theory and the American Musicological Society's AMS 75 PAYS Endowment, funded in part by the National Endowment for the Humanities and the Andrew W. Mellon Foundation.

Lastly, given my natural impulse to hide in a corner never to emerge, I have found profound forces of inspiration in the kind words of my mentors:

Christa Brüstle, whose presence always reminds me to think beyond what is at first apparent, particularly regarding gender; Andreas Dorschel—the ideal Doktorvater in every respect!—for providing unwavering support, financial, emotional, personal, and professional; Ellie Hisama, for her dedicated encouragement of any and all justice work; Darien Lamen, whose seminar Music and Labor saw the beginnings of chapter 6; Judy Lochhead, whose class on phenomenological approaches to music analysis led to the earliest draft of chapter 5 and whose unwavering praise remains a reliable constant; Erin Manning, a Deleuzian who convinced me to cut the Deleuze; and Christian Utz, who, despite only partial alignment of our music theoretical interests, always provided a platform for me to express my ideas in my own words.

To Erik and Amelia Scorelle: Thank you for hearing me when no one else would.

Introduction

People flock to sexy songs, songs to have sex to, songs that sound like sex. In the last year alone, headlines boasting the best sexy songs trailed through *Cosmopolitan*, *Time Out*, *Women's Health*, *Harper's Bazaar*—and those are just the top hits on Google.[1] But before debuting any of these playlists to a lover, a diligent host might consider some pressing matters: What sounds sexy (to me, to my partner)? Are sex sounds arousing? Are sounds enough to get off?

Assuming an editor's decision of which songs to include isn't random, we may have a few more questions beyond the initial perks. And this book offers answers. *Sex Sounds* seeks to find out where this musical obsession came from. Looking to music of the last seventy-five years, this book investigates what, if anything, sexy songs have in common—musically, lyrically, or otherwise. Given that the book focuses only on music since the 1950s, an overwhelming majority of music in this period could be considered electrified in some sense—music that is amplified, music with synthesized sounds, with or without a beat, and so on. Throughout this book, I therefore use the term *electronic music* in an open and inclusive way to mean any music employing electronics. But, historically speaking, that's not how the term has been used. In concept, my focus on *electronic music* appears to provide a simple constraint, but in reality the stipulation actually proves quite tedious when attempting to explain this project to other people, mostly on account of semantics.

For a long time, when referring to electronic music, academics concentrated on music emerging either from France in the 1950s, *musique concrète*—music with concrete samples, or from Germany in the 1960s and 1970s, *elektronische Musik*—largely synthesized. These mythic origins continue to be passed down to new generations of university students in music technology courses even today, when many efforts are being made to show electronic music's wide and diffuse developments in broadcast radio,[2] live concerts,[3] and through sound recording and production, either professionally or domestically in the home.[4] In her revisionist history of electronic music and the studio, Margaret Schedel explains that histories that focus on the provenance of the genre in France, at the hands of Pierre Schaeffer, willfully ignore music being made by electronic means outside of Europe and North America, explaining that the concentration of electronic music studios there is "likely due to reporting bias, for the most dominant nations tend to control history."[5] In her chapter in *The Cambridge Companion to Electronic Music*, she points out predecessors to Schaeffer's experiments with recorded sound, including Egyptian-born Halim El-Dabh's (1921–2017) pieces processed with wire recorders, as well as the prevalence of studios emerging throughout the 1950s in Japan and Korea. Similar endeavors occurred throughout the world, in Chile, Ghana, South Africa, and essentially every corner of the globe. And considering musical experiments with electronics arose in popular music production quite early on, we could also point to earlier initiatives.

Nevertheless, we cannot dismiss the early influence European composers had on disseminating both the formalization of electroacoustic practices as well as historical lore far beyond European borders. European composers who traveled to Latin America, for instance, Pierre Boulez (in 1954) and Werner Meyer-Eppler (in 1958), are often credited with the formal establishment of electronic music studios in Buenos Aires, Argentina, and Santiago, Chile, respectively.[6] Schedel points to Karlheinz Stockhausen's similar influence in Japan, where the electronic music studio at Japan's national broadcasting organization Nippon House Kyokai (NHK) was modeled after the Westdeutscher Rundfunk (WDR) in Cologne with help from Stockhausen who worked alongside Meyer-Eppler, a scientist with known ties to Nazi military research.[7] That Joji Yuasa, Toru Takemitsu, Hiroyoshi Suzuki, and Kazuo Fukushima had already been experimenting electronically with sound before, from 1951 and probably earlier, shows historical precedence that is frequently sidelined in official histories of the genre.[8]

At the same time, El-Dabh himself traveled, first to New Mexico on a Fulbright fellowship to study with Ernst Krenek, then to Boston, and later to New York, where he played a large part in founding and securing the prestige of the Columbia-Princeton Electronic Music Center in New York (CPEMC) (discussed in chapter 7). Likewise, producer Francis Bebey (1929–2001), born in Cameroon, brought his knowledge of mathematics to the Sorbonne and the Studio-école de la Radiodiffusion Outre-Mer in Paris and then to New York University, before continuing his research on traditional African music and his composing with electric guitar and synthesizers in Ghana in 1957.[9] This is all to say that electronic music and electroacoustic practices circulated both officially and less formally by many paths globally and therefore cannot be definitively traced to any one person or institution. Origin stories that outline either Schaeffer or Stockhausen/Herbert Eimert/Meyer-Eppler in pioneering roles are referred to by James Andean as the "grand narratives" of "electroacoustic mythmaking."[10] Even more critically, in isolating and bracketing influence to a few key players, Andean identifies a "national agenda" in this mythmaking.[11] Not only do the stories limit the geographic and cultural influences apparent in electronic music history, they also capitulate gendered exclusions by forefronting the experiences of men's music-making while, as Francis Morgan shows, forcing every "other" non-man, non-European creator to be discovered anew with each mention.[12]

As important as it is to dispel these electroacoustic myths, there are significant ways in which the dominant narrative, whether or not accurately representative of electronic music history, also served to advance particular ideologies surrounding electronic music production, especially as these relate to gender and sexuality. Or, more accurately, reiterating these myths demonstrates the omission of discussions interrogating the central role of gender, sexuality, and race in this music's creation. Collective silence on the topics of gender, sexuality, and race plays into the normalization of particularly destructive tropes regarding who can make electronic music and how.

Readers of this book may not be familiar with every artist but, depending on the reader's background, there will likely be at least one familiar name. Even readers who know every name contained within these pages have not likely heard the specific pieces analyzed and discussed. In most instances this lack of familiarity, at least in the case of "composers" (distinguished from "artists" or "producers"), arises from deliberate oversight built into scholarly electronic music historiography; put simply, their "sexy music" has long been overlooked and deliberately ignored.

Take, for example, Pierre Schaeffer and Pierre Henry, composers of *Symphonie pour un homme seul* (1950–1951), a piece frequently pinpointed in textbooks, blogs, and documentaries as one of the most influential works in electronic music history for its novel use of vinyl records, live sampling, and conceptual framing as a "symphony" for the human body.[13] Though the work secured much prestige for its two composers, few scholars have ventured to analyze this piece, and never have I come across another source apart from this book that analyzes the work's "Erotica" movement, the subject of chapter 1. Before writing this book, I considered ignorance of this movement a surprising oversight. Now that I've witnessed how frequently electroacoustic scholars and music theorists alike sidle away from sex, I see the movement's omission as part of a larger avoidance in particular corners of music scholarship. I have come to expect that scholarly writing by electroacoustic practitioners and music theorists will quickly gloss over "sex" as this-thing-we-all-know-so-why-should-we-name-it. As will become clear later, music theory's dedication to objectivity, which aims to retain its historical correlation to the sciences, offers one explanation, as exemplified by composers like Milton Babbitt in the United States or Pierre Boulez in Europe who shared a vested interest in the development of both music theory and electronic music.[14] Oversight of such musical circumstance and rhetoric likely also accounts for the high volume of inappropriate sexual conduct in these disciplines. Ellie Hisama attributes this higher ratio of inappropriate conduct to the ways in which music theorists assign value, where certain topics—numbers and objects—count more than other sociocultural factors, which in turn tends to effect *"who* gets to count."[15] Yet, without an established discourse of sex in music, how should we define sex as an act without instance? This book aims to establish a sonic profile for sex acts as they have been represented in electronic music by looking to what electronic music creators themselves attribute with sexual or erotic associations—that is, representations of sex. Collectively, I term this music *electrosexual music*.

ELECTROSEXUAL MUSIC

The music of this book presents the imagined collective ideal of how sex ought to sound electronically. Each chapter provides a case study exemplifying the particular habits of electrosexual music-making. Given its prevalence, the electrosexual genealogy of music is often eclipsed or ignored both

Features of Electrosexual Music

CANON
- Distance as a gendered trope
- Women's audible sexual pleasure as "evidence"
- Sexualized and racialized intramusical tropes

(FEMINIZED) VOICE
Gender perception and composer expectation

CLIMAX MECHANISM
- Teleology and/as musical form
- Mimetic exchange between voices and instruments
- Erotic vocalisation as a gendered and racialized sonic trope

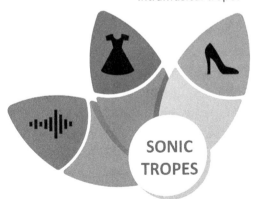

SONIC TROPES

Figure 0.1

in music theory and electroacoustic historiography alike. When it is considered, the subject is usually broached as an element of excess, as something "extra" or beyond music analysis. For this reason, it has proved crucial to demonstrate through multiple musical analyses that composers themselves took up the task of writing music that arouses associations to sexuality and eroticism. I aim, first and foremost, to show that electrosexual music is far more representative a collection than the typically presented electroacoustic figures—supposedly disinterested, disembodied, and largely white cis men from Europe and North America.[16] As an ideal that recurs through habits of mediation as much as compositional constraints, I find this music draws on two central characteristics, namely the feminized voice and the "climax mechanism." These features are *so* prevalent, as I show in examples throughout this book, their recurrence reveals an existing musical canon that shares electrosexual features but a canon that traverses established genre boundaries (figure 0.1). Second, the "feminized" voice—we don't know if the voice is a woman's, but it's meant to sound that way—is overly represented in this repertoire. And third, intersections of these gendered tropes with other racializing factors of association carve electrosexual music into a canon all unto its own.

Sexually explicit themes have paraded through electronic music since its inception, and yet, interviews with practitioners, feature articles, and historical surveys, repeatedly purge early electroacoustic musical experiments of its associations with sex and sexuality. Indeed, theories of sexuality appear to conflict with the disembodied and disinterested methodologies of early electronic music practitioners like Pierre Schaeffer, Pierre Henry, Toru Takemitsu, Milton Babbitt, or Karlheinz Stockhausen, who are often portrayed in musicological discourses as harbingers and pioneers. This book challenges this impression by investigating the reality of the repeated focus on sexual themes in electronic repertoire since the 1950s, while also throwing into question electronic music's electroacoustic origins by drawing connections between academic institutions, radio studios, experimental music practice, hip-hop production, and histories of independent and commercially popular music to identify the wider social and cultural impact sex has had on audible creative practices. That creators, producers, composers, and performers proclaim examples from each of these traditions as sexually connotative in some way highlights how common these themes are in music produced through electronic means. Such contributions point to a yet unacknowledged shared history across multiple musical practices, providing context for the natural progression to a global cross-genre pollination witnessed in recent electronic music.

Although this book investigates several mutual arenas of electronic music and sex, the possibilities of such expressions are far from infinite. As I explore in part 1, the electronic medium itself sets up certain creative limitations, particularly as electroacoustic philosophies veil the visible in amplification of the audible electronic medium. The book's initial chapters trace the historical fascination with electrified audible sex. These chapters establish the precedence of electrosexual music and justify investigations of sexuality as legitimate and obviously neglected in electronic music historiography. At the same time, the chapters examine reasons for electronic music history's pronounced exclusion of works created by women, people of color, women of color, and, in particular, Black composers. As Tara Rodgers observes: "Work that is not understood as significant may not even be saved, and therefore it can never reach a point where it would become the subject of scholarly analysis or historical accounts and would be recognized for its cultural value."[17] While electronic music's most recollected composers have often been white men, white women and non-white individuals have not only contributed as composers, performers, and listeners of electronic music throughout this music's history, but, as I argue, have been instrumental in shaping today's most common musical practices.

Historically, electronic musicians have been just as fascinated with new sounds as with exploiting and celebrating technological innovations. Andrew Hugill summarizes this impulse in *The Cambridge Companion to Electronic Music*:

> The origins of electronic music lie in the creative imagination. The technologies that are used to make electronic music are a realisation of the human urge to originate, record and manipulate sound. Although the term electronic music refers specifically to music made using electronic devices and, by extension, to certain mechanical devices powered by electricity, the musical

possibilities that these technologies have opened up are a recurring theme in literature, art, engineering and philosophy.[18]

With this framing in mind, electrosexual creators took to electronic music to explore how the medium could expand the limits of sex as we knew it, to modify, manipulate, and produce new sexual encounters. Different from acoustic music, electric possibilities include synthesized sounds, amplification, and broadcasting. At the same time, recording technologies allow for the exact replication and sonic displacement of sex sounds from physical sexual encounters.

Since the advent of sound recording with Édouard-León Scott de Martinville's phonautograph, the last 163 years of innovation in the technology of musical reproduction have fueled some of the most pressing and prescient discussions in recent music history. Already in 1934, Adorno described "The Form of the Phonograph Record" as the object that resembles nothing, least of all the sound of that which it reproduces.[19] Fifty years later, the replication of "real" sound caused an even greater severance of source from effect in what Michel Chion recognized as the "double meaning" of recorded sound, its referential potential continuously marred by contextual significance.[20] Composers have tools that *could* change a sound to such an extent that it would no longer be associated with its originating source or cause, and yet, they often do maintain "transcontextual" connections, especially with the recognizable human voice.[21] This discrepancy between sound as effect and its given or perceived source and cause has created a rippling divide across music-creative philosophies. One problem stemming from this divide is that, although a composer may intend either to retain or to sever the connection between cause and effect, an electronically produced sound's identity can be neither fully disclosed nor forever and completely veiled. This "spacing," as music theorist Brian Kane terms it, between a sound's effect and its source–cause relations, lies at the heart of electrosexual music.[22]

Ever since music theorist and philosopher Jean-Jacques Rousseau heard the cloistered adolescent nuns of Venice in 1743,[23] since the Jubilee Singers' North American debut from behind a curtain in 1871,[24] and even since the phonographic "blue disc" records popularized in the 1930s,[25] veiled sounds have stirred human curiosity with erotic inference. There is something about "sound unseen," as Kane calls such "acousmatic" phenomena, that forbids music from simply sounding.[26] The ritual of unveiling sounds has much in common with the scientific habit of hypothesis and theory, and these links were only made more apparent with sound's increasing electrification in the twentieth century.

Composers of electronic music—which in this book denotes any music incorporating electronic elements—have long tempted the limits of reproduction, and, reciprocally, many inventions made in service of music radiate outward to inspire and provoke new technological and also cultural, social, psychological, philosophical, and ideological developments. Indeed, technology has contributed to a surge of investigations into how and why repetition occurs and functions, and definitions and reformulations of what exactly is meant by repetition. Certainly, repeating cadences are crucial to creating sexual allusions through sound. This is not to say that technology sets an artificial standard of musical possibilities; rather, as this book shows, the kinds of technology, the sexual circumstances, as well as the creative and performative context all combine to construct a sexual reality specific to the particular work and to when and where it was created and heard.

SEX IN/AS THEORY

Electricity surges through today's most heard music throughout many parts of the world, but electronic music also builds on sonic imaginings from long before electricity. For example, Jean-Jacques Rousseau "confessed" his attraction to the young women in a "clausura" he visited in Venice in 1743, based solely on the sounds of their voices.[27] More than aural pleasure, Rousseau describes hearing these voices in sexualizing terms as "ravishment" (*ravissement*), "ecstasy" (*extase*), even "voluptuous" (*voluptueux*).[28] His desire is heightened by their young age—by music "sung in the galleries by girls only; not one of whom is more than twenty years of age"—even though he cannot see them. Rousseau recounts the girls' bodies veiled behind an "iron grate" that obscures and diminishes their visible presence. Aside from the prudence of sheltering the female physique from visual scrutiny, the veiling gate amplifies the girls' voices, which take on some form greater than human, as "angelic music." Certainly, Rousseau's tone is unsurprising for anyone familiar with the feminist backlash against his usual characterization of women.[29] More interesting for this book are the ways this anecdote has been recounted in musicological lore, especially how his words have fueled and been folded, seemingly uncritically, into electronic music discourses.

Rousseau's horror and appall upon discovering that the figures producing his beloved angelic sounds did not live up to his affected image of them, arises the moment the mute veil is lifted. It is clear from his account that the allure of the veil—the medium—is not merely one of curiosity, but that the veil's own erotic appeal is essential to its very existence as a separating divide—as *technê*. It is precisely the veil that lures. Rousseau's account appears in Kane's theorization of the *acousmatic*, the term applied today to the phenomenon of veiling sounds from their source and cause origins, much like the singers behind Rousseau's iron grate. Rousseau's confession is characterized as the "erotic drive to peer behind the grilles and behold the actual bodies fantasized by the nuns' voices."[30] As subjects of fascination, nuns are staged to arouse the spectator, but, according to Kane, if we, like Rousseau, uncover the ruse, if we point our attention to the fabrication of the experience and to that which *makes* the situation, the entire illusion is destroyed. In Kane's argument, our attention is drawn compulsively to the veil, the surface of the medium, while simultaneously failing to attribute it as the source of sound. The veil becomes a sensed but nevertheless invisible and inaudible boundary, and, in proscribing *technê*, Kane asks that we *distance* ourselves from our listening instincts to reinforce the rational reflection proper to Western philosophical inquiry. This distance is a metaphorical veiling gap reflexively filled by the substance of the electronic recording medium. For the electroacoustician, the electronic medium necessarily invokes distance, the divide between perceiving subject and object perceived. Made repeatedly conscious of the phantasmagoric veil as a (mere) device, the spectator's erotic desires are bracketed from sound and the electroacoustic genre therefore appears incapable of erotic signification.

Tracing the acousmatic concept from the time of Pythagoras in the fifth century BCE, through Native Americans in early North America, to nineteenth-century literature, opera, and contemporary popular music, Kane bridges the acoustic arts with the digital via the "electroacoustic" compositional philosophy that has dominated scholarly discussions of electronic music from the mid-1940s. For Kane, the "acousmatic situation is often treated as an aesthetic duty," as if "the traces of musical performance *should be* erased."[31] Indeed, Kane takes on this duty by faithfully relaying the acousmatic philosophy. However, he still has sympathy for the listener, Rousseau, who discovers that the girls were not only far from ethereal but "ugly" and "disfigured," "a cruel and misogynist joke" in Kane's assessment.[32] But *what* about Rousseau's story might be "cruel" or "misogynist," the bodily veiling or its subsequent revealing? Put another way, is the joke at the expense of the singing girls who have been denied their humanity by this veiling, or Rousseau, for being fooled into hearing these "ugly" girls as beautiful? Kane's *Sound Unseen*, though a faithful account of acousmatic tendencies, tacitly ignores the gendered and sexualized stakes of the acousmatic—one task of the present volume. This book revisits the acousmatic in its electronic manifestations to examine and interrogate sexual and sexualized assumptions underwriting electroacoustic musical philosophies.

But dwelling a bit longer in the acoustic, music retains some common sexualizing tropes that permeate the various styles of European and North American musical traditions. The popular blog medium shows how stereotypes of sexual gait and pulsation persist beyond academic ephemera. In "8 Songs That Parallel the Rhythmic Path to Orgasm," blogger Waylon Lewis assembles a list of musically unrelated clips, ranging from Aram Khachaturian's *Sabre Dance* to "The Battle," from Hans Zimmer's soundtrack to the film *Gladiator* (works that in and of themselves are hardly explicitly erotic) to demonstrate a common cliché of current listening practices: the pulsating *telos* of repetition.[33]

They start gently, methodically, but steadily, sweetly . . . they wander to and fro, but build all the while . . . they gather heat and form and storm and retreat and regather and build until! and then! And . . . yet . . . yes . . . and until. And until they finally crescendo, and collapse, and relax.[34]

Lewis's examples were not composed with the intention of paralleling a broadly conceived path to sexual climax, but nevertheless, in our current time and our contemporary hearing, a consensus of what is erotic in music seemingly abounds.

Thirty years ago, musicologist Susan McClary identified and developed a theory on the orgasmic teleology Lewis, above, dubs the "rhythmic path to the orgasm"—a trope that recurs later in electrosexual music. McClary's conspicuous analysis of Beethoven's Ninth Symphony identifies phallic intentionality in the closing of the first movement, referring coyly to the erected "beanstalk" of musical signification supplanted with violating oedipal symbolism. Importantly, like Lewis, McClary observes the "beanstalk" in music that was not expressly inspired by sex, rather, derivative of the male orgasm and a teleological expression of masculine aggression, McClary's beanstalk "marks the heroic climax of many a tonal composition," simultaneously on account of how music is composed and how music is heard.[35] She summarizes this prevalence as follows:

There is, to be sure, much more to classical music than the simulation of sexual desire and fulfillment. Still, once one learns how to recognize the beanstalk, one begins to realize how pervasive it is, how regularly it serves as a hook for getting listeners libidinally invested in the narratives of compositions. And when it turns violent (as it does more often and more devastatingly in nineteenth-century symphonies than in heavy metal), it becomes a model of cultural authority that cannot be exempted from social criticism.[36]

Given the prevalence of this perceived masculine aggression in music, we may be tempted to extend McClary's second-wave feminist analysis to other kinds of music, since her perseverance proved crucial in shaping the field of musicology. And there is no question of her instrumentality in instigating a necessary feminist intervention into musicology as a discipline. Yet, her beanstalk perpetuated three problematic generalities regarding sexuality in music: (1) that an ever-present erotic signification exists in much of "Western art" music, (2) that erotic signification in music is necessarily metaphorical, and (3) that gender is the primary motivation for sexual difference, that is, that universalizing differences exist between the sexual experiences of men and women. Music sociologist Tia DeNora reacts to this final point, writing, "To claim for example, that Beethoven's music is 'masculine' because it is 'powerful' (i.e. loud, emphatic *tutti* finales) not only skips a logical step, it simultaneously grounds itself on unwarranted assumptions about the 'nature' of the feminine."[37] McClary's missed step is an indication of a long-standing critical conflation of gender with sexuality, a confusion that arises equally at the hand of patriarchal conservatism as it does from opposing separatist feminists.[38] Although the second-wave feminism of the 1980s and 1990s thoroughly ruined the conventional representation of the phallogocentric male lineage, this dissolution simultaneously affirmed an opposing but equally biased alternative.

The conflation of sex with gender presupposes essential biological differences between male and female attitudes. To extend these attitudes to an assessment of sexual behaviors is not merely solipsistic but also reinforcing of a larger social stigma against sex, a stigma that brands sexual expression— especially certain kinds of expressions—as marginal and relentlessly transgressive. As cultural anthropologist Gayle Rubin explains, "Although sex and gender are related, they are not the same thing, and they form the basis of two distinct arenas of social practice."[39] Though lesbians must overcome the hurdles of women's oppression, these women also face "the same social penalties as have gay men, sadomasochists, transvestites, and prostitutes," thereby situating sexuality in a separate category from gender.[40] Where sexuality has been defined distinctly from gender, so too must eroticism, as an affective and emotional expression of human experience, be investigated through plurality and reexamined through categories that are not always-already delimited by gender. Rather, gender is only one of several avenues to interpreting sexual relations.

In her introduction to the collection *Musicology and Difference*, editor Ruth Solie praises the plethora of existing individual identities, yet is careful to also caution against the essentialism that threatens to delimit individuals according to singular identity categories. Moving to correct the *insistence* on difference, she writes, "Essential difference . . . fails to take into account the ways in which identity categories inflect one another."[41] Though the music explored in this book no doubt elicits many expressive variances among listeners, we can also find in this repertoire some experiential coincidences. This overlap, however, is not easily understood under the same logical deductions yielding to the "beanstalk" teleology. As frequently as the climax mechanism is heard and composed into music—a notion I interrogate further in chapter 6 —works employing this device as a musico-sexual means of expression differ dramatically. The drive to climax, as a common and simplistic representation of erotic impulses, fails to account for the variable minutiae of emotional expressivity. Several (if not all) of the works explored in this book could be distilled through analysis into a formal representation of the "rhythmic path to orgasm," and yet, beyond this feature, this repertoire displays wide variety, whether "sonic" (within the music "itself") or "contextual" (arising from the circumstances extrinsic to the formal constraints of the composition).

Over seventy years ago, Georges Bataille launched a critique against investigations, such as the experiments conducted by sexologist Alfred Kinsey, that delimit sexual expression only to orgasmic release.[42] Put simply, this project begins from the premise that sexual expression in music resounds similarly in the ears of most if not all listeners, and yet this expressive content is not necessarily experienced or interpreted in the same manner across all listeners. How one differentiates between the sexual expressions of each work is not only a question of individual psychology but also born of socialized aesthetic, as well as philosophical and enculturated ideologies. Distinguishing between music's sexual qualities requires an investigation of the ways in which music presents sex, of how composers envision the subject musically as well as how listeners understand and react to music. Further, at the heart of music's philosophical glimpse into sex is a question of what constitutes "sex." And, from there, we may also be curious about whether sex need be confirmed by sex *acts*. Indeed, how does music present sex acts and who en*acts* them?

To return to electroacoustic music's "aesthetic duty" to suppress sound's source–cause relations, although many examples of electrosexual music exist, few discussions of such works have taken place within institutional and academic settings. Indeed, while Schaeffer and Henry's *Symphonie pour un homme seul* (1950–1951) enjoyed remarkable exposure, meriting its composers a great deal of prestige as the earliest example of *musique concrète* in Paris (and arguably in Europe), the work's fourth movement "Erotica,"— the subject of analysis in chapter 1—is rarely if ever mentioned among the work's other movements. The same could be said of the popularity enjoyed by other creators aligned with the Western notion of "art music," Luc Ferrari, Robert Normandeau, and Annea Lockwood, whose electrosexual works are also discussed in this book. If, as philosopher Jerrold Levinson has argued, pornography's primary purpose is to achieve sexual release, and art is said to give rise to deeper aesthetic engagement,[43] the opposition of these two categories reduces pornography pejoratively to the triviality espoused historically by "popular" as opposed to "serious" music in comparative musicological discourses.

In my research into the electroacoustic, I have found the prominent puritanical resistance to pornography confronted by a no-less-prevalent creative fascination with the erotic, as exemplified by a statement made by Canadian electroacoustic composer Robert Normandeau in 2009: "music, unlike other contemporary art forms—literature, painting, cinema—has hardly dealt with eroticism as a genre."[44] As this book shows, Normandeau's statement is inaccurate at best and, at worst, contributes to the ongoing calculated erasure of electrosexual music. Given that a handful of the university's most celebrated twentieth-century composers concerned themselves with erotic electronic works, Luciano Berio, Schaeffer, Takemitsu, and Lockwood, to name a few, it is a wonder that electrosexual works feature so rarely in literature dealing with electroacoustic music. Sex is simply cast aside, as taboo as it is banal. This absence is so pronounced that it seems to have enforced an irreconcilable rift between such music and its reception. Normandeau's proclamation presents his piece *Jeu de langues*, analyzed in chapter 4, as uniquely innovative while his ignorance of musical precedents silently advances the common lore of the "pioneering" electroacoustic composer.[45]

Long before Normandeau entered the fray, new technologies have innovatively opened the door to sexual encounters in music, from sound recordings to fully realized digital compositions. An ocular bias in scientific work may have made it difficult to guarantee that sex, absent visible confirmation, *sounds*. Nevertheless, because electroacoustic sound is not motivated by visual "evidence" of a body responsible for producing these sounds, erotic nuance in this music takes on distinctive meaning for its listeners. As detailed in each chapter's analytical case study, the musical creators discussed in this book have all sought to evoke sex through overt use of its sonic envelope by appealing to and manipulating accepted norms of how human sexuality sounds. As I examine below, tension between the erotic and pornographic arouses the observer's intrigue because of a creator's intention but not necessarily in the manner intended. And, although the book explores philosophical, cultural, social, and scientific technologies of eroticism, sex acts, and sexuality through a musical lens, in this limited space I could not possibly hope to explore every instance of sexually inspired electronic music, let alone in music as a broader category. Instead, I present instances of music within social, sexual, religious, historical, and philosophical beliefs that construct musical sexuality at different places and times. It is my hope that readers will make their own inferences beyond the pages of this book.

The first chapter situates the earliest electrosexual music as a continued pathologization of women's sexual pleasure. Looking to the burgeoning psychoanalysis in *fin-de-siècle* Paris, I present the dominant attitude toward women's sexual pleasure as a symptom of disorder, an acousmatic curiosity as that which can be heard but not seen. However, thanks to medical and scientific documentation, women's climactic, periodic moaning was rendered abstractly and by the second half of the twentieth century—just as electroacoustic music reached the apex of its popularity—came to be the musical phenomenon we know today.

Recall from early in this introduction that in scientific study, an expert examiner seeks to understand its subjects—vocalizing women—while also maintaining a distance from them as objects of study. Feminist philosopher Luce Irigaray identifies this habit in a passage in Friedrich Nietzsche's *The Gay Science* entitled "Women and Their Action at a Distance." In Nietzsche's words, "The magic and the most powerful effect of women is to produce feeling from a distance, in philosophical language, *actio in distans*, action at a distance; but this requires first of all and above all—*distance*."[46] Distance ensures separation between same and different, between one and an-other. It is in his very desire to study Woman as an ideal construct, says Irigaray, that Nietzsche erects a distance from them. Yet, in his role as spectator, the (male) investigator—whether composer, philosopher, or psychoanalyst—moves simultaneously to approach, provoke, unwrap, and penetrate the veil in an effort to better understand her.[47] I find it easy to collapse any distinction between physician and psychoacoustician when in this very passage Nietzsche contrasts Poseidon's pained singing with what he observes as women's curious silence: "When man stands in the midst of *his own* noise, in the midst of his own surf of projects and plans, he is also likely to see gliding past him silent, magical creatures whose happiness and seclusion he yearns for—*women*."[48] In his resolve to broach this silence, man would allow her to speak, to betray her "Veiled Lips" (the title of Irigaray's essay), but only on his terms.

Such distances—gender gaps—are the difficulties one faces in telling the history of eroticism in electroacoustic music. The distanced subject, woman as represented by the man who examines, appears ever set-apart from the history that encapsulates her. Her story is recounted secondhand through the canon of representative musical compositions that capture and conceal her voice.

Like the composers identified in part I of this book, Nietzsche asserts a secure and singular distance between the controlling male examiner and his female object of study, but we see an alternative mode of inquiry in the writing of Marcel Proust as read by Gilles Deleuze.

In Deleuze's reading, Proust's writing hinges on a substitution of the sexes, on the possibility that gender is merely a courtesy, a placeholder or sign into which bodies are cast, but one by which bodies can also exchange places. Gender impinges on society as a sign all unto itself, permitting an exchange and *transposition* of identity.[49] As Deleuze notes, Proust, a lover of men, made many "transpositions" several times over to arrive at this theory, an intricate *abyme* of mirrored signs that inevitably adjoin the two opposing subjects.

> Insofar as the beloved contains possible worlds, it is a matter of explicating, of unfolding all these worlds. But precisely because these worlds are made valid only by the beloved's viewpoint of them, which is what determines the way in which they are implicated within the beloved, the lover can never be sufficiently *involved* in these worlds without being thereby excluded from them as well, because he belongs to them only as a thing seen, hence also as a thing scarcely seen, not remarked, excluded from the superior viewpoint from which the choice is made.[50]

By this logic, Deleuze argues, Proust's eroticism arises through a deliberate tension between the visible and the invisible, between sense—the *sensed*—and the non-sense of the "virtual" world. Deleuze summarizes Proust's sense of suspense—this "resonance" of a memory—as "real without being actual, ideal without being abstract."[51] What one envisions in the electro-acoustic experience is, as far as it is virtual, very real indeed: "The virtual is opposed not to the real but to the actual. *The virtual is fully real in so far as it is virtual.* . . . Indeed, the virtual must be defined as strictly a part of the real object—as though the object had one part of itself in the virtual."[52] In Proust's male-centric theory, Deleuze observes, men who like men sleep with "women who suggest young men." In a reverse transposition of this theory, Irigaray's erotic sociality hinges on relations among women, not in the gap left by women's absence or in women's seeming out-of-place-ness, *ec-stasy* (see chapter 7), but in a space outside this context, a place wherein women exist on *their* own terms.[53] Though queer, Proust–Deleuze contra Irigaray reinforces a gendered binary that is only queered to the extent that men and women are interchangeable and transposable. Such transpositions are a point of departure for some electrosexual composers like Canadian Barry Truax.

As I recount in chapter 9, in the 1980s and 1990s, Truax emphasized his desire to compose homoerotic music to project his identity as an openly gay composer through his works. In doing so, Truax says he fulfilled the need for representation from a yet underrepresented group of queer composers. Hannah Bosma's 2003 article "Bodies of Evidence, Singing Cyborgs and Other Gender Issues in Electrovocal Music," was the first to point out a gender disparity in the overwhelming majority of electronic music composed by men featuring women's voices.[54] On this account, despite their privilege, men who compose music—even homoerotic music—face an impossible hurdle to overcome: we urge them to include a greater diversity of voices to dispel the "master/slave" dynamic—with all its racist baggage, as I explore later—that potentially results from a (white) man "fixing" the voice of a girl/woman. On the other hand, if men "correct" the disparity by composing for *more men*, then, once again, women are excluded from the creative processes of electroacoustic music. Alice Shields's woman-centered performances throughout that time, such as her electronic opera *Apocalypse*, explored in chapter 7, fill a similar role regarding gender, albeit in a heterosexual context.

Musicological queries into eroticism often premise their investigations on the assumption that music is, first, (only) a representation of real, lived experience, and second, that eroticism is (only) evoked in music through imitation. In his chapter on "Erotic Representation" in music, musicologist Derek Scott asks three questions in his investigation of "how gender difference is constructed in music":

1. How does a composer represent sexuality?
2. How does a performer convey sexuality?
3. How does a listener interpret sexuality?[55]

Scott's crucial use of the word "represent" presumes that erotic inference, what he sometimes terms "sexuality," exists somewhere outside of "the music" proper. Sex is only *re*-presented in music.

Scott's assessment resonates with feminist philosopher Alice Jardine's definition of representation, which states, "Representation is the condition that confirms the possibility of an imitation (mimesis) based on the dichotomy of presence and absence and, more generally, on the dichotomies of dialectical thinking (negativity)."[56] Sexual representation can be defined as much by what is there as by what is not. She further extrapolates that, "The process of representation, the sorting out of identity and difference is the process of analysis: naming, controlling, remembering, understanding."[57] Scott's three questions not only assume that eroticism is imitable, meaning identifiable and thus replicable, but, since music seems "itself" incapable of eliciting its own sexual criteria, erotic music is therefore limited *only* to imitation. His questions imply that sexuality is secondary to music, as if music remains unencumbered by sex when unmarred by any one person's erotic intentions.

Given Jardine's definition of "representation," I would argue that listeners likely *do not* experience music *merely* as representation, because sexual stimulation would be difficult if not utterly impossible to identify or qualify under such definitive terms. In fact, as I see it, the main philosophical problem of studying eroticism through the lens of any claims that music *expresses* emotion is to assume that eroticism, or really any feeling, is readily identified by finite and universal traits. Eroticism is not identifiable in discrete terms and, for this reason, is not adequately "represented." Eroticism arouses complex physical sensations and emotions that emanate as a response to certain sensed qualities, and although listeners may agree that some music sounds erotic, what makes eroticism sound is neither uniquely individual nor entirely universal. That listeners *may* hear music differently, even from the composers whose music they are responding to, does not account for the many moments when expression is similarly experienced by both a composer *and* her various listeners. Such consensus confirms some expressive qualities of music as recognizable and describable, but these are not necessarily discrete emotions that can be "named by an emotion word" such as sad or calm.[58] If eroticism arises through a consensus of multiple listeners, what then qualifies some works as erotic and not others?

Without regard to medium, it would be difficult to consign all erotic artworks to a single category. If we recognize erotic artworks on the basis of visible subject matter, for example a painting of a nude body,[59] then such content separates the painted medium from the musical one. One cannot know from sound alone whether the performer is clothed—that there is even a body is up for some debate. In short, whereas symbols may be shared among the arts, the *form* and *content* of representing those symbols in music differs from that of the other arts.

In music absent text, identity and representation—both functions of association—are complex matters. Associations may abound, but can we determine if all listeners share the same associations? In terms of associations, the source of sound would likely matter less in music organized by certain rules or systems that function independently of instrumentation. In this case, we would be forced to consult either the work's creator to find out their intentions, or the spectator to confirm their response. Here we arrive again at modular categorization, this time within music as a medium. We may accordingly wish to divide up tonal works that rely upon the same syntax from nontonal works that may rely on a system devised specifically for each work. Instrumental works absent text are considered by most to be abstract or no more than ambiguously referential[60] and should therefore be distinguished from electronic works, which may employ "concrete" samples of sounds recorded from the real world. Whereas instrumental works merely allude to the subject, a concrete electronic work uses actual utterances (whether human, bird calls, or mechanical sirens) to present the "subject" it depicts. And yet, this is not entirely true across the board with electronic works. We can further divide electronic compositions into works that use sampled sound and works with sounds synthesized by either analog or digital generators. But even between these two categories, there are many crossovers. Erotic musical works, therefore, do not easily categorize into a single genre, and for this reason this book collects examples of several such categories side by side. This section alone introduced several dialectical pairings: gender/sex, serious/popular, art/pornography, sight/sound, men/women, queer/straight, texted/untexted music, instrumental/concrete music, sampled/synthesized electronic music. As in Kierkegaard's doctrinal *Either/Or*, music, which in essence is sensual and erotic, elicits especially polarizing dichotomies that require not either pole but both, through ethical *and* aesthetic deliberation.[61]

PUTTING THE "ART" IN ART MUSIC

I have chosen to restrict the selection of works in this book to those deemed erotic or sexual by their creators and, therefore, begin from the assumption that the works collected here were intended to arouse either erotic sensations or, more plausibly, elicit allusions to sexual scenarios. This stipulation does not, however, exclude the possibility that these works were intended as either erotic artworks, *or* pornography, or both by their creators. As I explain in chapter 3, to avoid the lofty connotations of high art, Luc Ferrari's self-proclaimed pornographic framing in his various "anecdotal" works aimed to present situations as they really happened, with limited processing and hence minimal interference from himself as their creator. This intention does not, however, preclude Ferrari—assistant to *musique concrète* inventor Pierre Schaeffer—from being classified among notable "art" composers of the time. Similar to John Cage, it was precisely on account of Ferrari's appeal and resistance to art as an institution that his works gained prestige *as* art. Again, the intended purpose does not dictate use. Nevertheless, the method of composition seems to affect the relevance and significance of certain symbols within the music as well as the music's connection to external associations. Each chapter in the first part of this book turns its attention to a particular compositional attitude, beginning with the phonograph samples used in *musique concrète*, and progressing more or less chronologically through analog tape realizations, the Musical Instrument Digital Interface (MIDI), digital granular synthesis, and digital audio workstations (DAWs). Such points of focus provide each work its own conceptual framework, but, taken together, these works shed light on the broader category of electrosexual music.

Whereas artworks gain institutional and academic recognition, such institutions generally position pornography in a separate category or, in the case of silent omission of electrosexual works from the electroacoustic canon, ban it outright.[62] Hans Maes enumerates many artworks that suffered such fates, such as J. M. W. Turner's pornographic drawings that were burned by John Ruskin and "several of the explicit frescoes of Pompeii and Herculaneum" destroyed or banished to the storerooms of the

Naples Archaeological Museum.[63] But I would point, on the contrary, also to notable works that were not transported, destroyed, or censored, for example, to the plethora of erotic sculptures lining the outer walls of the Lakshmana Temple (Khajuraho, India) and decorating the Javenese-Hindu Candi Sukuh or to the writings in Vatsayana's *Kamasutra*, to emphasize that the criteria for praise or persecution are culturally determined. As I examine in chapters 7 and 8, any determination of either art or pornography rests in the controlling hands of those with deciding power in the geographical region and historical time period in which the work is assessed. Where certain practices of Hindu art, dance, and music were at one time accepted simultaneously as art *and* pornography—sexual education being an intrinsic facet of art—the British occupation of India raised some questions about the merit of certain traditions, and subsequently these practices took on new, and often damaging, significance in the hands of Indian and non-Indian creators alike. The conflict between sex as either pornography or education crops up again in the educational pornography, or edu-porn philosophy of pornographic actress, sex educator, and former prostitute Annie Sprinkle. In chapter 8, I analyze Sprinkle's *Sluts & Goddesses* video workshop, which features a soundtrack by influential electroacoustic composer Pauline Oliveros. Edu-porn uses visuals and sound to instruct sexually curious participants and viewers, parodically. That is, sex sounds—whether constructed musically or otherwise—have the potential to teach listeners how to hear sex in real-life situations. Therefore, it is important to articulate the qualities of these sounds, how they are created, for what purposes, and, in turn, how sounds are interpreted and understood.

In a report to the Committee on Obscenity and Film Censorship in the 1970s, Bernard Williams defines pornography as the explicit depiction of sexual material, that is, "organs, postures, activity, etc."[64] Any argument, however, that, on the one hand, pornography concerns explicit depictions of sexual acts, while, on the other hand, erotic artworks convey only sensations either by allusion or association, depends very much on what one considers sexually explicit representational content. Arguably, the plastic arts, film, and narrative prose all allow for express representation, but music does not. The distinction between content and medium requires one first to ask what the characterizing elements of each artistic medium are and then to extrapolate on the specific manner of representing content in each. Yet one can find exceptions to even the most fundamental distinction between the visual and sonic arts when it comes to the issue of representation.

On the subject of representation in music, philosopher Susanne Langer argues that it is easy to attain organic unity (*Gestalt*) in artworks that represent the human form (visibly). She writes in *Philosophy in a New Key*: "Even when we would experiment with pure forms we are apt to find ourselves interpreting the results as human figures, faces, flowers, or familiar intimate things."[65] However, unlike the visual arts, whose forms—line, color, and shape—correspond (traditionally) to the object they depict, music's forms—pitch, rhythm, contour, and timbre—are not generally representative of some other object of reality. Susanne Langer warns us that we should be wary of reducing the value and import of an artwork to merely its recognizable models. The model's "interest as an object" to the artist, says Langer, "may conflict with its pictorial interest and confuse the purpose of [the artist's] work."[66] In any case, music is not directly representational, though specific artworks may conceive of a symbolized subject.[67] For the most part, Langer does not attribute art with expressive or arousing properties. As Stephen Davies summarizes, "she emphasizes the idea that music symbolizes not occurrences of feelings but the concept of them."[68] It is perhaps for this reason she raises the possibility "that erotic emotions are most readily formulated in musical terms."[69] That she identifies "erotic emotions" with the forms brought about in musical terms shows that Langer imbues music with properties that distinguish it from the other arts, but, more importantly, that she nominally isolates "erotic emotions" over all others points also to eroticism's unique expressive potential.

The term *emotion* here requires some qualification, since for Langer music is not *e*-motive; by emotion she means simply that music conveys some semblance of eroticism—an erotic essence, to foreshadow my first chapter. For many composers of erotic works, music need not arouse the audience sexually, nor ought it to express erotic sentiments. Rather, a musical work itself is the erotic object, a "presentational" symbol, to invoke Langer's term for the symbols of art as opposed to the discursive symbols of language: eroticism is alluring without representing anything alluring. However, this position, which Langer proposed in the 1940s, no longer serves to oppose music from the visual arts.

In the twentieth century, the plastic arts were increasingly drawn toward abstract forms; conversely, electronic technology has brought to music the sound samples of real life. Changing technologies of the last seven decades have altered the face of musical composition, such that what was once strictly referred to as electroacoustic music has since blossomed into a rich soundscape of artistic possibilities, the sounds of which extend even beyond the previously delineated borders of music.[70] Technology has changed the very material of music, the manner in which music is preserved and shared (if at all), and its execution in performance. Technology compromises *phonopoiesis*—making sound—both as it affects the quality of sound and in the manner of accessing sound, whether as creators or receivers. It is on account of these new developments that I have chosen to isolate music with electronic components from "purely" instrumental or vocal music. Although music at one time may have been opposed to painting or drawing on the basis of how subjects are presented in each, we can no longer separate music from the other arts on these grounds. Similar to photography, electronic possibilities grant music the semblance of representation, though the represented object, both in art and music, is necessarily mediated. Mediation transforms the object into a symbol with constituting elements that may differ from some originary model as I later examine in the context of sampling. Electronic media affect the resulting relationship of the sonic and the visible. Sound and vision mediate one another such that neither is entirely determined *by* the other nor is one separable *from* the other—with the stipulation that *how* something is presented relies as much on the manner in which it came to be. The creator's craft is as significant as the views of those who encounter the work.

Reception certainly factors into responses to the question, "What is pornography?" For philosopher Michael Rea, "x is pornography if and only if it is reasonable to believe that x will be used (or treated) as pornography."[71] The determination of the work's quality is, therefore, beholden entirely unto the observer. Except that people use pornography differently. Where one may be using electrosexual music to get off, another person may be using the music for a more analytically removed study. Then again, the former and the latter may in fact be the same person in different instances.

Let's turn here briefly to German Dadaist Erwin Schulhoff's *Sonata Erotica* for solo female voice from 1919—the first musical work to notate a woman's orgasm. In addition to her visible presence on stage,[72] the sounds of Schulhoff's moaning woman resemble and therefore recall a woman experiencing sexual pleasure. According to film scholar Linda Williams, in the context of pornography, men's sexual pleasure is often visual—men's satisfaction is graphically conveyed with the crowning "money shot"—while women's satisfaction is equated, according to John Corbett and Terri Kapsalis, with the "quality and volume of the female vocalizations."[73] In a comparable instance, then, for example when listening to a sound record-ing of Schulhoff's work, listeners may conjure a woman to mind, even absent the visual aspect. This inferential association is what I refer to in the title of part I as the "Electroacoustics of the Feminized Voice," because the sounds of ecstasy draw on feminizing tropes. Throughout the book I refer to the "feminized" voice when we have no record of the performer. We cannot know if the recorded voice originally belonged to a woman, but it's meant to sound that way, and that's what the composers tell us in many cases. The work's originary (or its intended) scenario delimits neither how the work will be used nor its purpose. While an erotic artwork may very well represent the human figure, its intended purpose does not lie solely in such allusion. The composition may invoke certain formal or idiomatic expressions independently of the perceived woman's sexual pleasure—I may become aroused by the sounds I hear or become repulsed and dis-gusted by their implied objectification of the female form—and this is as true of art as pornography.

The Oxford English Dictionary defines pornography as "the explicit description or exhibition of sexual subjects or activity in literature, paint-ing, films, etc., in a manner intended to stimulate erotic rather than aes-thetic feelings. . . ."[74] By this definition, the moment that art treads into the realm of sexual arousal it risks being ousted from its lofty seat among the artworks of the world. In his article "Why Can't Pornography Be Art?" Hans Maes raises just such an objection to pornographic content, which he summarizes in two points: "One could say that a pornographic representation is (1) made with the intention to arouse its audience sexually, (2) by prescribing attention to its sexually explicit representational content."[75] The example of the moaning woman, above, only satisfies the second stipulation. The woman's vocal "organ" exhibits the quintessential lewd posturing of the sexual act, which is meant to arouse the audience, but while she may *sound* arousing, she may not actually arouse any of her listeners—think of straight women or gay men, if one can appeal to generalities—and this argument extends equally to pornographic works in which sexual intercourse is not the quintessentially arousing act.

Take another example: the case of sexual fetish. Listeners may find themselves aroused by some action other than "the" quintessential sex act—examples include the clicking of high-heel stiletto shoes, which draw attention to the feet, or the plucked string, recalling the tactility of the plectra, whether material or flesh, such as the sounds of Juliana Hodkinson's and Niels Rønsholdt's *Fish & Fowl*, analyzed in chapter 5. These examples may satisfy the first criterion, by arousing some listeners, but they are not bound to the second, that sexual arousal in pornography consists of sexually explicit representations. It is not sufficient to deny the label of pornography solely based on content, since it is the work's use that ultimately determines its satisfaction of both criteria. Yet, as Langer proposes, it is neither the content nor the use but the *purpose* of the artwork that deems it as *art*. As will become clear in this book, media's sexual aspects are not beholden solely unto a creator, the object itself, or to its reception: electrosexual intentionality lingers in a consensus of these varied constitutions. Whatever their intended use, erotic artworks are prone to mishandling by users, and therefore appeal equally to pornographic censure. It will therefore be important in this book to consider the threats of representation, namely exploitation and objectification, issues I take up more thoroughly in individual chapters but briefly introduce here.

PORNOGRIFYING MUSIC

Beginning in the 1960s, feminism's second wave urged resistance to the lewd and disrespectful portrayal of women's sexuality in the pornographic films of the previous decades. Antipornography feminists argued that the women in such films serve no other purpose than to perform graphic and explicit sexual acts in order to please heterosexual men. Absent character development or any particular identity, these women are typically insatiable, and, at the same time, their pleasure is made present only in the service of men. In the 1970s, radical feminist activist Andrea Dworkin helped shine a light on the industry's habitual mistreatment of women, who, in many cases, were coerced into performing in pornographic films or were filmed and publicized without their consent.

Where many readers are probably familiar with this history of pornography in the cinematic context, electroacoustic extensions of these abuses are likely unknown. Celebrated composers of the electroacoustic canon and contemporary music more generally have also objectified women toward pornographic ends. Ferrari's *Presque rien avec filles* for dancer and memorized sound (1989) and Normandeau's *Jeu de langues* (2009), the subjects of chapters 3 and 4 respectively, both employ women's voices without their knowledge in the composers' attempt to arouse "intimate" and "erotic" associations in listeners. Likewise, the "Erotica" movement from Schaeffer and Henry's *Symphonie pour un homme seul* (1950–1951) also employs an uncredited woman's voice—at least she's meant to sound like a woman, but we cannot know for sure who performed the eroticized sounds. Each of these examples raises concerns about sexual consent because of how the men use these voices and intend their reception. As I have written elsewhere regarding such "Specters of Sex": "In this sense, cutting, pasting, sampling, and thus manipulating a woman's voice to produce a piece that elicits sexually connotative meaning or erotic overtones could be likened to sexual coercion," a claim explored more thoroughly in the chapters of this book.[76] Such cases recall objections advanced by scholar and lawyer Catherine MacKinnon during the Women Against Pornography (WAP) movement in the US in the 1980s.

MacKinnon argued that pornography was inherently exploitative because women's sexuality became a form of commerce that did little to benefit its primary laborers.[77] In this regard, women were made into objects construed for the pleasure of a class to which they did not belong and, indeed, were presented in spaces that habitually excluded them.[78] Similarly, electroacoustic spaces have long excluded women's contributions as equal creators to men, who are more typically touted as composers and therefore compensated with prestige in the form of academic positions or board nominations.[79] Together, MacKinnon and Dworkin launched and defended several lawsuits against the pornography industry on allegations of distributing films that were captured without prior consent of the participants. Not least of these were filmed acts of rape caught on tape and distributed for profit. Consequently, the WAP group became motivated to dismantle and altogether abolish the pornography industry, determining that its films were inherently violent toward, and biased against, women. Electroacoustic institutions have never experienced comparable censure, most likely because electrosexual music has hardly been given significant attention for it to receive the equivalent public scrutiny.

Despite their important work to secure legal measures against women's mistreatment in the industry, many subsequently criticized MacKinnon and Dworkin for claiming that all pornography necessarily incited violence. Many women were concerned that a complete and total rejection of pornography would be a disservice to women who derive pleasure or profit from it. Excluding a female audience is, therefore, just as dismissive of women, once again discounting them solely on account of gender while also undermining their ability to assert their own subjectivity. In other words, WAP succumbed to the very accusations MacKinnon launched against the pornography industry. Part I, "Electroacoustics of the Feminized Voice," therefore includes several examples that work to mitigate against blanket refusal of the electrosexual project. In some cases, a creator's transgressions against women can be reframed contextually within the performer's own experiences, as for example in actress Rio's comments about performing in Ferrari's staged pseudo-lesbian encounter *Les danses organiques* (see chapter 3).

In the early 1980s, questions of moral integrity surrounding pornography eventually attracted academics to the debate, and a conference with the title "Sexuality" was held at the Center for Research on Women at Barnard College in New York City. Organized by Carol Vance, along with Ellen Dubois, Ellen Willis, and Gayle Rubin, the event was meant to broach the topic of sexuality as an issue central to feminist concerns rather than a marginal or transgressive topic. As Carol Vance outlined in her invitation letter, the conference highlighted an array of issues concerning women's sexuality to address sex as "a social construction which articulates at many points with the economic, social, and political structure of the material world."[80] Despite its wide reaching aims, antipornography groups identified the conference solely with issues of pornography, S&M, and the butch/femme dichotomy, which they viewed as profoundly violent and particularly objectifying of women. Antipornography groups pleaded with the university to cancel the conference and, when the university did not heed their requests, a small collective rushed the venue in violent protest.

A cornerstone event in the "feminist sex wars," the conference made important scholarly contributions to academic discourses on feminism and sexuality, including Gayle Rubin's workshop "Concepts for a Radical Politics of Sex,"[81] which sowed the seeds for her important essay "Thinking Sex." Rubin was not content with the simplistic antagonism between women and men and attempts to codify pornography as a cruel and improper industry. She accused antipornography campaigners of vilifying all sex "acts" that did not align with a severely limited and conservative idea of sex (see chapter 7). Later, pornography scholar Linda Williams observed that the art/pornography divide more drastically impinged on women's work than men's: "Performance artists, especially women performers whose gendered and sexed bodies serve as the basic material of the performance, are often vulnerable to vice squads, or to censorship by the National Endowment for the Arts (NEA), because their art and thought occur through the body."[82] Equally present now, following President Donald Trump's 2020 executive order banning government institutions from funding scholarship involving "race-based ideologies" including "critical race theory," even those who write about these topics struggle to secure research funding, perhaps another reason for the absence of discourses engaging with either pornography studies or critical race theory from electroacoustic historiography.[83] Critical considerations of race are crucial to my analysis of electrosexual music and its thematic constants.

WHO MAKES ELECTROACOUSTIC MUSIC?

Despite electronic music's reputation as "fixed" media, this music often arises in situations that are improvisational, experimental, iterative, performed, realized, patterned, and expressive, all terms encapsulating a sense of feminist craft—and it's hardly coincidental that *craft* (arts and crafts, textiles, weaving, etc.) is a term frequently applied to women's work and to the work of communities minoritized and marginalized from academic (musicological) study, communities ignored or exploited (or both) precisely on account of their craft. For example, Lisa Nakamura has written about the instrumentalization of indigenous women's craft in computer engineering in the United States. Her article "Indigenous Circuits: Navajo Women and Racialization of Early Electronic Manufacture" details how the geometric patterns woven by Navajo women into rug designs became integral to the electronic design of computer chips manufactured by Fairchild Semiconductor, "the most influential and pioneering electronics company in Silicon Valley's formative years."[84] As Nakamura explains, although "women of color . . . are almost never associated with electronic manufacture or the digital revolution," Fairchild's entire operation was founded on the labor of Navajo women, because of how the company envisioned parallels between the intricate and delicate work involved in the design of both rugs and computer chips—so much so that the resulting chips ultimately even looked visibly similar in design to the traditionally woven rugs. This is but one instance of how women of color's craft is embedded in technology, of how the presence of women and especially women of color in the history of technology is at once both implied and denied. That is, like Bosma's electrovocal music, we are left with a trace of women's presence without explicit acknowledgement that their labor is also innovative, even productive.

Until now, I have referred abstractly to "the voice" and significances attributed to its gendered aspects. Still another crucial vantage for the critical analysis of how people create, circulate, and perceive electrosexual music is how "the voice" becomes racialized. Electroacoustic composition insists on the virtue of transcendental authority, thereby elevating analytical investigations of musical aesthetics above bodily concerns, which somehow always remain relegated to the purview of women and those otherwise minoritized on account of sexual behaviors/orientation/identity, and/or race/ethnicity. I take up this racializing impetus in each of the chapters in part I but define how racial othering becomes generalized more broadly and in effect naturalized through music in part II.

Today, aside from soundtracks to actual pornographic films, sexually explicit content in recent music is most commonly heard in (by which I mean associated with) popular music and, in particular, hip hop—the most dominant musical genre in US listening habits according to a 2017 report by the global measurement and data analytics company Nielson Music.[85] Hip-hop music often boasts sexually explicit lyrics that, historically, have been either critiqued as demeaning to women, or, when performed by women, painted as a way of reclaiming feminist agency from men, whose claim to power is given and assumed. Hip-hop scholar Tricia Rose summarizes this historical polarity:

> Those who take on sexism in hip hop can generally be divided into two broad groups: (a) those who use hip hop's sexism (and other ghetto-inspired imagery) as a means to cement and consolidate the perception of black deviance and inferiority and advance socially conservative and anti-feminist agendas; and (b) those liberals and progressives who are deeply concerned about the depths of the sexist imagery upon which much of hip hop relies, but who generally support and appreciate the music, and are working on behalf of black people, music, and culture.[86]

But even in the earliest days of the genre, when hip-hop scholars were cultivating categories of gendered and sexualized conduct in these discourses, Rose cautions:

The political differences between these two groups is not absolute . . . Although the language of disrespect, the emphasis on degradation of women (because it thumbs its nose at patriarchal men's role as protector of women), has roots in white conservatism, it also has solid roots in black religious and patriarchal conservative values . . . Black female resistance to this perception ["that black women were sexually excessive and deviant as a class of women"] encouraged a culture of black female sexual repression and propriety as a necessary component of racial uplift.[87]

Part II, "Electrosexual Devices," examines Black women's "repression and propriety" more in depth, in the respective stories of Donna Summer and the musical group TLC. In each case, the performers projected a sexual aesthetic that did not necessarily or entirely align with their personal ethics.

The burgeoning electrosexual critique I introduce in this book owes much to hip hop—the music as well as its thorough discourses on musical borrowing and digital processing and its now trenchant analytical tools for sexual(ized) performance both visible and audible. As these discourses are much more developed than in the electroacoustic context, there is more rudimentary material to present through electroacoustic music in part I, "Electroacoustics of the Feminized Voice," before turning in part II, "Electrosexual Devices," to the complex iconography of popular music and hip hop and how electroacoustic music borrows and capitalizes on established tropes in Black music, clinging to sexualization while abandoning deeper engagement with the layered subtexts advanced by such tropes.

A parallel discussion to Nakamura's archeology of circuits arises in the ethics of an archeology of sounds, or the colonialist impulse to forage for sources without properly crediting one's creative inspiration. In contemporary art music, this practice is referred to as using "found sounds," but usually omits identifying where the sounds were found. Similarly, electronic sampling—the effect of weaving prerecorded music into a newly produced work—often engages in appropriative practices because of diverging expectations of how source material should be identified, habits of shared knowledge-building that are spelled out in some forms of musical engagement and not others. For example, both hip hop and electroacoustic music employ sampling through similar technological means—whether as recognizable or disguised references, and yet, the mostly exclusive receptions of hip hop and electroacoustic traditions tabulate the practices into discrete categories. In his article "Was Foucault a Plagiarist?" Mickey Hess likens sampling to the way Foucault tends to cite his sources—without formal acknowledgement but with assumptions of his readers' existing knowledge.[88]

In his 1977 interview, "Prison Talk," Foucault explains:

> I often quote concepts, texts and phrases from Marx, but without feeling obliged to add the authenticating label of a footnote with a laudatory phrase to accompany the quotation. As long as one does that, one is regarded as someone who knows and reveres Marx, and will be suitably honoured in the so-called Marxist journals. But I quote Marx without saying so, without quotation marks, and because people are incapable of recognising Marx's texts I am thought to be someone who doesn't quote Marx. When a physicist writes a work of physics, does he feel it necessary to quote Newton and Einstein? He uses them, but he doesn't need the quotation marks, the footnote and the eulogistic comment to prove how completely he is being faithful to the master's thought.[89]

In Foucault's version, citation is an author's way of forming alliances while also rupturing bonds and therefore asserting boundaries between academic disciplines. Hess draws attention to this quote from the well-regarded French philosopher to show that even recognized scholars—or *especially* scholars who presume to have proven themselves intellectually—employ a kind of citational practice akin to how sampling works in hip hop, as paying dues and recognition. As I show throughout part I, electroacoustic composers also use sampling to signify a sound's implied contextual meaning, though they rarely flag inherited meaning overtly by acknowledging the music's original creators, and this is where electroacoustic sampling borders on appropriation.

In my article "Categorising Electronic Music," I performed search queries of music databases for the term *electroacoustic* in search of non-white non-men composers. Not finding these composers, my search instead revealed that, according to these databases, those representing electroacoustic composition fit into a hegemony of white men of European backgrounds. This hegemony arises primarily in how keywords are designated for this music in the specific databases I searched, including one of musicology's most valued databases, *Répertoire International de Littérature Musicale* (RILM) Abstracts of Musical Literature (1967–Present Only). The hegemonic identity of electroacoustic composers also perpetuates in the way electroacoustic spaces recognize influential composers, as I showed in the case of the dedicated electroacoustic website and database, *electroCD*.[90] I argued that, although many composers of color work in electronic music, the search term "electroacoustic" remains exclusionary because of who declares themselves as an advocate of this music, and not necessarily in how their music is made. This study showed that a search of "electroacoustic" in the RILM database returned 2,814 results, and yet only one of these results examined music written by Black composers.[91] Turning to the *electroCD* database, I found several composers of color listed alphabetically in the browse function—including Toru Takemitsu, George E. Lewis, and Nam June Paik—but, of those appearing in the "featured" section on the website,[92] I could recognize no composers of color: "The site lists a total of 4,364 electroacoustic artists. A featured category, indicated on the website by a star icon, boasts headshots and additional details for 61 composers . . . , none of whom appear to be Black or even people of colour."[93] Looking to yet another database, the Composer Diversity Database (CDD),[94] I could isolate musical qualities as well as identifying qualities for the composers such that I could search uniquely for "Black" composers of "electroacoustic" music, which returns sixty-eight results, thirty-five of whom are women. However, since performing this research, many individuals included in the CDD have expressed concerns with how their information appears in the database, since the database publicly identifies queer composers who may not be out and categorizes artists's music by uncertain conventions (for example, in 2020 Stevie Wonder appeared as an electroacoustic composer).[95]

Crucially, one of the main claims of electroacoustic philosophy clings to the "aesthetic duty" of concealing source–cause relations that in turn amplify sexual suggestion and closely related associations to gender and race.[96] This becomes particularly problematic as many other forms of electronic music engage in unconcealed intertextuality, which is generally discouraged and can even become ridiculed in electroacoustic circles as a way of more strictly delineating between "serious art" music, in which electroacoustic music so desperately wants to be included, and what in music aesthetics has traditionally been the assumed frivolity of popular music. Indeed, these differences were inscribed on musical objects. Media scholar Jacob Smith reminds us, "The phonograph industry made clear distinctions between its high- and low-culture products, distinctions that were physically inscribed on records: from 1903 Victor's operatic recordings bore a 'Red Seal,' in contrast to the 'Black Seal' of popular records."[97] Though recordings no longer bear these physical markers—indeed, physical formats have become obsolete—racialized divisions between musical categories live on in the public consciousness.

Whether objectification or reclamation, vocalizations and sound effects associated with sex acts are mainstays of the electrosexual repertoire. Indeed, because of this common sonic denominator, I have taken precautions in framing electrosexual music's historical origins and subsequent developments. This book is in two parts, the first half dedicated to contextualizing electrosexual music within the dominant cis white racial framing in electroacoustic music, because much of this music works to reinforce this dominating habit. The second half of the book, while still attentive to this framing, turns to examples that break the electroacoustic mold either by consciously objecting to its narrow constraints or by emerging from, building on, and, in a sense, competing with a completely different historical trajectory. In order to better explain this bifurcation of the book's structure, I now turn away from defining the limits of sexual expression to briefly expound on notions of "human" expression and the trouble of locating origins in electrosexual music.

Musicologists of the early aughts frequently discussed electronic manipulations of the human voice in "posthuman" terms, referring repeatedly to N. Katherine Hayles's investigation in *How We Became Posthuman*.[98] Positioning her critique of the digital turn within the Cartesian duality of the mind/body split, Hayles was concerned with understanding "how information lost its body," why "in the posthuman, there are no essential differences or absolute demarcations between bodily existence and computer simulation, cybernetic mechanism and biological organism, robot teleology and human goals."[99] Alexander G. Weheliye's timely intervention into posthuman musicological discourses returned to Hayles's analysis to illuminate her text's uniformity of examples: "Even when giving examples of paradigmatic posthumans, Hayles falls back on white masculinist constructions by citing the Six Million Dollar Man and Robocop as avatars of the posthuman condition; at the very least, the category might have been expanded by including the Bionic Woman."[100] Weheliye goes on to interrogate limitations of the concept Hayles takes for granted as "human," building "her definition of subjectivity from C. B. Macpherson's classic, *A Theory of Possessive Individualism*. . . . Hayles asserts that the liberal subject was thought to predate the market by virtue of wielding ownership over the self (the natural self)." This observation leads Weheliye to determine: "Though careful to stress that the liberal version of selfhood is only that—one particular way of thinking about what it means to be 'human'—and not wishing to resuscitate this rendition of subjectivity, in the end Hayles unwillingly privileges this modality, for it serves as her sine qua non for human subjectivity. Put differently, Hayles needs the hegemonic Western conception of humanity as a heuristic category against which to position her theory of posthumanism."[101]

At issue here is that prioritizing the "Western"—white European cis men's—perspective of human subjectivity diminishes any supposed divergences from this model as secondary and always-already proximal to this model. It is because of this differing notion of human subjectivity and its relation to the "posthuman" that I divide this book structurally in two parts, and for this reason, this organization could very well be flawed in the same ways as is Hayles's critique.

Cautious as I am of reinscribing the hegemony Weheliye outlines above, part I, "Electroacoustics of the Feminized Voice," nevertheless reiterates the electroacoustic origin story that similarly centers heterosexual cis white masculinist tropes of electrosexual expression precisely because these tropes have not yet been explicitly identified in electroacoustic discourses. "Electroacoustics of the Feminized Voice" defines and interrogates the assumed norms of electroacoustic sexual expression in works that represent women's presumed sexual experiences via masculinist heterosexual tropes, even when composed by women. Part II, "Electrosexual Devices," then presents examples that depart from such assumed norms, beginning from musical examples that actively respond to and disrupt these norms, though, in so doing, some end up unintentionally doubling down on the norms. Again, it is important to show the stereotypical ideal of women's sexual expression in order to illuminate how these ideals become "naturalized," self-assumed, underexplored, and thus tacitly ignored by the electroacoustic community.

Popular music features in part II as an agglomeration of broad musical categories that developed in parallel to the electroacoustic, though somewhat in distinction—even segregated—from that practice and its aesthetic philosophies.[102] Case studies in the book are arranged in loose chronological order, adjusted slightly for topical connections. Where part I provides examples that typify the electroacoustic sexual canon, the extent to which sexuality has been belabored in popular music—which is arguably driven by electronic and digital media—means it is less necessary for me to rehash old arguments. Regrettably, neither is there space to present thorough histories of feminist and queer interventions therein, though I revisit these discourses where relevant and in my introduction to the second part of the book. Instead, chapters in part II represent unique interrogations of sexuality and therefore require more extensive analysis and theoretical detail. After presenting the electroacoustic stereotype of the electrosexual, I pick up the threads of a presumed universal subjectivity and its resulting exclusion in my introduction to part II, where I present examples that do not fit into this model of electroacoustic sexuality because of who created the music, their aesthetic or philosophical outlook, as well as how an artist's personal identity factors into their views of sex and its musical representation. Of course, listeners are just as crucial in these analyses, and where possible I aim to contextualize the music within fandom and scholarly reception alike.

Electrosexual music exists everywhere music exists, and it exists in droves. In the music of both part I and part II, white men have often received credit as creators of sexuality in the ideal or abstract, territorialized in a space far removed from their own personal proclivities, whereas women of color are simply not afforded that privilege, neither in "popular" nor in "classical" framings. Because classical music has a long history of privileging white narratives, music classified as somehow aligned with "serious" or "elite" (read *elitist*) institutions like universities and scientific communities more broadly—as electroacoustic music is—frame sex often as an "extra" musical trait, as opposed to the popular/dance musical heritage commonly associated with a minoritized underclass of non-white, non-straight, non-men, which, in the dominating academic discourses, is a group emblazoned with inherently sexualized traits. As feminist musicologist Susan Cook explains emphatically, and I quote her here liberally to convey the full weight of her critique:

> Now "the popular" has become my passion, because for me the most troubling legacy of twentieth-century modernism perpetuated by twentieth-century scholars regardless of their historical foci has been the creation and maintenance of hierarchical—and largely fictitious—dichotomies of all kinds. One of the most fiercely believed in draws a distinction between "classical" and "popular," or "serious" and "popular," or "cultivated" and "vernacular."
>
> Like so many hierarchical categories, the "popular" and the "classical" are imaginary, and yet they are powerfully imagined. A great deal of energy goes into keeping these categories circulating and often in simplistic and uncritical ways. We use the labels easily, yet it is rarely clear what we do indeed mean, what constellation of socioeconomic contexts, practices of consumption, or criteria for inclusion or exclusion we seek to identify—in short, what kinds of value we are assigning and why, what kinds of "popularity" we either celebrate or debase. These categories, like patriarchy itself, were historically created, and every time we use the terms, we call into being something that does not necessarily exist except as a kind of shadow. And through naming it without defining it, we give it extraordinary discursive power to shape us and our larger cultural landscape. . . .
>
> What bothers me about our fictional categories like "popular" and "classical" is that they are set into tension with one another. They don't simply exist as a pair of labels, resting comfortably side by side, but are almost always set up in inequitable relationships of power and prestige wherein "the popular" gives "the classical" its worth; the "classical" is worthwhile only if the "popular" is worthless. And the hierarchies keep replicating internally so that within

the "classical" you uncover the further delineation of populars that can similarly be dismissed or discounted."[103]

The bias Cook characterizes as an ill-defined "shadow" ties in with the psychoanalytic "shadow" composer Alice Shields seeks to address through her music in chapter 7. Indeed, many racialized and gendered uncertainties linger in the abstract non-definite notion of "sexuality." Where psychoanalysis seeks to confront and banish the shadow, Shields shines a light directly on sexuality to show its discrete and finite geographic, historical, and gendered borders.

This book situates electroacoustic music alongside many forms of electronic music, from esoteric experimental examples to the most common household names in disco, R&B, hip hop, and pop—genres commonly traced to their roots in Black music. Indeed, comparably to electroacoustic music, Weheliye locates Afro-modernism precisely in the uncertainty afforded by electronic music's acousmatic dependence on source–cause relations. Yet, unlike how electroacoustic music denies origins, Afro-modernism does not stop there. The philosophy (which is not tied to any single musical style) simultaneously insists on humanity, especially regarding sonic barriers—boundaries insisting on race as many concepts that are constantly being (re)constructed. Weheliye writes:

> We should understand this disturbance of the alleged unity between sound and source not as an ordinary rupture but as a radical reformulation of their already vexed co-dependency, which retroactively calls attention to the ways in which any sound re/production is technological, whether it emanates from the horn of a phonograph, a musical score, or a human body. . . . This interplay between the ephemerality of music (and/or the apparatus) and the materiality of the audio technologies/practices (and/or) music provides the central, nonsublatable tension at the core of sonic Afro-modernity.[104]

Similarly, Guthrie P. Ramsey locates musical "Afro-modernity" at the intersection of improvisation and technology, concerning "the creation and, certainly, the reception (the political and pleasurable uses) of musical expressions . . . articulated attitudes about [African-American people's] place in the modern world."[105] Thus, we note that electronic technology affords possibilities for rescripting mediations that might at one time have been considered demeaning or appropriative.

Tying the racializing aspects of intertextuality also to gender, in the words of philosopher Sara Ahmed:

> Living a feminist life does not mean adopting a set of ideals or norms of conduct, although it might mean asking ethical questions about how to live better in an unjust and unequal world (in a not-feminist and antifeminist world); how to create relationships with others that are more equal; how to find ways to support those who are not supported or are less supported by social systems; how to keep coming up against histories that have become concrete, histories that have become as solid as walls.[106]

A feminist electroacoustic theory affords a retooling of antifeminist and/ or racist figureheads. Looking to the composers whose work is explored in part I, neither Schaeffer and Henry, nor Ferrari or Normandeau could be described as feminists. In fact, their respective methods of cutting up the "feminized voice" has become a canonized feature of electroacoustic music since—and not just the sexy stuff.[107] Nevertheless, there is value in centering the voices of the women they sample and curate.

Electroacoustic composers desire to assert control over how women conduct themselves in electrosexual works by cutting up, displacing, and intermingling their voices with various other sounds. Whether they succeed in ultimately *taking* control is up to how we listeners interpret the works, how we correspondingly either echo, confront, or contradict this desired narrative/ narrative of desire. For Ahmed, "Many feminist figures are antifeminist tools, although we can always retool these figures for our own purposes," by which she means to expose racism and sexism, while simultaneously throwing into the question the very mechanisms by which oppression and suppression are institutionalized.[108] Whereas, historically, critics favoring the so-called "serious" forms of classical music in the West dismissed "race music"—jazz, blues, and other genres of Black music—as belonging hierarchically to an inferior "vernacular," for example, John Cage or (arguably) Theodor Adorno, examples of electronic music valued in academic circles demonstrate that the sexually suggestive was far from coded only for particular audiences. Indeed, owing to these cross-genre manifestations "academically" coded genres also circulate racializing and gendered tropes between modes of artistry.[109] The task of this book is to illuminate the mechanization of sex alongside its accompanying interlacing tropes of sexism, racism, misogyny, queerbaiting and straight-bashing, to simply show how pervasively sex sounds in all contemporary musical currents.

I ELECTROACOUSTICS OF THE FEMINIZED VOICE

1 SCHAEFFER AND HENRY'S "EROTICA" (1950–1951)

Man is an instrument that is too seldom played.[1]

—PIERRE SCHAEFFER (1952)

Electroacoustic music has a long history of taking sounds from one place and time and using them toward alternative ends altogether; this phenomenon of disruption was given the name "schizophonia" in the 1970s by R. Murray Schafer.[2] As a characteristic facet of electroacoustic music, the schizophonic habit is frequently cited in regard to Pierre Schaeffer's (1910–1995) philosophy of listening and, as we will see, tends to be exploited to the extreme by composers creating electrosexual works. Schaeffer's *musique concrète* was conceived as somewhat of an antidote to "naïve" listening, to hearing music merely as a collection of identifiable sounds from the real world.[3] Like diligent phenomenologists, Schaeffer's listeners are asked to confront their commonly held beliefs about sound and to approach each piece—and each sound in each piece—anew, without prior associations, convictions, persuasions, or considerations of a sound's source and cause. The Schaefferian approach, with an arguably more challenging objective than classical phenomenology,[4] forces listeners/investigators not only to confront what they take for granted as *common* sense, but the sensical in the context of this music conflicts in many ways with the sensical of the extrinsic, everyday actuality. In short, electroacoustic listeners must make sense of what, for all intents, is auditory illusion.

In 1943, Pierre Schaeffer founded the *musique concrète* studio at the Radiodiffusion-Télévision Française (RTF), where he had been a radio technician since 1936. Pierre Henry (1927–2017) joined Schaeffer in the studio in 1949 until 1958. Both Schaeffer and Henry took interest in cinematic effects. Schaeffer's philosophy of sound organization was inspired by British documentary films of the previous decades,[5] and in a more recent interview Henry expressed how closing one's eyes at the cinema created a similar effect to his compositional approach with recorded sounds:

> . . . il me semble d'une manière plus générale que les moyens d'enregistrement du cinéma ont permis une écoute nouvelle, donc un concert de haut-parleurs. Quand on va au cinéma, si on ferme les yeux, on assiste à un concert de haut-parleurs. On entend une musique symphonique ou dansante et c'est de la musique concrète si l'on veut. C'est comme les "aberrations" dans les arts plastiques, des choses un peu cachées et mystérieuses que l'on dévoile. La musique concrète, c'est amener le mystère devant les gens et qu'alors ce mystère se déploie, s'écoute, s'analyse et se critique.[6]

> . . . it seems to me more generally that the means of recording the cinema allowed a new listening, as a concert of speakers. When you go to the cinema, if you close your eyes, you are attending a concert of loudspeakers. We hear symphonic or dancing music and it's concrete music if you want. It's like the "aberrations" in the plastic arts, things a little hidden and mysterious that we reveal. *Musique concrète* brings mystery to people and only then this mystery unfolds, is listened to, is analyzed and critiqued.

This cinematic current runs through many of the examples in this book, including Luc Ferrari's notion of a "cinema for the ear" (chapter 3) as well as the *film noir* inspired *Fish & Fowl*, the piece analyzed at the end of part I.

Reciprocally, Schaeffer and Henry's *musique concrète* influenced the soundscape of French New Wave cinema. Their studio even served as a hub for the audiovisual experiments of directors like Jean Cocteau.[7] But differently from cinema, in which recorded sound occurs in tandem with the visible, the acousmatic necessarily keeps sound at a distance from sight—excepting the respective imagined visions of any listener. The composers' respective involvement in radio further bolstered their intrigue for divorcing sounds from sight. In this chapter, I identify how early sampling and mixing techniques served to advance scientific (and scientistic) experimentation that, perhaps inadvertently or at least unremarkably at the time, brought a colonizing impetus to electrosexual music, starting with its earliest instance: Pierre Schaeffer and Pierre Henry's "Erotica" (1950–1951).

The 1950s were a significant time to experiment with the sounds of human physiognomy. Notably, it was around this time that scientists were getting serious about researching the pleasures of sex, for example, in the United States, sexologist Alfred Kinsey founded the Institute for Sex Research at Indiana University on April 8, 1947, followed closely by publication of his *Sexual Behavior in the Human Male* in 1948 and *Sexual Behavior in the Human Female* in 1953, books that were wildly popular and widely translated.[8] The two-year period 1948–1950 saw a series of legislative amends regarding obscene content, specifically regarding "obscene phonograph recordings," or what were known then as "blue records," which had been circulating since the late 1890s, first on wax cylinders and then, from the 1910s, on discs.[9] According to Jacob Smith, in the US "There seems to have been a legislative and cultural sea change in regard to obscene records in the mid- to late 1940s. At this time there was an increase in the coverage of obscene records in the national press, which led to federal legislation in 1950."[10] France, around the same time, witnessed a number of relevant publications: in the popular press two special issues of the satiric magazine *Le Crapouillot* were dedicated respectively to *La sexualité à travers les âges* ("sexuality through the ages") (July 1950) and *La sexualité à travers le monde* ("sexuality throughout the world") (April 1951); diagnostic examinations from a medical perspective, for example from Pierre Simon; and Georges Bataille's notable philosophical investigation of *Eroticism* from 1957.[11] One must also wonder whether the Nazis' famous destruction of breakthrough research from the important German sexologist Magnus Hirschfeld—one of the earliest to advocate for homosexual and trans identities as we know them today—served as an impetus to further the militant shock-effect of sexual research at this time.

By this time, recording technologies were well developed, as were tools and techniques for manipulating perceptions of those recorded sounds. The voice, in particular, served a pointedly curious role, and Schaeffer and Henry's *Symphonie pour un homme seul*—a "symphony" for the human body—made great strides to explore the extent of these techniques. Though the composers experimented and improvised with sound according to a new philosophy, they were simultaneously adapting tried-and-tested approaches to sound that remain sedimented in their practice. Where, musically, *Symphonie* broke apparent barriers of representing the human body, sonically, their "Erotica" movement, the fourth of the piece, drew on existing radio and cinema techniques, retaining gendered, sexualized, and racialized aspects those techniques enforced. More specifically, the composers minimized women's presence by manipulating a feminized voice toward erotic ends without crediting the performer in either the recording's published material or their comments about the piece. Erasing an unknown woman's performance has since had a profound impact both in how Schaeffer and Henry's larger piece *Symphonie pour un homme seul* has been portrayed in subsequent accounts as well as how it has served as an example for the simultaneous erasure of women's performances in electroacoustic music and more broadly in electrosexual music. In later chapters in part I, I show that many electroacoustic composers drew on Schaeffer and Henry's "concrete" cinematic philosophy and even repeatedly used women's voices without their knowledge or consent toward erotic or sexual ends.

SYMPHONIE POUR UN HOMME SEUL (1950–1951)

The live premiere of *Symphonie pour un homme seul* arrived at Paris's *École Normale de Musique* on March 18, 1950, alongside other notable *musique concrète* works, *Études de bruits* Nos. 2, 4, and 5 (1948) and excerpts from the *Suite pour quatorze instruments* (1949). The tone for the live *École Normale* concert presented the composers as pioneers of sonic discovery, with introductory remarks by composer and musicologist Serge Moreux comparing the practitioners to Robinson Crusoe's pioneering domestication of the elements: "Listening to Schaeffer's scores has nothing to do with juvenile and honest musical civility. It is, perhaps, about discovering a sound continent as virgin as Robinson Crusoe's island," a description to which I return shortly.[12] Schaeffer and Henry joined artistic forces to create in sound something people had never before experienced. They had the idea of conjuring the experience of being human in sound—actually many sounds, hence the title *Symphonie*, *sum* + *phony*, or sounding together. The work emerged gradually throughout the 1940s from sound experiments Schaeffer and Henry conducted in their collaboratively founded *Groupe de Recherche de Musique Concrète* (GRMC) in Paris. That is to say, of the recordings captured for the purpose of *Symphonie pour un homme seul*—and there were many—those that survived were curated from spontaneous hands-on play with microphones, record players, speakers and speaker arrays, keyboards and other musical instruments, variously gendered actors, and anything else sounding in the vicinity.

Conceived in the late 1940s as "an opera for the blind," with its final version appearing in 1966, *Symphonie pour un homme seul* is divided into twelve scenic movements, (1) Prosopopée I, (2) Partita, (3) Valse, (4) Erotica, (5) Scherzo, (6) Collectif, (7) Prosopopée II, (8) Eroïca, (9) Apostrophe, (10) Intermezzo, (11) Cadence, and (12) Strette.[13] *Symphonie* uses a combination of sounds that are, as Schaeffer says, either "interior to man" (various aspects of breathing, vocal fragments, shouting, humming, whistled tunes) or "exterior to man" (footsteps, etc., knocking on doors, percussion, prepared piano, orchestral instruments).[14] Taken together, the sounds are meant to paint a complete sonic picture of man, however idealized. Schaeffer explains,

> The lone man should find his symphony within himself, not only in conceiving the music in abstract, but in being his own instrument. A lone man possesses considerably more than the twelve notes of the pitched voice. He cries, he whistles, he walks, he thumps his fist, he laughs, he groans. His heart beats, his breathing accelerates, he utters words, launches calls and other calls reply to him. Nothing echoes more a solitary cry than the clamour of crowds.[15]

In other words, the sounds of the human body resonate.

Around the time Schaeffer and Henry composed *Symphonie*, they gained access to various sound manipulation tools at the GRMC—audio signal generators, filters, and also the first tape recording device arrived at their studio. This is to say that Schaeffer and Henry *could* have transformed or distorted their source sounds into so-called objective entities that no longer serve as referents to actual, musically extrinsic objects, and yet they chose to retain some such identifying features.[16] 1951 would have been the first year Schaeffer had (officially) come to work with tape, though he was still using the technique of "closed-groove looping," as it is now regarded, using a sticker or other objects to block the record player's needle from advancing along the spiral grooves and forcing it to skip back repeatedly over the same groove (as when the record is scratched). As the story goes, Schaeffer first became occupied with looping through a happy accident brought on by a scratched vinyl record. The scratch caused a segment of a bell (absent the attack) to repeat, giving the impression of a sustained sound, and this technique came to be widely employed throughout *Symphonie pour un homme seul*.[17] In both tape and vinyl looping, the length of the looped fragment is dictated by the size of the record, though a looped sample may be slightly transformed depending on the speed and direction of the record's playback. On the importance of Schaeffer's happy accident in the formation of his *musique concrète* philosophy, Simon Emmerson notes that "such regularly repeated sound rapidly loses its source/cause recognition and becomes 'sound for its own sake.'"[18] Schaeffer furthered the technique by way of the various tools and electronic "instruments" he developed and employed at the time, including the three-head tape recorder (1952) and the morphophone (1953), for example.[19] That Schaeffer did not interfere with the material in *Symphonie* enough to change the audible identity of the feminized voice and piano in "Erotica," that he used vinyl records and not tape, and that he maintained extrinsic connections with real-world sounds, points to his desire for listeners to *recognize* these sources. As Schaeffer himself says, at least of the first version of the work, "The objects of the initial series were perfectly recognizable."[20] Furthermore, the work's performance "made use of the *pupitre d'espace*," a spatialization system that could route monophonic *objets sonores* to four loudspeakers positioned around and above listeners, thus enhancing the distantiation of a transcendent feminized voice in the "Erotica" movement.[21]

In the case of the "Erotica" movement, listeners are confronted with laughter, breathing, and sighing that, as the title insinuates, are cried out in a sexual encounter. Here closed-groove looping produces a regular rhythmic cycle within a temporally delimited framework, a periodicity reminiscent of the pulsating sexual organs in *the act* together with the vocalizations that typically accompany it. That the movement is entitled "Erotica" and not "Rapport" (as in, *les rapports sexuels*), or something comparably overt, attests to its mere suggestiveness, to the movement's purpose of arousing erotic desire but of not *re*-presenting sex. It is through repetition that the temporality of the "Erotica" movement itself emerges, and this constant provocation through repetition is meant to arouse an object-like, or generalizable erotic quality.

THE "EROTICA" MOVEMENT

Structurally, "Erotica" consists of six successive looping background tracks made primarily from short vocal snippets (table 1.1), and atop this looping one hears the alluring and cooing laughter of a mysterious and unidentifiable feminized voice. The first background loop (0:02–0:05 and again 0:07–0:11) features a crescendo, ascending vocalization on the syllable "oua"; the second (0:15–0:32 with fade out) alternates TA-ta on two notes approximating the pitches A and B and is accompanied by a vamping vaudeville-style um-pah um-pah; the third loop enters directly after the second fades out and is a three-note cycle rotating between pitches f–Ā–g–f–Ā–g;[22] after that, a fourth loop (0:39) sounds as a continuously sustained pitch (much like the bell in the happy accident); in a fifth loop (0:48–0:57) the voice returns; and the movement culminates in one last stream (0:58–1:15), a repeating figure

Table 1.1

Structure of the "Erotica" movement from *Symphonie pour un homme seul*

Loop Description	"oua" <<<	TA-ta (A–B) and vamping vaudeville-style um-pah um-pah	cycle rotating between pitches f–Ā–g–f–Ā–g	Overlapping "sustained" pitch	Vowel-consonant (E–F#)	repeated descending pitch
Time	0:02–0:05 pause 0:07–0:11	0:15–0:32	0:32–0:39	0:39–0:48	0:48–0:57	0:58–1:15
Duration	:09	:17	:07	:09	:09	:17
Loop No.	1	2	3	4	5	6

descending in pitch comparable as a same-but-opposite fragment of the first loop.[23] Atop these loops, laughing and sighing repeats through mechanically induced looping, though less periodically than is heard in the background. The voice's periodicity offsets the continuous background looping with curiously tumbling laughter, which rises progressively in pitch over the course of the brief movement until, at long last (1:05), a short, codetta-like return of the opening laughter signifies the movement's relaxation and conclusion.

Schaeffer's characterization of "Erotica" as a "man's voice [seeking a] woman's voice" is curious since there is no distinguishable man's voice to hear.[24] We hear only what we imagine is a woman's voice from its relatively higher frequency rage, a voice that serves as a trace of human presence. According to philosopher Adriana Cavarero, part of our pleasure in hearing a voice arises from "[imagining] how this person might be different from every other person, as the voice is different."[25] It is not just anyone's voice, but *this* woman's voice. And who is this woman you might ask? Quite frankly, I do not know. We have no record of the performer. We don't even know if the recorded voice originally belonged to a woman, but it's meant to sound that way (and that's what the composers tell us). The "Erotica" movement was included in its entirety already from the earliest versions of *Symphonie*, and the movement was part of the 1955 Maurice Béjart production available on YouTube (starting at 4:27).[26] The voice is, however, remarkably similar to Brigitte Bardot's 1964 hit *"Moi je joue"* ("I Play")— foreshadowing a "game" that would later become common to electrosexual music, as I explore in chapters 3 and 4.[27] The sample is especially comparable to Bardot's outro solo, as she refrains the words *"plus fort"* (harder) interlaced with moaning and the exclamation "ay yai yai," words we hear also in "Erotica" (at 0:19 and 0:41).[28] Samples for *Symphonie* were first assembled in 1950, when Bardot was only fifteen. Though, not coincidentally, this was the same year she appeared on the cover of *Elle* and was first discovered as a performer. I have come across no evidence Bardot performed for Schaeffer or Henry, but *Symphonie* surely captures some erotic sentiment from the time, and, who knows, maybe Bardot heard it somewhere.

Suspending our disbelief, we might imagine the composer hears the man's "voice" metaphorically; after all, who should we presume is there to pleasure the woman if not a man? Unlike the woman's cooing, the man's presence is not confirmed by his voice, then by what? Working diligently to layer together many sounds to create the movement's intimate atmosphere, we envision the composers at the phonograph as human samplers controlling the volume, cadence, and repetition of the many simultaneously looping layers of a woman's voice.

Mixing recorded sounds without accompanying images, Schaeffer and Henry purposefully shed sounds of their originating source and cause, allowing listeners to attend to sounds on their own terms. Some would later say this approach facilitated a kind of listening to sounds *as objects* for their own sake, and many composers since Schaeffer and Henry have gone to great lengths to disguise source–cause associations in their work, though listeners have not been so keen to let origins go. And what if the voice in "Erotica" was never a woman's at all? What if the man seeking a woman's voice instead describes a process of transformation? By recording the voice and altering the frequency, either slowing or quickening its periodic motions, what if this voice is a *man's* voice merely projected as a woman's? (What if this voice is *merely* a man's?) Does the voice still remain encoded within an enforced gendered binary? (Or does this only happen when we hear the voice as a woman's?) And if it retains gendered markers, how so?

As an early example of Schaeffer's philosophy put into practice, the source sounds in "Erotica" border on the abstract while also maintaining concrete references. Though the vocal sounds originate from an actual body, they are transformed beyond any human being's ability to perform them. The piece walks a fine line between record as "fixed" evidentiary representation and the schizophonic uncanniness described by R. Murray Schafer, "how a voice or music could originate one place and be heard in a completely different place miles away."[29] Given the movement's identifiable human voices, Schaeffer's "Erotica" did not (or had not yet) successfully realized his aspirations for "acousmatic music": an organized collection of sounds for which no origin is determinable.[30] But, to challenge an assertion that Schaeffer's *musique concrète* relies undoubtedly on *a priori* objective forces, we must remember that the composer did in fact modify sounds, however simplistically, to such an extent as to be producible *only* by electronic means. It is for this reason that Schaeffer finds written notation unsuitable for *musique concrète*.

In contrast to "ordinary music," which is conceived first mentally as an abstracted concept, proceeding then through theoretical notation to be executed in performance, *musique concrète*, conversely, originates in material, preexisting elements, and proceeds *toward* the abstract.[31] While the voices in "Erotica" vaguely manifest transcontextual associations among listeners, one cannot point to existing objects of the extrinsic world outside the piece that produce these sounds. In this sense, Schaeffer's sound object *is* distinctively objective; and his and Henry's delineation of *Symphonie* into two proportionate "Prosopopée" further supports this approach.

The word *Prosopopée* stems etymologically from two others. *Prósopon*, the mask or face in ancient Greek theater revealing a character's emotional state to the audience (while shielding the face of the actor), an identity however stilted, and *poiëin*, meaning to make, the verb from the same root as the noun *poiesis*. Schaeffer and Henry's *Prosopopée* present the appearance intended by the composers, shielding or veiling the origins of a sound only to the point that a listener is left questioning their beliefs. The composer's contribution then aims to sidestep the question of *who* or *what* makes sound, and of who or what makes sex sounds. Whether or not sexual imaginings are aroused in listeners by what is heard or even if evoked by the musical constitution of "the work" as such, the "Erotica" movement is meant only to capture or bottle up the essence of the erotic. Given common erotic sonic tropes, such as the moaning woman—though repetition of any kind can be enticing to varying degrees—the erotic then becomes the shared referential point of access to the erotic musical artwork, but neither the source/cause of the originating sounds nor the effects of the erotic essence can be declared as definitively intersubjective. The object, in this case, is not an instance of individuals engaged in sexual intercourse; rather, it is nothing other than "Erotica" *itself*.

EXOTICIZING, EROTICIZING, OTHERING ONE'S SELF

Schaeffer and Henry arranged recorded sound samples in the way they imagined was most likely to conjure erotic sentiment, not according to preexisting musical forms (for brief descriptions of the samples, see table 1.1). Still,

Schaeffer and Henry's method was not really galvanized by spontaneous epiphany; the movement's inspiration, in Schaeffer's words, was to create a duet in which a "man's voice [seeks a] woman's voice" with "light percussion accompaniment" and a "Tahitian record."[32] Schaeffer does not expand on this description, but it is easy to imagine how he envisioned the context for such an erotic encounter. In colonial memory, Polynesian culture is commonly associated with sexualized ritual dance.

As described by French conquerors, the Tahitian 'upa'upa dance was an indecent sexual performance enacted by a boy and a girl for an enamored audience (such infantilization is denoted in many Western descriptions).[33] From accounts of the music's rhythmic impulse to the gestures commonly employed—the boy rhythmically scissoring his legs behind a girl's "grass"-skirted undulating hips, the French invaders viewed the dance as a spectacle to be witnessed passively, or at most with critical skepticism. Revised accounts, such as that given by Marist ethnologist Patrick O'Reilly, describe the important role of nondancing participants in terms of their active drama, a role those unfamiliar with the practicalities of the dance situation—its culture, its history—could not deign to fulfill.[34] The realities of the 'upa'upa dance are certainly fascinating in themselves, yet Schaeffer and Henry's work was not premised on any realistic portrayal, instead they fashioned the Tahitian image as a vessel for their unshackled sexual fantasy in a comparable manner to Antonin Artaud's image of Balinesian culture explored in chapter 7. In essence, the quip about the Tahitian record simply perpetuates an eroticized and exoticized colonial phantasm with which Schaeffer and Henry's French listeners were already likely familiar. Using prior associations is one way music captivates sexual curiosity without making sex anew. It would be a futile exercise to trace similarities between this fantastic illusion and the realities of actual Tahitian dance practices.

Early modernist Linda Phyllis Austern identifies the European exoticizing of ethnicity and gender with the formation of Europe's geographical boundaries at "the cusp between the Renaissance and modernity . . . marked by the first major wave of European imperialist expansion."[35] According to Austern, physical proximity weighs on the intangible sonic dimension intertwining gender, race, musical aesthetics, and sexual proclivity with exotic women's submissiveness—all stereotypes embedded and entrenched in Europe's very musical language. In a more contemporary context, ethnomusicologist David F. García's *Listening to Africa* examines an exoticizing project in recorded music:

> Listeners of music as performed by bodies racialized as black, whether on the streets or on stage in Havana, as recorded in the anthropological field in Trinidad, Brazil, and Cuba or in studios in New York City and Havana, and listened to in the laboratory or domestic spaces, were disposed to localizing that music in temporally and spatially distant locations, the effect of which was that the listener remained anchored in, or at least aware of, their physical modern present.[36]

Distance, therefore, preserves a European austerity in recorded musical practices, and electroacoustic practice is no exception; it is perhaps even responsible for reinvigorating a colonial posterity in contemporary music as so many examples in this book follow this pattern. Austern and García push against the Eurocentric narrative of a history of Western Art Music, while possibly reinscribing authority to musical creators. Does it then matter whether we ascribe such agency to white men or, conversely, to non-white non-men? To investigate possible answers to this question, let us examine more closely the relationship of Moreux's apt likening of Schaeffer to Robinson Crusoe.

Daniel Defoe's story tells of the "heroic" trials of Robinson Crusoe as he procures slaves from Portugal to West Africa. On a voyage to Brazil, he is castaway and arrives stranded as a lone man on a desert island near Trinidad. Realizing he is not alone, Crusoe fears the land's inhabitants as "savages and cannibals" and takes it upon himself to domesticate one such mutineer, whom he names Friday, by introducing him to "other flesh," teaching him English and introducing him to Christianity.[37] Many authors have taken up retellings of Crusoe's story to draw attention to the novel's masculinized Eurocentric Enlightenment ideals. Michel Tournier's 1967 novel *Vendredi, ou les limbes du Pacifique* (*Friday, or The Other Island*) emphasizes the perspective of the prisoner-become-servant, while J. M. Coetzee's *Foe* is told from the perspective of Susan Barton, a woman also stranded on the island inhabited by Crusoe and Friday. Further removed from these retellings, literary critics and philosophers have illuminated the tale's constructed notions of national identity, individual subjectivity, gendering stereotypes, and colonialism apparent via these secondary sources.

Alice Jardine's investigation of Tournier's novel shows how, during a depressive episode following his shipwreck, "Robinson decides that he must henceforth master both himself and the island if he is to survive."[38] Envisioning the island itself as a female companion and naming her Speranza, "now, in time and mastery, she is his slave. Woman is, therefore, no longer absent from Man's adventures, even though he remains outside of intersubjectivity."[39] In this sense, Woman appears through his eyes, and this second subjectivity, according to Jardine, causes Robinson to turn his investigative voyage inwards to "his Self, his Man-hood." Seeing Woman in terms of his own construction of her affords him an understanding of the Other as no other than one's self, thus breaking down the distinction between subject and object. In the same way, Schaeffer and Henry's objectified Woman sounds by way of their intersubjective construction of the "lone man" at his phonographic helm—indeed, as I show in other chapters, many composers explicitly confess as much outright.[40] Still, more than this, Tournier's Robinson eroticizes Speranza-become-Woman, not only making love *to* her but also to himself-become-her. This is the point where Jardine identifies the origins of Gilles Deleuze and Félix Guattari's (in)famous theory of "becoming-woman," summarized below.

Like the relationship she recognizes in Robinson/Speranza, Jardine understands erotic intersubjectivity in Robinson's encounter with Vendredi (Friday):

> And so arrives Vendredi—an Indian whom Robinson has unwittingly helped to escape from his tribe (where he had just been condemned to death by "a witch"). He and Robinson quickly become brothers, "lone brothers" who "form the ideal, sterile, eternal couple; other couples are only satellites; they live in the midst of vicissitudes."[41] Robinson at first tries, of course, to master this "savage" just as he had tried to master Speranza. But Vendredi operates according to another logic completely. He tries to obey his master, to whom he thinks he owes his life, but, whenever possible, laughs, sings, plays, dances.[42]

Just as Schaeffer and Henry fix the "feminized voice" on their phonograph records, she, as Vendredi, continues to become mobilized, untamed, through the composers' *phonopoeisis*—their improvisatory play, whether autoerotic, with one another, or with each performance anew. To generalize even further, through this lens the entire colonial mission could be characterized as driven by autoerotic gratification.

It is this homoerotic colonizer–colonized relationship that Gayatri Chakravorty Spivak homes in on in her critique of Coetzee's *Foe*. Spivak marks Crusoe's island as "a 'pure' example" of "the colonization of other spaces," where the term "pure," drawing on Samir Amin, refers to how "colonization by European forms, as a whole, part of the gradual formation of a periphery."[43] That is to say, in prioritizing the point of entry into the perspective of "Other" as originating in the cis, white European man, Robinson Crusoe's story inscribes otherness as peripheral to some more dominant originary experience, whether Crusoe's or, more dialogically, Defoe's Enlightenment-driven project. More complexly, in Coetzee's novel, writer Daniel Foe approaches Susan Barton to tell her story, to convey in writing what she, apparently, cannot. While recalling her past to "Foe," Barton simultaneously embarks on a search for her missing daughter who, unbeknownst to her, has been taken to (enslaved in) the New World. Where she sits below Foe in terms of prowess and power, she aspires in her own narrative to give voice even to those more oppressed than she is, like the captured Friday.

Pointing to Coetzee's *Foe*, Spivak underscores a colonizing trajectory from "the terms *colonialism*—in the European formation stretching from the mid eighteenth to the mid twentieth centuries—*neocolonialism*—dominant economic, political and culturalist maneuvers emerging in our century after the uneven dissolution of territorial empires—and *postcoloniality*—the contemporary global condition, since the first term is supposed to have passed or be passing into the second."[44] Certainly, there are a number of parallels between the white South-African novelist Coetzee's instrumentalization of Susan for his Crusoe portraiture, and Schaeffer/Henry's notion of the erotic expressed through an anonymized "feminized voice." For Spivak, "the critical frame" is necessary to deconstruct these colonizing habits, but also to call attention to the process of philosophical interrogation in the first place—or, in my case, to musical analysis.[45] Beyond any reprimand facing Schaeffer and Henry, what role do I as analyst have in pointing out the parallel historical context between *Symphonie*'s construction of the "feminized voice" and Spivak's reading of Coetzee reading Defoe? Drawing on Jacques Derrida, Spivak articulates this problem via the philosopher, "philosophiz[ing] in the margins," where the margin remains peripheral, "wholly other" to the primary text, or rather, marginal.[46] This notion is what Spivak understands as "*marginal* in the *general* sense."[47] While turning focus toward the object of study, the studying subject positions herself in the periphery, and in this way minimizes her own beginnings, how she first came to encounter the object, and how she now draws conclusions that cannot in themselves be conclusive endings. In a more "narrow" sense, Spivak says "marginal" pertains actually to a broader notion of peripheral, to "victims of the best-known history of centralization: the emergence of the straight white Christian man of property as the ethical universal," to the ways in which Schaffer and Henry benefit from using the "feminized voice" as I identify it above, and how it benefits neither the woman recorded nor, presumably, the one (myself) who recognizes and articulates "centralization" in this context.[48] In other words, Spivak's critique is important because she does not benefit from occupying the marginal in general, since, like myself, her identity presupposes an always-already marginal role in proximity to canonized European forms.

DESIRING WOMAN, BECOMING OTHER

We can trace Deleuze's later notion of desire back to his reading of Proust, first in *Proust and Signs* from 1964, and later expanded in *Difference and Repetition* from 1968.[49] Proust's theory of "transposition," as Deleuze interprets it, results in constant and continuous exchanges between and among various identities. Identity as a contextually defined concept emerges through this continual transposing of signs, where definitions of concepts such as gender or sexuality, which rest on tensions strewn among many identities, also incite instability and perpetual *re*-definition. Proust's slippery signs leave identity and relationships among identities somewhat open, according to Deleuze, who employs Proust's example as a homosexual norm from which to extract a generalizable theory of the erotic not premised in an antipodal gender binary. The point here is not to show homosexuality as forming an alternative relationship to the dominant heterosexual norm but rather to admit that such substitutions or transpositions are an inevitable facet of loving in the space of a philosophy such as Deleuze's, within a constantly changing, temporally dependent, perspectively determined reality. But who is Deleuze to make such determinations of desire?

In their monumental *Thousand Plateaus*, Deleuze together with Guattari imagines a philosophy of becoming, wherein the world is an assemblage of continuous plateaus imminently unfolding. Deleuze and Guattari outline a number of possible realizations of this philosophy, from the abstract, "Becoming-Intense, Becoming-Animal, Becoming-Imperceptible," as is the title of their chapter detailing this philosophy, to more concrete or worldly instantiations of becoming-rat, becoming-dog, becoming-whale, becoming-child, becoming-woman, becoming-vegetable or -mineral, and becoming-minority.[50] Their point with these many becomings is to envision the imminent possibility of all things and the simultaneous interconnectivity among things.

In Jardine's summary, "D+G want to denaturalize Bodies of all kinds—and especially the 'human' one. To do that means denaturalizing sexuality and especially its polarized genders, a process the philosophers began already in *Anti-Oedipus* with the terms 'desiring machines.'"[51] While, in *Anti-Oedipus*, Deleuze and Guattari insist on the gendered dichotomy embodied by the psychoanalytic metaphor of the oedipal mother, in their later conceptions, they come to term this assemblage the Body without Organs (BwO), after Artaud's "war on the organs."[52] The BwO is a denaturalization of the human body; it is "what remains when you take everything away."[53] The "osmosis," as Jardine calls it, "maintains no identities, no images. For example, to be caught up in a 'becoming animal' means not that one will resemble either Man or the Animal, but, rather, that each will 'deterritorialize' the other."[54] And yet, the absence of identity—actually, simply the possibility of denuding identity, or in other words, the potential for *deterritorialization*—implies that there exists already a territory from which to take away.

Deleuze and Guattari aspire through disembodiment toward a gender-inclusive philosophy, writing that "all becomings are molecular [as opposed to molar], including the *becoming-woman*, it should also be said that all becomings begin and pass through the becoming-woman. It's the key to all other becomings." But, if always in a constant state of becoming, according to feminist philosophers, Irigaray, Jardine, and others, the BwO prioritizes the white and masculine as the state *from which* to become.[55] After all, there is no becoming-man or becoming-majority, for man is assumed. Thus becoming-woman has practically nothing to do "with women *per se*. . . . 'Woman' . . . is the closest to the category of 'Man' as majority and yet remains a distinct minority."[56] This is what Simone de Beauvoir meant when she wrote, "One is not born, but rather becomes, a woman," perhaps a phrase of inspiration for Deleuze and Guattari.[57]

When Woman is delimited by the properties of an ideal or when defined in relation to what she is not, the being that emerges is not of this world, though she is made real by collective imagination. Indeed, her existence, her image, her presence as the "protagonist"—which I examine further in chapter 5—emerges from a collective desire. But desire *for what?*

Were we to envision desire not as a Lacanian yearning for the unattainable or non-object, but as the longing for an effigy of the Woman, her laughter and cooing, alongside the fragmented "Tahitian" accompaniment, then, although the guise construed is of the individual's mind, such an appellation is not outside of music; it is neither removed nor reactionary. The guise is then necessarily shaped by the constraints of the music. Desire comes, therefore, *in excess* to the sounding "material," and not from a compulsion of something lacking. Having a space or void requires something to have been taken away in the first place, but with music there is nothing *a priori*. Sounds are material, and sex is composed through this material's fundamental inextractable matter.

In Lacan's framework, we yearn for the Woman we cannot quite grasp and for a tradition of music "Erotica" does not quite meet, simply because of its erotic designation of which it can never be stripped. Lacan's definition of desire rests on the "primacy of identity," to borrow the words of Deleuze, for whom "identity" is always linked to a foundation, a ground, in other words, to a hierarchical constitution.[58] Lacan would have us believe that the object of desire exists outside one's self, as a mirrored other (*autre*); the object, therefore, a projection of one's desires but irreconcilable and hence secondary to the primary existence of the perceiving subject. Lacan's understanding of desire as reactionary rests on an imposed duality of the given and the constructed, on the age-old distance imparted between nature and culture, between nature and man. The psychoanalytic model would have us believe that the subject, as given, is not in control of impulses and thought. Hence, what is desired by Lacan is always at odds with the subject. This is precisely the definition of the drive, a concept Lacan borrows from Freud, which always takes its aim toward a certain goal of being fulfilled.[59] When the goal is unmet, desires remain unfilled, always in lack. But this is not how Deleuze and Guattari envision desire.

Psychoanalysis has historically purported *jouissance* as the insatiable desire to fill a void, to voice silence. In contrast, Deleuze's later philosophy *a*-voids the psychoanalytic emphasis on lacks and veils. For philosopher Elizabeth Grosz, Deleuzian desire is productive and connective:

> Instead of aligning desire with fantasy and opposing it to the real, instead of seeing it as a yearning, desire is an actualization, a series of practices, bringing things together or separating them, making machines, making reality. Desire does not take for itself a particular object whose attainment it requires; rather, it aims at nothing above its own proliferation or self-expansion.[60]

Famously, however, Deleuze's rhizomatic desiring machine begins as if from nowhere. Perhaps it is always-already becoming determined, but determined by *whom*? Do we accept all experiences of desire equally or unequivocally (un-equi-vocal-ly, as in equal voice)? When Schaeffer and Henry say, "The lone man should find his symphony within himself," does this include the woman who rejects their notion of symphony because of its detrimentally constructed representation of women's (sexual) pleasure? From another angle, I, a queer person, may hear something alluring in this "feminized" voice, queering my hearing in a way that tangentially—marginally—relates to the composers' own targeted hearing. Another critic may even agree that this woman-loving-woman hearing falls within the bounds of the composers' intention, thus once more centering their "centralized" perspective.

One recognizes the beginnings of posthumanism in the philosophy of Deleuze and his joint collaborations with Guattari, where the posthuman is typified by a move away from the mere human and progressively toward something beyond the individually subjective. Thus, momentous acts of becoming invigorate man toward new outcomes—new lines of flight. And yet, in Deleuze, Defoe, and Schaeffer/Henry, man's construction of reality, of which women and other minorities were a part, weds the marginalized minority to the masculine majority in ways that prevent minorities from ever gaining independent agency. Within the progressive posthuman outlook, which I pick up again in the introduction to part II, white cis men from Europe and North America retain a primacy of identity that always-peripheral minorities could never hope to overcome. But the search for the organic seed to deliver us from ever plummeting into the spiraling rabbit hole of marginalization creates its own fictitious paths. Indeed, the virtual goes both ways, it extends both to the future and the past.

So, what of the real woman whose voice we hear in "Erotica"? Is her true identity now irrelevant, since we are invited—any one of us—to hear even ourselves in the role?

In many ways, the case of Schaeffer and Henry's "Erotica" is representative of all the electrosexual music I've come across. Electrosexual music—like the genre of more general "electroacoustic" music—frames listeners' expectations by playing on preexisting "extramusical associations." Charles Darwin, with his sound studies hat on, recognized a sound's source or cause as an evolutionary imperative. Historian of electroacoustic music Simon Emmerson summarizes Darwin, writing, "Not knowing the cause of a frightening sound is (in the first instance) a severe disadvantage if we feel it poses a threat, though one substantially reduced in the 'cultural' confines of a concert hall or in private listening."[61] Sound helps situate our surroundings, to remain grounded and *emplaced* in an environment rather than floating aimlessly in space. Our surroundings (context) shape how we hear and interpret sounds (content). Jardine's search for the origins of man's drives parallels Kane's admittedly futile search for the source and cause of acousmatic effects.[62] That is, we may never be able to find the origins of the sounds, of becomings, and yet, bracketing out such origins is neither neutral nor singular. Context is important for how we interpret content and vice versa. By ignoring gender, sex, and race—indivisible from and integral to the previous two—critics reinscribe and reinforce the premise that composers were themselves not concerned with aspects of identity.

Where electronic music's origins cannot be singularly traced to a white, European male lineage—because, in fact, electronic music was not "invented" by white male Europeans—as will become clear, the way electroacoustic music is celebrated (by a lineage of composers), created (with singular attribution), and categorized (as "academic" in nature), conflates any actual origins of this music's nascence by redrawing historical lines according to a false, neatly aligned trajectory. Rather than diffuse this mythical lineage, a task I perform in later chapters, the next chapters reemphasize this canonized trajectory, pointing to celebrated composers and their most regarded compositions as a way of illuminating the habitual erasure of women's instrumental(ized) role in this music's creation. My return to Schaffer and Henry emphasizes the canonicity of their work to revise and disrupt the predominant narrative coursing through electroacoustic historiography, one that labels electroacoustic practice as disembodied and sexually disinterested. My analysis performs a similar disruption and revision to the way Shanté Paradigm Smalls disentangles a presumed heteronormative masculinity in the Sugarhill Gang's monumental classic "Rapper's Delight" (1979), with the aim of illuminating embodied and cultural ways of understanding music that are already accepted as foundational in a genre, whether electroacoustic or hip hop (more on this in later chapters).[63] My hearing, like Smalls's, finds sex an apparent and unapologetic subject in music the canon already celebrates and in this way makes room for sexuality studies in existing historical and aesthetic framings of the genre. In so doing, there are ways in which this analysis cannot escape reinscribing the centrality of white men's notions of sexuality, since to deny this threatens to misrepresent this work by moving too far adrift of its situated, contextually dependent history.

2 ANNEA LOCKWOOD'S *TIGER BALM* (1970)

Annea Lockwood (b. 1939, Christchurch, New Zealand) is best known for her work with nature. Her electronic documentation of the world's rivers, various works sharing the title *River Archives*, was inspired by her childhood trips to the Waimakariri River.[1] She trained in electronic music, first in Holland in 1963 with Gottfried Michael Koenig (b. 1926) at Bilthoven Electronic Music Studio, then in London in Peter Zinovieff's Putney Studio, which later became EMS,[2] as well as studying psychoacoustics at the University of Southampton from 1969 to 1972. In 1973, she moved to New York, where she formed close personal and working connections with two other composers, Pauline Oliveros and Ruth Anderson. As Elizabeth Hinkle-Turner observes: "Additionally all three of these women have been relatively open about their lesbianism and have helped to foster a comfortable creative atmosphere for women composers in an academic setting regardless of their sexuality" (I return to this in chapter 8).[3]

Lockwood's interest in sounds of and in nature led her to early provocations with *Piano Transplants* (1969–1972), four installations ranging in concept from relatively short-term performances to less probable even "impossible" projects.[4] Around this time, in 1966 or 1967, Lockwood started talking about collaborating with choreographer Richard Alston, and from 1972 their talks bore a number of projects: *Heat* (1966–1967), *Tiger Balm* (composed in 1970, choreographed by Alston in 1972), *Windhover* (1972), *Headlong* (1973), *Lay-Out* (1973), and *Soft Verges/Hard Shoulder* (1974), all performed at The Place by Alston's Strider ensemble, the first independent dance group to emerge from the London School of Contemporary Dance.

Heat, Lockwood and Alston's first collaboration, involved experimenting with fire and informed her fascination with the sound of fire evident in later works like the *Piano Burning* (1968) installation involving an upright piano set on fire and played until it burns up. The composer recalls these associations in an interview with Rebecca Lentjes:

> An English choreographer, Richard Alston, and I were thinking of cooking up a dance work called "Heat," in which we would warm up the auditorium or performance space. I needed to record fire, so I had experimented with my fireplace, and with a bonfire in a courtyard, but none of it quite worked. It didn't sound hot enough. Then it occurred to me that there was a piano graveyard in London where the garbage disposal people would dump decrepit pianos that people didn't want anymore. And I thought, "Well, let's burn a piano." So we wrapped a mic cable in asbestos, set the mic in a reel-to-reel portable tape recorder, and set the piano on fire with a twist of paper sprinkled with lighter fluid so it would start slowly.[5]

Regarding *Piano Burning*, Lockwood recalled that she could usually play the blazing piano for about forty minutes before it got too hot, "but throughout the piano's destruction audience members could hear the sound of strings breaking, wood burning, and microphones melting."[6] The thrill of such spontaneous risks resultantly infuses Lockwood's music with an apparent variety and sonic richness that seems to have compelled Alston to work differently with her than any composer before. According to dance historian Angela Kane, before Lockwood, Alston's choreography usually juxtaposed various musical styles through borrowing, even electronic sampling, and overlay of multiple works. Lockwood was one of the few composers with whom Alston collaborated as a sole contributor.[7] This focused collaboration becomes significant when seeking to understand what choreography could have added to Lockwood's *Tiger Balm*, a work in which the composer herself vocalizes in sexualized moaning while (presumably) engaged in an autoerotic encounter.

TIGER BALM (1970)

In *Tiger Balm*, loud, repetitive, low-frequency purring conjures a cat to mind almost immediately. Listeners familiar with a cat's reasons for purring can intuit its pleasure.

After three minutes of purring, an ostinato in gongs enters as an acoustic backdrop soon emphasized rhythmically by a now syncopated groove against the purring. Before long, a twanging jaw harp enters to propel this groove. About a third of the way through the twenty-minute piece, a woman begins to moan in the same rhythmic cadence as the purring. Lockwood explains, "The sounds [in *Tiger Balm*] flow in a transformational process, often merging and emerging based on shared characteristics which are evocative of the tiger's presence—cat, mouth harp, tiger, woman, plane; all variants of the same sonic energy."[8] Our ears are drawn to more than just a medium: the purring cat (rumored to belong to visual artist Carolee Schneemann) and moaning woman are intertwined like Schaeffer and Henry's Tahitian percussion and looping laughter, drawing the streams, both sonic and sensual, seamlessly together.

The breathing of a living animal *is* for all intents evidence of that animal. That the purring cat is nowhere in sight does not contradict this sound's identity as a meaningful referent. Lockwood says as much about her pieces in an interview with composer John Young:

> One of the things that, to me, is seductive about sounds is that this very immaterial substance conjures up "a cat"—a cat's purring conjures up "a cat"—with a real kind of concrete presence. When the cat coughs in *Tiger Balm*, everyone responds—I think it's the concreteness of the feeling "here is a cat" that people are responding to, so that link between the sound and its source is very much a part of the piece on one level—as it is in many of my pieces. I think 'Musique Concrète' is a very nice term—a very real term! But quite apart from that, if I'm thinking of myself as working with phenomena then I like to work with the real phenomenon *itself*.[9]

The ambiguity left open by extrinsic associations is also what spurs on the intrinsic associations among sounds within the piece.[10]

That one associates a real (but not actual) cat with the sound of its voice does little to explain how the cat's purring relates to the woman's breathing in *Tiger Balm*. This relation depends on how we "bracket" the streams, in phenomenological terms, or, to use music theorist Christopher Hasty's term, how we "segment" the piece.[11] "Musical streams," drawing on Stephen McAdams and Albert Bregman, refers to "how the auditory system determines whether a sequence of acoustic events results from one, or more than one, 'source.'"[12] Rather than segment the woman and cat into separate "streams," we can bracket in both cat and woman to imagine them sounding on the same plane. According to Dora Hanninen, a music analyst's associations are what define segments and orientations—the way listeners orient themselves. These associations are therefore important for understanding "the motivation or rationale for particular segments and segmentations" in music analysis.[13] The cat's cough (0:38, 1:35, 3:09) in *Tiger Balm* is notable particularly because of what happens later in the piece. The regularly lilting purr stumbles momentarily as the force of the cat's breathing increases, causing it to emit a glottal vocalization, an involuntary alimentation much as Schaeffer's laughing woman in "Erotica." Caught in the cat's throat, the "cough" Lockwood mentions surely arouses empathy in human listeners (and additionally, my cat's curiosity was also piqued by the purring of my computer speakers). Young elaborates in his analysis of the piece that at the moment of the cough, the cat's purr moves from "static sign" to something more; the cat's cough not only initiates the game but triggers with it a series of possible plays.[14]

When the woman joins the purring tiger (10:49), she too elicits the rhythmic cadence of breath, inhale–exhale, weak–strong, an inverse to the Western classical impulse of strong–weak in duple meters, but that breath as a musical stream nevertheless relies on the perceived alternation of strong and weak beats, like so much of music (I pick this thread up in chapter 5).

A struggle ensues as the woman's formerly anacrustic inhale rushes ahead of the cat's dominant exhale, confronting, challenging, and syncopating that dominance as the breathing drives intensively toward an impending convergence (13:00–13:15). To an attentive listener, a new improvisatory stream joins the dancing breath of cat and woman, the breathing listener.

This would be Lockwood's ideal performance of the work: "I really hope that people are going to allow themselves to be enveloped by the sound and taken over by it—have it, in a sense, streamed through their bodies and be absorbed by it."[15] *Tiger Balm* is an experiment in audience participation, suggests Lockwood. "I was concerned with how our bodies respond to sound, and with the concept of sound as a primal energy and nutrient."[16] Composer–performer–listener entwine.

LOCKWOOD'S VOICE

Whereas eroticism is generally thought of as a subjectively arousing feeling, composers of electrosexual music intend their art to be consumed *collectively* as a common sensory experience, whether in the concert hall or merely in an imagined collective. Though music may be conceived with an "ideal type" of listener in mind—one who listens in a certain way or who is attentive to certain features—a composer does not usually write with a single individual in mind.[17] The collective attitude pertains to the contract between and among listeners and composers, a contract very much like the "rules" Don Ihde identifies as required for two individuals to engage in an argument: there must be some common premise from which the two proceed.[18] The rules of the game are continuously redefined on the basis of an intentional common ground enacted even without any explicit verbal agreement, but "Fortunately, in a very general sense, the phenomenologist can rely on a certain latent 'phenomenological' ability on the part of others. . . . Thus I can rely preliminarily on the other to have such and such experience and on the other to be able to detect whether such and such may or may not be the case."[19] For the electroacoustic game to work, part of the composer's job in *the erotic contract* is to predict which musical parameters will encourage particular reactions from listeners and to implement these parameters within some structural formulation to trigger the listener's imagination toward particular imagery.

Lockwood ensures such real-world associations by attempting to maintain something of the natural environment of the sounds she captures. As she explains to her interviewer, DJ and electroacoustic theorist, Tara Rodgers:

TR There are these very different trajectories in the history of working with sound—on one hand, there's the spirit of musique concrete, with composers like Pierre Schaeffer who would catalogue and classify sounds and their properties so extensively. And then there is this other way of working, which you describe which involves discovering as many sounds as you can, but mostly accepting them and leaving them as they are.

AL Or not needing to *fix* them. 'Cause I think they're essentially not fixable. But in their natural state, *sounds in their natural state*—that's a concept I sort of like—are not fixable, are they?[20]

One could interpret Lockwood's perceived "natural state" as reliant on the hive of collectivity through which the electroacoustic soundscape is construed. Sound, as heard in electroacoustic music, is never mere duplication; magnetic tape skews the sonic image captured by the microphone, and the loudspeaker shapes that sound according to the spatial dimensions of the place in which it is transmitted and broadcast. Like the psychotherapist who diagnoses the maladies of the ecstatic patient, the electroacoustic composer isolates the uncontrollable "abnormalities" of sounds from their natural environments—yet, what is construed as *abnormal* is certainly *contextual*.[21]

Electroacoustic composers seem particularly concerned with the ideological implications of sounding contexts, as they are constantly preoccupied with the relation between a work's intrinsic details and its extrinsic significance. Although Lockwood was probably not thinking directly of *Tiger Balm* in the above discussion of "sounds in their natural state," and more likely referring to her life's work of documenting sound maps of the world's rivers, her statement provokes an interesting question for this book.[22] Let's, for a moment, imagine that Lockwood *is* speaking of *Tiger Balm*. What, we might ask ourselves, is the "natural" soundscape of this erotic circumstance? The question of a constructed soundscape of sex points importantly to certain limitations of the ideological thrust of the soundscape movement. For soundscape enthusiasts, "the" environment, or "natural state" refers only to certain conceptions of environment and only to some types of engagement with "natural" contexts. What is intriguing about the works discussed so far in this book—Schaeffer and Henry's

"Erotica" and Lockwood's *Tiger Balm*—is each composer's ambition to normalize the erotic soundscape, to capture an erotic essence as a completely legitimate part of the audible human environment. And certainly, the normalization of exoticized sex does not go unnoticed—including the Tahitian in "Erotica" and Lockwood's reference to the Chinese herbal remedy "tiger balm," presumably employed favorably in the eroticized encounter between woman and pussy cat. Positioning Lockwood in the ultimate authorial role, Young hypothesized that Lockwood recorded herself masturbating, maybe with the aid of the Chinese rub.[23] *Tiger Balm* is an artistic project that invites such conjecture, whether true or merely a disturbing hiccup of an inseparable and inevitable "voice-body," a lingering phantasm of the source.[24] Incidentally, Young's inference supplies an ideological connection between Lockwood's erotic work and its historical precedent in Schaeffer and Henry's *Symphonie pour un homme seul*, which, as its title suggests, is intended to be performed by a single individual whether by a composer or by Schaeffer's stipulated "lone man," should *he* be able to find a symphony within *himself*.

The common setup for these performances, with Schaeffer surrounded by a slew of phonographs tailored for his particular rendition, ensures that the emergent erotic soundscape is always-already stipulated by the person at the controls (in this case the controlling composer). Such is the composer's desire to participate, and this participation is never passive. With the composer at the desk, the gaps imposed by the uncertainty of a sound's source or cause become filled with the composer's expressive intent—whether these intentions are met by listeners or not. But narcissistic exhibitionism is not the sole purpose for creating the experience. Another purpose is to provide a place for audiences to submerge themselves in the music jointly.

Decades after composing *Tiger Balm* (possibly following Young's presumptions), Lockwood confessed publicly: "[*Tiger Balm*] is part of my blending with myself as a woman—making love with one's essence, merging with it."[25] One need not necessarily fulfill—that is, remain faithful to—what the composer intends listeners to identify as source and cause to arrive at a composer's desired effect, the point is rather to engage collectively with the music, each on their own terms. We do not know if the voice is indeed

Lockwood's, but if it is, it is to her credit that she owns and names the performance. If the voice is not hers, perhaps such false attribution typifies the electrosexual lineage.

Music theorist Edward T. Cone hypothesized that composers want listeners to give them credit by hearing the composer's "voice" in their music even when no vocalizations sound (like the absence of a man's sounding voice in "Erotica"). In her critique of Cone's theory, Marion Guck paraphrases him, writing: "Together, the personae of a vocal work or the agents of an instrumental work comprise the complete *musical persona*." She later quotes directly from Cone, "all roles are aspects of one controlling persona, which is in turn the projection of one creative human consciousness— that of the composer."[26] Yet, Guck argues Cone places too much weight on the composer's authoritative role in the interpretative hierarchy, and I believe Lockwood would sympathize. As I read it, Lockwood's suspicion of *musique concrète* in the interview with Rodgers above points directly to electroacoustic music's presumed elevation of the composer's "voice," as if, in using her own voice, Lockwood's creative authority over *Tiger Balm* is assumed in irony. Looking to the work's performance provides yet another layer of interpretive flexibility.

As mentioned, Alston took up choreographing *Tiger Balm* in 1972 with six dancers of his Strider company. Prior to his collaborations with Lockwood, he had a reputation as an experimental interdisciplinary artist. In an interview with Stephanie Jordan in the context of "British Modern Dance," Alston claims he aspired toward "radical juxtapositions of the new sensibility," as he is recognized for "his enthusiasm to overturn convention, to make an artistic virtue of plunging into the unknown and attempting the extraordinary for its own sake."[27] It seems likely that the incongruities of "radical juxtapositions" between the purring cat/tiger and Lockwood's sexual expression fit within this aesthetic, though it is unclear in terms of influence whether Lockwood's work appealed to him because it exhibited these criteria outright or whether he discovered such elements in her music and brought them out through choreography. Significantly, though Lockwood's is the only human voice we hear, Alston's troupe reimagines the scene, not as a singular autoerotic experience but as a group ritual.

Andrew Porter, a writer for the "Musical Events" column that ran in the *New Yorker* from 1972 to 1992, describes the performance at the Place Theatre on August 14, 1972 (Ross 2015):

> The first section of the score, the snarling and snoring, is laid out spaciously, first with "magic pictures," glowing and fading, or a ritual group, five white-robed figures, who then perform a slow ritual dance. The central episode, the vocalised orgasm, is done by Christopher Banner naked; a dance of tensions, of energy posed and then released in a body stripped so that the play of forces can register directly. In a quiet close which continues in silence, after the music is done, he is robed too and joins the patterns of the others.[28]

Funnily enough, the review makes no mention of any incongruity between the quality of Lockwood's audible feminized vocal orgasm and Banner's nude male form. It even appears from many reviews that audience members were focused solely on the shocking sight of a naked man, and, outside of ascribing its orgasmic flight, completely ignored the feminized qualities of the voice altogether. That is, the sexual quality is given primacy and heard more readily than the gendered aspects resonating through the voice. It may be that by this time audiences were quite desensitized to the sounds of women's orgasmic moaning, but naked men on the other hand still held some shock value, as one headline boasted: "Audience roars for more as man dances in the nude." Still another describes the event as a "Nude dancing . . . roar'n'snore bore."[29] Knowing this, and positioning this supposition alongside Lockwood's invitation for listeners to absorb the sounds "through their bodies," how might we understand Alston's choreography?[30] Does Banner dance in collaboration with Lockwood's own performance? Or does he somehow embody this feminized voice, transgressing gendered norms which assign feminized voices to feminized bodies? Is Lockwood streaming an authorial compositional "voice" via the dancer or does she expressively penetrate and intertwine with his performance? And how do such questions allow us to reframe the patrilineal history her later work seeks to disrupt?

Certainly, one *could* listen to each of the pieces featured in part I of this book with the passive acceptance of an assumed compositional intentionality, an intention that requires listeners to get *turned on* by a composer's self-proclaimed sexy music—that is, if the composer tells us how to listen, and she says her music is sexy, then a "good" listener will get turned on. We might also assume that accepting this fate is to (over-)appreciate the composer's "voice" in terms of its mastery over the feminized voice—as a woman's voice or as an ambiguously gendered voice produced with the intention of taking on token feminized traits, such as the erotic, exotic, submissive, and so on—manipulated by composers as they see fit.

On the other hand, despite the desires of even the most dictatorial composer, they ultimately hold very little claim over the audience and therefore cannot predict or, worse, dictate the ways in which listeners hear and interpret their music. One crux on which this debate hinges, as Guck points out, is that rather than an audible persona, music can convey audible dramatizing *acts*, which I read to include also *sex acts*.[31] That is, musical acts cannot be reduced merely to an *actor*, especially in electroacoustic music which, on account of its interpellation of concrete and synthesized sounds, typically has no identifiable performing agent.[32] Such interpretive flexibility is precisely what allows Alston to imaginatively choreograph the work. The term "interpellation" is commonly borrowed from Louis Althusser in hip-hop theory, as a way of explaining how meaning is co-constructed by artists and listeners via knowledge inherited or articulated through the electronic interpolation of samples. Interpellation, in Althusser's designation, is ideologically constructed by the subject, whereas hip-hop scholar J. Griffith Rollefson recognizes colonizing imbalances of power and appropriation owing to the kinds of associations listeners bring to the experience.[33] (More on this later.) That is, though the original source of a record may have been captured by way of one person's ideology, later samples or choreographed and staged realizations of that record might monopolize or alter the original intent.

This more open-ended hearing acknowledges the weight of a listener's inference, whether listeners simply take joy in the music or whether they then manipulate the music toward their own creative (or erotic) ends. In this hearing, we can imagine championing the vocalists, who are primarily women— real and imagined, simply because of reflexive associations between orgasmic vocalizations and the feminine—and whose contributions to these works are minimized in a prioritization of *the* composer's "voice." For Schaeffer and

Henry, this means elevating the composers who script the voice in certain ways, but in Lockwood's case power appears to be legible under a different subtext that allows listeners to attribute sexual performance to her and simultaneously ignore her voice in the choreographed performance. Indeed, when, in a public lecture, I offered the supposition that the voice in "Erotica" could belong to the composers, the suggestion was met with laughter and taken as a joke. I heard no such response when presenting Young's interpretation of *Tiger Balm*. Why the difference? Is it because in "Erotica" the voice sounds feminized and therefore could not possibly align with the men who composed the movement, or is it because the composers themselves—Schaeffer/Henry and Lockwood, respectively—are presumed to embody different sexual potential, whether sonic or not, on the grounds of their gendered identities?

Either way, what remains evident of women is much *more* than their bodies. Their literal and metaphorical voices remain a persistent and uncompromising presence, and I cannot emphasize enough that without these women vocalizing, electrosexual music would not exist, nor could it exist. Women's voices are a necessary staple of this music. Men's voices, whether Schaeffer and Henry's metaphorical "voice" or the implicit lover's, if and when they appear, sound only *in response to* women's—as ornate complements to the primary canonized facet of this music.

3 LUC FERRARI'S *LES DANSES ORGANIQUES* (1973) AND *PRESQUE RIEN AVEC FILLES* (1989)

In 1958, Pierre Schaeffer and Pierre Henry arrived at a great disagreement based on artistic (and personal) differences. After many momentous collaborations, including the several-year-long experiments collectively titled *Symphonie pour un homme seul*, the two went their separate ways. Henry took with him many friends, composers, and researchers, from the *Groupe de Recherches de Musique Concrète* (GRMC) and Schaeffer was forced to repopulate the studio anew.[1] That year he attended a concert of original works by Luc Ferrari (1929–2005) with the composer at the piano, and Schaeffer, thoroughly impressed, invited Ferrari to join his newly titled *Groupe de Recherches Musicales* (GRM).[2] In 1962, Ferrari became director of this group, at a time when composer Beatriz Ferreyra, singer Simone Rist, engineer and composer Guy Reibel, and physicist Enrico Chiarucci were working to test Schaeffer's concept of electroacoustic ear training by designing a *solfège* for electroacoustic music. This research eventually culminated in Schaeffer's *Traité des objets musicaux* (1966).[3]

While Ferreyra and Schaeffer moved toward the goal of some generalizable typology of musical parameters, Ferrari did not share this vision. Setting himself apart from Schaeffer's research-oriented practice, Ferrari aspired to engage with the real world around him, with sounds as they occur in their environment. Ferrari called this practice "anecdotal music."[4] Eric Drott explains,

> Unlike music that derives its meaning from the play of abstract forms, anecdotal music has the advantage of not requiring any specialized knowledge of musical

syntax or style to be deciphered. And insofar as anecdotal music fashions messages out of the quasi-universal code of everyday sonic experience, it is within the grasp of any potential listener, from the most naive to the most educated. Ferrari thus describes his anecdotal works as "an attempt to find a music that is at the same time simple and unfamiliar, and thereby suitable for mass dissemination."[5]

Ferrari's "anecdotal music" aimed to capture and record sonic environments in all their peculiarity. Aware as he was of the microphone's immanently mediating presence, he aimed for the recordings to stand on their own with minimal interference from the hand wielding the microphone.

Ferrari's project conflicted with Schaeffer's ideology in its basic premise. Schaeffer's concern for devising a specialized approach to sound—a *solfège*—did not match up with Ferrari's altruism to encourage listeners from all walks to record their own electroacoustic music. Schaeffer's approach required a guide with certain skills to cultivate listeners. Ferrari sought to relieve the composer of his lofty status, aligning himself conceptually (but not musically) with John Cage,[6] showing a passive Cagean attitude in works like *Presque rien no. 1, ou Le lever du jour au bord de la Mer* (1967–1970).

Over the course of several months, living in a seaside town on the Croatian coast, Ferrari would place a microphone outside his window to record between three and six in the morning every day, capturing the passing fishermen, the sounds of the coast, and the world just awakening. These recordings were then condensed, with supposedly minimal interference from the composer, into a unified scene or environment. Ferrari's series of four works with the title *Presque rien* are mediated, according to the composer, mostly by the microphone, not the composer's hand.[7] The works present musical materials without stipulating their aesthetic ordering, not unlike Cage's chance scores (for example *Music for Carillon no. 1* from 1952), which leave the organization of graphic segments to the hands of the performer. According to Ferrari, *Presque rien no. 1* thus anticipates the electroacoustic practice that became known under R. Murray Schafer as soundscape composition (though in reality these philosophies are not as similar as Ferrari purports, because Schafer's works and those of his followers are usually subject to quite a bit of after-production).[8]

We might better understand Ferrari's compositional approach through Deleuze and Guattari's philosophy of art, which recognizes that an artwork's greatest value is in its reception: "The artist creates blocs of percepts and affects, but the only law of creation is that the compound must stand up on its own. The artist's greatest difficulty is to make it *stand up on its own*."[9] The work, as Deleuze and Guattari conceive of it, is preserved in the sensation conjured by the material of the work. "We paint, sculpt, compose, and write with sensations. We paint, sculpt, compose, and write sensations."[10] Another offshoot of philosophies that promote a work's success independently from its creator emerges from the study of material semiotics in the humanities, especially in sound studies. This thrust attributes agency to sounds themselves to underdetermine the sensations of whimsical observers.

Music Theorist Brian Kane's book *Sound Unseen* follows this trend, arguing that what truly sets apart electroacoustic music from other genres is sound's mediation via the capturing medium, "the recorded character of the recording," a perspective akin to Drott's attribution above that Ferrari believed there could be a "quasi-universal code of everyday sonic experience."[11] In this chapter I examine Kane's analysis, which centers on Ferrari's *Presque rien no. 1*, from three perspectives: (1) the human voice as characteristic sound with its own (exclusive) meaning; (2) the medium of delivery; and (3) "hearing presence," the twofold presence of one who hears as well as any presence the hearer perceives in music. I challenge generalizations drawn from Ferrari's first *Presque rien* with comparisons to his electrosexual works, *Hétérozygote* (1963–1964), *Unheimlich Schön* (1971), *Les danses organiques* (1973), and finally *Presque rien avec filles* (1989), to show that although Ferrari's works could be collected indiscriminately by bracketing out sexuality as merely attending to the recorded medium as such, Ferrari's attention to "intimate" music-making very much relied on listener inference. His compositional philosophy raises ethical suspicions regarding consent and sexual harm to such an extent that theorists and historians ought to consider suspect ethics even in works where the composer does not state outright sexual intentions.

1 THE VOICE

Unlike the oboe or piano, which are usually heard in a musical context, the voice sounds in many contexts. The resonating voice of an invisible source and cause has been described in psychoanalysis, particularly by Freud, Lacan, and recently Mladen Dolar, by its allure because of the desire to disclose the body it seems to conjure. And, as explored by Roland Barthes in his famous essay "The Grain of the Voice," it is precisely one's sounding voice, distinct from the language it sometimes speaks, that marks the individual.[12] Philosopher Adriana Cavarero has explained that the particularity of the voice as signifier is unique to its identity as *sound*. She turns to the example of Italo Calvino's story "A King Listens," which tells of a king who falls in love with the voice of a woman he has never met. Calvino describes the King's fascination as follows:

> What attracts him is "the pleasure this voice puts into existing: into existing as a voice; but this pleasure leads you to imagine how this person might be different from every other person, as the voice is different." In short, the king discovers the uniqueness of each human being, as it gets manifested in the uniqueness of the voice.[13]

In Cavarero's reading, the king accepts the voice as such, not as an indication of its speaker but as an object and end unto itself. "No appearance of a face corresponds to the phonic emission. Sight does not even have the role of anticipating or confirming the uniqueness captured by the ear.[14]" When it comes to the voice as evidence of human presence, Cavarero argues, sound trumps vision.[15]

Yet, where philosophy traditionally jumps straight for sound as signifier or to its lexical "articulation," both Kane and Cavarero advocate for separating two facets of the voice—its sounding quality and its function. Like any musical instrument, the voice is beholden to the "institution" of music, but its resonances are also entangled in "extra-musical" significance—its socialized identity—namely aspects pertaining to the source and cause of the sound, that is, the vocalizing body, and whether this body belongs to a real-life speaker, or one imagined by the listener.[16]

2 MEDIUM OF DELIVERY

Absent visual evidence, Calvino's king is perplexed. He cannot distinguish whether the voice is a figment of his imagination or if it comes from else-where. The illusory voice incites as much allure as paranoia. Cavarero's analysis insists on perception to argue for the voice's identity as convened aurally as its most spectacular feature. Kane, conversely, dismisses listener perception, since the King cannot even distinguish between reality and his fictive imagi-nation. Instead, Kane turns his focus to the sounding medium. According to Kane, it "makes no difference whether the voice is real, remembered, or imagined," because it will incite the same response from the king, but in each case the source and medium differ.[17] Only once one determines this source, can they determine the sound's effects: the identity, function, and meaning of the voice.

Kane turns to Ferrari's anecdotal music to evince this separation between a sound, its source, and its reception. He explains that even when a composer aspires to convey a complete scene or acousmatic soundscape, the fact that the sound is recorded makes audible the processing of the sounds, which, in turn, causes the electroacoustic *medium* to also become audible. For Kane, Fer-rari's anecdotal music—and possibly even the entire electroacoustic project—fails at realism because it always sounds mediated. The electrified acousmatic environment is not faithful to the *real*, since the manner of *presentation*—the recording—takes precedence over any one listener's experience. Far from a *re*-presentation, the anecdotal soundscape is only a *vestige* of the original environment. Roger Scruton might argue that it is precisely this inevitable mediation that enforces musical representation, because, as he writes, "Rep-resentation requires a medium, and is understood only when the distinction between subject and medium has been recognized."[18] But, unlike Kane, Scru-ton takes for granted that a listener is responsible for instituting this distinc-tion. Kane's Ferrari dispels the myth of the great composer, and, since anyone can prepare music this way, the observer isn't remarkable; there is no such presence. At least none worthy of investigation.

Focusing on the medium as he does, Kane's Ferrari avoids any discus-sion of whether sounds might gain contextual significance in relation to other sounds, because his analysis does not prioritize different sounding qualities. By undermining the listener in favor of the medium of presentation, Ferrari ultimately equates the voice with musical instruments—with *any* sounds for that matter—"but not in any *way* whatsoever."[19] Hence the medium becomes the central factor in determining a sound's presence. All recordings retain a trace of the medium of recording.

At this point, it becomes tricky to discern whether Kane is reading Ferrari or advancing his own electroacoustic philosophy. Because of what Kane terms (after Derrida) the "spacing" of sounds from source and cause, which occurs due to mediation—on account of having been recorded—sound loses all semblance of identity, including gender, sexuality or race, which are notably absent in this discussion. He privileges mediums in a way that transcends even the work of performers and composers, the recording environment, and, perhaps, most strikingly, the listeners who hear music (*as music*)—*including himself*. It's as if Kane accepts the uniqueness of the voice but would rather bracket out the listener from assigning this status. We therefore have a paradox: if sound is of special significance, and the voice is a notable example of this significance, for *whom* is it significant?

In bracketing out human presence from his analysis, Kane aligns himself with the recent material turn in the humanities, with a movement that prioritizes the independent ontology of objects within the field of sound studies.[20] He unravels the chain of significance by pointing our attention to the distinctiveness of the veiling medium. The veil is of course of central acousmatic concern. However, in reducing an electroacoustic work to its medium, privileging that medium over sound's physical properties, Kane erases any distinction between original and copy, which are both in and of the same medium. But in emphasizing this static result, Kane seemingly privileges all sound over any particular sound, collapsing the difference between one voice and another, for example, the girls whose voices Ferrari stole without their knowledge or consent to create his "intimate" anecdotal works. Certainly, understanding a recording as *music* recognizes the recording in its interpretation as music *with inherited structures and meanings*. The twenty-first century is hardly the time to reinforce semantic gradations that bracket out formalized organizational schemas at the expense of those that are socialized—and Ferrari would hardly agree to this framing.

3 HEARING PRESENCE

Kane's attention to the medium overlooks one very important aspect of the recording, and that is its *content*—what has been recorded. As soon as we accept the work as musical, we admit a performing and/or creative presence as well as listeners for whom there is perhaps no distinction between the real and the virtual, between hearing and *hearing as*. In the words of philosopher Anne Sauvagnargues: "Art is not imaginary doubling. . . . Art is reality itself."[21] The unique sounds of electroacoustic music, wherein new sounds are generally produced for each piece, provide a space in which to explore perceptual potentials that resonate (*re*-sonate) beyond the limits of tangible objects, and sometimes even beyond what makes sense. But this doesn't stop us from trying to make sense of what we hear, nevertheless. Otherwise, why worry about any medium at all? What purpose does mediation serve without a perceiving presence or receiving object (whether computer, amplifier, tree, water, or air)?

Steven Connor argues, in contrast to Kane, that the recorded voice always maintains its link to the body, that the voice can *never* be disembodied, that the sound of the voice always retains an irrevocable "'voice-body,' the body implied by or intuited from the voice."[22] Although the virtual medium as a veil of anonymity severs the voice from the original speaker, still some distinctions are to be made between voices once one is willing to admit questions of identity and, hence, representation. Indeed, one ignores such things at a great risk. Ignoring identifying criteria is the first step to removing race, gender, sexuality, and other factors from the discussion, subverting these onto the level of secondary or ancillary points of discussion. It then becomes very easy to decide that such points belong to a niche of scholarship rather than within the main philosophical debates.

Kane's analysis portrays Ferrari's *Presque rien no. 1* as nothing more than sound recording, that is, unremarkable replication. Against the composer's claims of his own creative detachment, Kane urges that the implicit hand of technology, forced by Ferrari, alters the soundscape to such an extent that it is no longer compatible with reality.

> In a soundscape recording, the listener relies on aural cues for the reconstruction of spatial relations, evaluating distances according to their volume, reverberation, and spatial attenuation. A well-mixed soundscape can give us the illusion of depth, and we will hear *through the recording* the intended spatiality: the distant water lapping, the closeness of footsteps on the floorboards, the passing of the motor, a singing voice reverberating off the hard surfaces of the street. In other words, we receive *an image*. *Presque rien* does not present this kind of soundscape. If one listens closely to the mix, the listener may notice that everything is pressed up to the surface and presented with nearly equal audibility and clarity. . . . When have you ever experienced an auditory environment in which motors, insects, and lapping waves are all *equally audible?* Ferrari's mixing resists a realistic reconstruction of the environment, effacing the difference between foreground and background.[23]

Putting aside the sheer limitations of the electronic compositional medium, where the nuances—the type of dimensional depth and overshadowing—he desires from Ferrari may not have been available to such an extent at the time this music was composed, Kane argues that because the presented environment is not faithful to the real experience, our ear is drawn to the manner of presentation—to the medium of recording—forcing us to recognize the piece (or the pieces of the piece) *as recorded*. From this Kane concludes that *Presque rien*, far from representation, retains only a *vestige* of the original environment, a mere "trace." The recording provides a hollow trace or replicable mold absent reality's unmistakable details.

Kane proposes that the "nothing" of Ferrari's *Presque rien* (almost nothing) pertains to its very essence as artwork, to the generalizable ontological status of art in the current collective consciousness. Summarizing Jean-Luc Nancy, Kane writes:

> The challenge is to think about the trace not as something that leads us back to the source or idea that produced it or would subsume it. Rather, it is to try and think of the trace *as a trace*, as surface, as being right there at the surface and opaquely present in all of its sensibility. This cluster of Nancian terms—vestige, exposure, surface, and trace—spurs us to think of the artwork

differently: not as the artwork intended to represent nothing, but as the artwork that has nothing to represent.[24]

As I read it, Kane's theory comes as a justification for what he recognizes as Ferrari's failed attempt at capturing reality. This resolve rids the music of its context completely, raising *Presque rien* to a representative example of "the recorded character of the recording," no more, no less.[25]

> Perhaps one could state the paradox like this: While trying to meet the transcendental condition of *recording whatever*, the recording is also stuck in the immanent condition of always being a recording of some particular thing.
>
> This is the paradox of *Presque rien*. It is *almost nothing* because it is *almost anything whatsoever*. And to be *almost anything whatsoever* means that while the recording records this particular morning at this particular seaside in this particular Croatian seaside town, it is also indifferent to this fact. It could be replaced by something else, by some other recording, but this replacement would still encounter the same paradox.[26]

Kane's point is that unlike Schaeffer's vision of *musique concrète*, which asks listeners to focus on sounds in themselves, Ferrari's anecdotal soundscapes demand that listeners disregard the aural details of the medium's sound. Ferrari's unrealistic rendering draws attention away from music.

Within its own paradox, Kane's analysis seems to privilege recording above the specificity of any one work. The medium overshadows the musical form, the work as monument (to recall Deleuze and Guattari), and any listener's perceptions. Yet, reducing the recording to nothing more than a recording reads like a resignation, a capitulation to objectivity. Perhaps it is true that Ferrari, like Schaeffer, did not refer to himself as a professional musician, as he identified with grassroots movements that sought to bridge the presumed cultural divide between the Parisian elite and the "deprived" provinces.[27] Instead, Ferrari identified with the concept of a *réalisateur* drawn more to "games" above organized musical works.[28] Still, he abided by nomenclature that would suggest independently identifiable works and himself as author, which is how we distinguish between the *Presque rien* of the seaside and his later work *Presque rien avec filles* (1989)—with girls. By ignoring the source or cause of sampled sounds (whether imagined or real), in focusing merely on the surface of the recording and the superficial medium of the object, Kane threatens to erase and thus silence the specific captured voices and their *musical identities*. Of course, such identities are not universal or static but rather imposed from outside by listeners and various listening constraints; it is precisely this variability that Kane wishes to abandon, maybe identifying too much with Ferrari's own claims about himself.

While I agree that musical realism is unique among the arts, especially as realism is conveyed within the electroacoustic tradition, I do not agree that the capturing medium is more apparent than or even separable from the captured sound and its *perception*. Indeed, the purpose of such a distinction remains unclear and raises further questions about what else Kane silently omits from his analysis to benefit his subject, Ferrari. In one reading, we might understand Kane as privileging Ferrari's approach to reality as unique; *Presque rien* is pressed up to the surface, just like any other electroacoustic work. But Kane ignores Ferrari's favorite artistic subject matter—sexual intimacy—and in so doing misses an opportunity to align Ferrari with a comparable contemporaneous creative practice: pornography. As I examine in later chapters, pornographic films tend to record the visible scenes in silence. Later, the actors (or other actors) go into the studio and perform sounds that can be dubbed onto what we see. Unlike Kane, who finds flatness a deterrent to realistic portrayals, pornography scholar Linda Williams determines that sacrificing spatialization in fact enhances the spectator's sexual fantasy: "In hearing the sounds of pleasure with greater clarity and from closer up, auditors of hard core sacrifice the ability to gauge the distances between bodies and their situation in space for a sense of connectedness with the sounds they hear."[29] Because of their close audible proximity to the action, porn viewers can more easily envision themselves in place of the performers at the site of sexual performance.

SURVEILLING PRESENCE

Electroacoustic music aligns itself with nonfiction and photography as an art that captures the semblance of actual things and also of actual people. The music calls for citation practices that have rarely been implemented.

Surely, the ethics of surveillance have been on the minds of those wielding sound recording equipment since the advent of telegraphones and the first wiretapping.[30] And we should not be so naive as to assume that recording, even in its beginnings, did not plunge into the murky waters of voyeurism.

Given Ferrari's affiliation with Schaeffer in the years before composing the first of four pieces with the title *Presque rien*, it would not be a stretch to assume that he was influenced by Schaeffer's compositions, though he claimed to avoid falling into the hunt for *concrète* objectification often attributed to Schaeffer. I would further like to draw out Schaeffer's influence, apparent already in Ferrari's earliest works precisely in their conceptual bond between sound source and recorded result, the very relation Schaeffer is said to have dispelled. Kane dismisses Schaeffer's theory of reduced listening for its "objective" approach, but emphasizes that, despite our ability to recognize the voice in the *Symphonie* as a voice, we are still unable to pin down a definitive identity for that sound's source and cause. Thus, listeners attuned both to Ferrari's soundscapes and to Schaeffer's concrete creations are trapped in ambiguity—in the spacing—between effect, cause, and source. But this space is nevertheless negotiable.

Ferrari acknowledges being influenced by *Symphonie* toward realizing an earlier tape piece, *Hétérozygote*, but says that this connection was not observed by Schaeffer, who was harshly critical of the piece. Like the "Erotica" movement of *Symphonie*, Ferrari's *Hétérozygote* features relatively few source recordings with "practically no manipulations."[31] What links this work with the later series of *Presque rien* works is the presence of unadulterated human voices speaking in identifiable languages, though the exact identities of the performers, not credited on the album, are veiled from the majority of the work's listeners. Unless one recognizes the voices, the performers of Ferrari's *Hétérozygote* or *Presque rien* or Schaeffer's "Erotica" movement remain anonymous. Significantly, Ferrari's self-proclaimed erotic pieces often feature women in a central role (as do most electroacoustic works featuring the voice).[32]

This trend also continues in Ferrari's later "intimate" electroacoustic works, *Unheimlich Schön*, a *Hörspiel* (radio drama) done at Südwestfunk (SWF) Baden-Baden for a "respirating young girl who thinks of something else," and the third piece of his anecdotal series, *Presque rien avec filles*, described thus by the composer:

> *Dans des paysages paradoxaux, un photographe ou un compositeur est caché, des jeunes filles sont là en une sorte de déjeûner sur l'herbe et lui donnent, sans la savoir, le spectacle de leur intimité.*

> In paradoxical landscapes, a photographer or a composer is hidden, young girls are there in a kind of *déjeûner* on the grass and give him, without they know (*sic*), the show of their intimacy.[33]

Ferrari's recorded girls do not know that they have been documented or captured by the recording, and we can suppose further that he never informs them. However naively one may want to read the "intimacy" in the composer's description of *Presque rien avec filles*, Ferrari's repeated emphasis in interviews and in publications on the sexual allure of recording as a form of veiled observation—*sur-veil(l)*-ance (with valance?)—maintains a hard edge.

If, as per Kane, the electroacoustic *Presque rien* is indifferent to its environmental source, then the recorded content would have little consequence for listeners, but I argue that sound *is* of consequence precisely because of the ethics of citation and surveillance. Not only does Kane's analysis undermine listeners, but he reclines on the tradition of abstract music, positing electroacoustic music as one mere inconsequential extension of instrumental music. But listeners do not hear *Presque rien avec filles* as a mere recording, and it is unethical to collapse qualitative distinctions between Ferrari's seaside recordings in his earlier *Presque rien* and his intimate surveillance of young girls.

Ferrari's electroacoustic music flirts with the boundary between source and representation, between documentation and creation. His many erotic electronic works build on his notion of "anecdotal music," of taking sounds from the world and resculpting them toward particular artistic aims—a familiar idea today, when we still lack appropriate ethical acknowledgment. As the composer explains, by recording a sound and then listening to it intently again and again in a studio, he has an opportunity to approach sound in his own intimate and sensual manner. By this attitude, Ferrari maintains some connection with his mentor Schaeffer: in supposing that the recording, once divorced from its originating environment, acquires anonymity, sound becomes sound as such with no previous ties to a source or cause. When confronted by his interviewer Brigitte Robindoré about recording a woman unbeknownst to her, Ferrari proclaimed that this sly manner of recording is simply part of the game.

Ferrari Take a woman encountered in a German market who plays with her voice when buying a kilogram of potatoes. Something absolutely extraordinary happens in the vocal sounds at that moment. How that woman buys the potatoes is a mysterious and profoundly human thing and a profoundly sensual thing, too. This observation opened up my preoccupation with sentiments. Naturally many find this approach to be so incongruous to music; they feel it is almost pornographic. Yet I discovered this. It is almost as if I were a musical psychoanalyst. How does the psychoanalyst hear his patient? Once I have recorded the woman buying the potatoes, I feel the same intimacy as the psychoanalyst in discussion with his patient.

Robindoré The difference being that the woman in the German market may not know that you are recording her, while the patient is usually conscious of his relationship to the psychoanalyst.

Ferrari There is another facet to the game here. Indeed, I preserve bits of intimacy, like stolen photographs. Naturally, she does not know, and it is just this aspect that makes it even more remarkable. I have captured something on tape, I bring it into my intimate world—my home studio—and I listen to her again. And here an extraordinary mystery is revealed. As ordinary as it appears on the surface, I am discovering this act in the studio as a blind person, as there are no more images. There is only the sound. And what happens in the voice at this moment? Something extraordinary. I learned this primarily when working with foreign languages. When I heard Algerians speaking of their life—I do not understand Arabic—I came back to my workspace and was faced with a language that left me without reference points. So I listened to the emotion; it is not so much the melody that I heard. With time, you can begin to feel what is being said, thus it is not possible to cut just anywhere in the conversation. Speaking is so intimate. It comes from the deepest part of us: from both the head and the sexual organs, from the heart and from all that we can imagine. Speaking is a place where everything comes together."[34]

This account confirms Ferrari's affinity for games. As biographer Jacqueline Caux notes, "[Ferrari] shows . . . a clear fondness for playfulness. For him, the act of making music is a game: a game of chance, an irreverent game, a serious and at times critical game, but more often a perverse game."[35] And, of course, in the context of erotic art, this perverse game is played at high stakes.

Elsewhere, Caux documents Ferrari's intentions when speaking of "intimacy." When asked of his increasing focus on the "intimate" beginning in the 1980s, Ferrari explains: "It starts with *Histoire du plaisir et de la désolation* (1981), which marks my first encounter with the symphonic orchestra. I wanted to explore the harmony of an orchestra while focusing on the intimate." Defiantly, Ferrari proclaims: "And what is more intimate than sex?"[36] Though orchestral, Ferrari acknowledges that his work on the "history of pleasure desolation" invokes intimacy through sexual allusion.

THE ELECTROACOUSTIC GAME

Acousmatic uncertainty is a ruse at the core of the electroacoustic impetus. Composers invite listeners to have their own associations, introducing familiar text or sounds while simultaneously toying with these convictions. Such is the "game" electroacoustic composers are fond of playing. The game arises at the hands of the composer, but the players can be anyone from listeners to performers, or, when left to chance, a composer may even pull the wool over their own eyes. Since sound is a particular effect (in the most technical terms, an impulse set in motion—thus sound*ing*), the game is dependent on the persisting ambiguity of the source(s) and cause(s) of that effect. The object of the game is to construct a soundscape from the implicit circumstances of these co-occurring effects (e.g., laughter, moaning, inhaling, exhaling, kissing, or any other sound). Sexual allure arises with similar mystery and uncertainty, and throughout this book many artists invoke both sex and humor in surprising ways. The sexual thrill of the pieces explored until now, *Symphonie* and *Tiger Balm*, rests primarily on the voice, the sounds vocalized and the voice-body's imagined environment (which the composer helps invoke).

Electroacoustic composer and theorist Simon Emmerson insists that even when heard musically as a complex of organized sounding relations, listeners still "analyze" electroacoustic music to assign meaning to sound. In the course of the electroacoustic game, listeners are constantly seeking to place sound within a context *beyond* what they hear.[37] Emmerson distinguishes a *game* from *narrative* in that the latter is concerned only with the manner in which a piece unfolds, while the former necessitates listeners and composer(s) negotiate relations reflexively, concurrently, and cumulatively. The game becomes a competition in which players, preoccupied with the actions of an opponent, take part and strategize to negotiate the piece—regardless of historical proximity,[38] whether or not the composer is there in the flesh. Although, the composer's contextual and historical situation may inform listeners of the musical "style" or idiom and thus help listeners interpret a work's structural organization, as Leonard Meyer says, "The constraints of a style are *learned* by composers and performers, critics and listeners."[39] In other words, the "gap" left open between effect, source, and cause in electroacoustic music is sculpted in part through a mutual agreement between composers and listeners, by way of their shared imagining.[40] To put it into the philosophical terms of John Searle, electroacoustic music is defined by way of "collective intentionality."

Intentionality is generally conceived as an attitude or an attention to something in the minds of individuals, while collective intentionality, as Searle points out, involves *cooperation* among individuals. Searle's "collective intentionality" therefore, involves two (or more) players *believing* that they share a collective goal and each in turn working independently toward achieving this goal.[41] Yet some individuals are given special status by virtue of the deontic power they are perceived to possess—for example the president of the United States, as Searle suggests, or, in my estimation, the all-mighty composer.

Physical buildings also stand to gain from collective intentionality. This is how buildings become institutions.

> I imagine a tribe that builds a wall around its cluster of huts, where the wall performs the function of restricting access in virtue of its physical structure because it is too high to climb over easily. We then imagine that the wall

decays until nothing is left but a line of stones. But let us suppose that the inhabitants, as well as outsiders, continue to recognize the line of stones as having a certain status: a status that we could describe by saying it is a boundary. And they continue to *recognize* that they are *not supposed to cross* the boundary unless *authorized*. I want that to sound very innocent, but in fact it is momentous in its implications.[42]

Searle concludes from this example that the line of stones comes to impose "an obligation on those who recognize it as a boundary," and we recognize a similar boundary in the confines of the concert hall, an institution wherein certain behaviors are practiced as part of the ritual of listening (sitting still, quietly, and unimposing; applauding at the conclusion of a performance—but not between movements; not crossing the stage's threshold, etc.).[43] The boundary, as I examine later, becomes apparent also in the construction of musical categories, of genres and distinctions between different kinds of electronic music on the basis of *who* makes the music (their race, gender, sexual orientation) and not in *how* they make their music. In short, "The intuitive idea is that the point of creating and maintaining institutional facts is power, but the whole apparatus—creation, maintenance, and resulting power—works only because of collective acceptance or recognition."[44] In the case of the electroacoustic "performance," the simple act of listening to the musical work is to participate in the agreement, to sign the contract, and to share in a desire to hear electroacoustic music as a genre—which is also an institution.

As Searle explains, "Human beings and some animals . . . have the capacity to cooperate. They can cooperate not only in the actions that they perform, but they can even have shared attitudes, shared desires, and shared beliefs."[45] In Western Art Music, for example, we acknowledge the institutional entity of "the composer" to whom we attribute a certain respect accorded in the terms by which we *attend to* their music. Philosophers Deleuze and Guattari, in this regard, give the example of a painter. "No two great painters, or even oeuvres, work in the same way. However there are tendencies in a painter. . . ."[46] The institution of "artist" is premised on such individual tendencies, and the composer is no exception. We recognize Ferrari because of his proximity to other figures, such as Schaeffer, Messiaen, and Stockhausen. Together these figures form an institution, and despite Ferrari's unethical practice, he gets a pass because he holds a powerful status on account of this proximity.

Similarly, the philosopher Searle himself gets a pass. Because of his status, he believed himself above the rules of social conduct each time he sexually harassed or assaulted students and colleagues.[47] His virtue extended by other individuals to enclose him in a protective blanket, the sheath of an institution—the philosophical collective, UC Berkeley, and so on. Institutions, too, gain status through mutual recognition among multiple players. The lowly woman complains to Searle's institution, but she does not have access to him. His power, upheld by recognition and status, extends also to the institution.

Feminist philosopher Sara Ahmed writes of institutions in the context of discrimination, harassment, and complaint. She describes walls as hurdles, barriers to women, especially women of color. Walls stand in the way of those who do not quite fit institutional norms. They do not quite fit and a woman cannot (nor should she) change the color of her skin, or her beliefs or gender in way that would make her a better fit—that is, an unlikely target— though she may work to change institutional norms. The rules of the institution are upheld by the same collective intentionality, by the majority. The minoritarian invited to the institution as a foreign, mysterious, "diverse" member, unlike the institution's founders, becomes a cliché. She cannot change the rules on her own to fit her needs, and as Ahmed determines: the cliches of diversity work become futile, immovable, like "banging your head against the wall."[48] How can the young girls be expected to change Ferrari's *Presque rien avec filles*? After all, it is only a work of fixed media. In repeatedly valorizing the medium, we abandon those girls, trapped in the work as any other sound object. The girls envelop Ferrari and anyone else who seeks to bask in their voices. We find ourselves again confronted by the gendered "gap" between listener and heard, whereby, according to film scholar Kaja Silverman, "the female voice becomes the receptacle of that which the male subject both throws away and draws back toward himself. . . . On the one hand, he is obsessed by the desire to establish his complete control over the sounds emitted by others—to over-hear, much like the disembodied voice can be said to over-speak. On the other hand, he is strongly and irrationally attracted by the female voice, which activates in him the desire to be folded in a blanket of sound."[49] But, any sound?

Turning specifically to the voice, Kane declares that electroacoustic music situates the voice as an object of desire. The sound of the voice in electroacoustic music is an object deserving—*requiring*—our attention. It is rumored, and commonly reiterated in the canonical texts of electroacoustic theory, that the term "acousmatic" comes from a story of Pythagoras teaching his disciples, the *akousmatikoi*, from behind a veil so as to separate the sound of his voice from the manifestation of his visible, physical form.[50] In Kane's view, when the veil is pulled aside to reveal Pythagoras the master, his students may not be disappointed by the demythicized body, but may in fact come to laud the erudite, flesh and blood man for his ability to construct such a believable ruse—a hypothesis surely evidenced by the acclaim Pythagoras still enjoys today. Lest we forget, Kane reminds us that it is not the speaker that allures us but only his voice. Once the spacing between source and sound is overcome, "the acousmaticity of the sound is gone," and thus the tension, the desire to know, resolved.[51] The veil becomes enshrined as the source of desire. Pulling aside the veil does not satisfy our desire to reveal the source of a voice, since the body of the speaker lingers as yet another veil that hides the mechanism by which sound is produced. One can lift the veil of deception, reach into the living speaker, and into the mouth and throat of the voice-producing body only to be overcome with disappointment in finding that the producing mechanism hardly corresponds to the meaning with which voice, sound, and language have come to be defined. For this reason, Kane concludes, "Whether we focus on the Pythagorean veil or the veil that is the speaker's body, the voice is always an emblematic object of desire—in Lacan's terms, an *objet a*."[52] Pointing to the famous example of Nipper the dog from Francis Barraud's now classic painting *His Master's Voice*, Kane writes: "We are always lured in by the voice with a source, regarding it—like Nipper—as our Master's voice."[53] Turning our attention once more to *Presque rien avec filles*, folded in a blanket of girls' voices, listeners (men in both the philosophical and musical default) become erotically aroused by the valance of the medium. Lifting the veil dampens the sexual charge because instead of the exquisite female form, there stands the humbling figure of the Master, the composer, creator, the philosopher and intellectual *par excellence*.

Every composer predicts certain listening behaviors in order to calculate possible future responses, and listeners meet composers with mutual expectation. But in electroacoustic composition this prophecy aspires, in the best-case scenario, to mandate what Kane describes as an ungroundable "flicker[ing]" between source and sound.[54] "The privilege that Schaeffer gave to reduced listening in the theory and practice of *musique concrète* has set the terms of a great debate within sample-based electronic music ever since: to refer or not to refer?"[55] But listeners do not merely seek a *sound's* source; in sound art and acousmatic traditions they even seek to arrive at the attributable source *intended by the composer*, whether what is recognized is the actual originating source or merely a projection that abides by the rules of the dis-acousmatizing game. The Pythagorean myth feeds into the permanence of the speaker's historical recollection, raising not only the image and sound associated with the thinker, but also valorizing our collective recollection of the one whose reputation becomes marked by certain status. When read through the lens of a music-compositional philosophy, the master who constructs the game is not necessarily the "performer" producing the sounds on the recording. The compositional "voice" flickers between the identities of a composer and performer. Given this substitution of voice for Voice, I further propose that the composer's elevated status is an inherent aspect of listening to electroacoustic music—arguably more than in other musics. This would then be composer's Voice as Institution, or what Kane terms, after Heidegger, the "voice of conscience."[56] The practitioner who studies his subject, who captures and liberates her voice, remains undisclosed, absent, taken as always-already given. The authoritative voice (of the composer or philosopher) must remain silent in order to maintain its power.

The phenomenological account of intentionality is distinct from the classical Cartesian conviction ("I think, therefore I am") because in phenomenology one cannot take oneself as a reliable narrator.[57] Ihde summarizes, "All experience is *experience of——*. Anything can fill in the blank. The name for this shape of experience is intentionality."[58] While intentionality is itself directed, the state of being directed also occupies a phenomenological category with its own intentionality.[59] "The implication—again quite properly 'anti-Cartesian' in the phenomenological radical alternative—is that I do *not* 'know myself' directly in Cartesian fashion. What I know of myself is 'indirect' as a reflection *from* the world. This also applies to others: I know myself as reflected from others.[60] Not only do I know myself through reflected experience but the means by which objects, or institutions, acquire meaning for me only by way of cumulative reflection. That is, I do not define for myself and others every "thing" each time I encounter it, rather there exists some already-agreed-upon knowledge that I share with others. Such, explains Searle, is the power of language: "I have my intentional states of hunger and thirst, for example, regardless of what anybody else thinks. But the intentionality of language, of words and sentences, called 'meanings,' is intentionality-dependent. The intentionality of language is created by the intrinsic, or mind-independent, intentionality of human beings."[61] Contrary to common opinion, argues Searle, though meaning is founded in language, "meaning itself restricts the possibilities of what one can do with language."[62] We can extend this requirement of meaning also to concepts, where concepts like the "erotic" require intentionality. The erotic is itself defined through a collectively intentional belief of how one performs eroticism, filtered, of course, through the voice of conscience—who we accept as the authority on the matter.

LES DANSES ORGANIQUES (1973)

Two women meet for the first time in a dark recording space (a studio? a bedroom?), their would-be lesbian encounter recorded and set electronically atop a kind of cheesy background of ritualistic bongos. This is the scene constructed by Ferrari in his 1973 *Les danses organiques*, a phallogo-centric fantasy, a pseudo-lesbian encounter that realizes and entertains a male composer's own erotic desires.

The piece is very much like the soundtrack to a pornographic film, complete with minimal dialogue, something like a narrative arc, and even a simplistic tribal-sounding background (ritualistic bongos introduce the women, occasionally processed with reverberation or sudden panning left and right, and as the atmosphere intensifies quick-rhythmed flutes enter in homo-phonic accompaniment). Somewhat static, the same sounds of kissing and breathing repeat unceasingly, without any apparent organization, for close to forty minutes. Presenting as "unmediated" yet obviously staged, the sex act makes for a stale and somewhat unnerving misophonic nightmare, sounding mainly of slurping in very close proximity, almost inside the ear—"pressed up to the surface," to draw on Kane's description of Ferrari's music.[63]

My impulse, as a queer woman of color frequently asked to "perform" for men in this way, is to reject and retract any artistic merit from Ferrari's fantasy, but when asked about the composer's voyeurism, one of the per-formers, Rio, responded, "Tout depend de l'attitude de l'homme. Cela peut être au contraire une complicité de sa féminité intérieure avec la connivance de deux féminités. Alors, à ce moment-là, c'est tout-à-fait autre chose: c'est du domaine de la participation et non de la récupération. [It all depends on the man's attitude. This may be on account of the composer's will to par-ticipate in the collusion of two women. So, at that point, it has to do with something else: it has to do with participation and not appropriation.]"[64] Rio seeks to understand, to empathize with the composer whose *Cinéma vérité* (we imagine a shaky handheld recorder) appeals to a more "true" version of reality than his own.[65]

Does Rio give Ferrari too much credit because of his stature as a com-poser? Or, given the work's pornographic nature—the sexual narrative, its deliberate production, a recorded instance of actual sex—might we accept Rio's testimony as authoritative?

4 ROBERT NORMANDEAU'S *JEU DE LANGUES* (2009)

Montreal composer Robert Normandeau's (b. 1955) electroacoustic signature is premised on the genre's inherent game, as described in the last chapter. The concept of a "game," plays directly into electronic music's philosophical underpinnings—electronic music of every kind plays with overt and disrupted associations, whether simulated, synthesized, or concretely captured. Normandeau, who rose to prominence in the 1990s, embraces the French concrete lineage from Schaeffer through Ferrari, often relying on sampling and production in his works.[1] He even composed an electroacoustic composition focusing on the electroacoustic game released on his first album with the opportune title *Jeu* (1989).[2]

The word *Jeu*, not easily translated to English, takes its dual meaning from the noun "game" and the verb "play," as in, to play a game, but also to play a musical instrument or a musical piece. *Jeu* invokes all of these different connotations in a strategic cat-and-mouse chase between the perceived source, cause, and effect. The work is divided into the following five movements: I. *Les règles du jeu* (Rules of the game), II. *Mouvement d'un mécanisme* (Movements of a mechanism), III. *Ce qui sert à jouer* (Things to play with), IV. *Les manières de jouer* (Playing styles), and V. *Les plaisirs du jeu* (Joys of playing). And Normandeau provides a description of the work:

> La règle du jeu. Jeux du cirque, du stade. Jeux d'adresse. Jeu de massacre. Jeux de société. Être hors-jeu. Mettre en jeu la vie d'un homme. Aimer le jeu. Se ruiner au jeu. Faites vos Jeux. Les jeux sont faits, rien ne va plus. Le jeu d'un verrou, d'un ressort. Donner du jeu à une fenêtre, à un tiroir. Jeux d'orgue.

Jouer prudent. Jouer dangereux. Jouer double-jeu. Un jeu brillant, nuance. Des indiciations de jeu. Jeux de mains, jeux de vilains. Jeux de prince. Jeux de l'imagination, de l'esprit. Jeux de mots. Un jeu d'enfant.

Rules of the game. Circus games, Olympic games. Games of skill. Wholesale massacre. Parlor game. Out of Play. To gamble with one's life. To like to play. To ruin oneself at gambling. Make your play. The die is cast, "rien ne va plus." The play of a bolt, of a spring. To loosen a window or a drawer. The game of skittles, of bowls. To have every opportunity. To hide one's game. The big play. Organ stops. To play carefully. To play dangerously. Double play. A brilliant, nuanced manner of playing. Stage directions. "Stop fooling around or it will end in tears." Plays of Prince. A game of imagination. Play of words. A child's play.[3]

In *Jeu*, Normandeau draws inspiration from Ferrari's view of electroacoustic music as a "cinema for the ear," which determines that a sound's "meaning is as important as the sound." The "cinema for the ear" aims to arouse the minds of its listeners. As Normandeau explains: "The sound of a train will trigger the imagination of the listener in such a way that they are reminded of their own train—and not a train that can be viewed by everybody, like in a film. It is like reading a novel, where everybody imagines their own landscapes and characters."[4] Leaning on the surrealism of painter René Magritte, Normandeau exclaims, "the sound of a train is not a train, it is the sound of a train."[5] Thus, sound is neither representative of tangible, actual objects nor is it reduced to a free-standing form; rather, the sound of the piece is representative of sound in some conjured mental reality. Normandeau conceives of this "cinema" as a technique of abstractable concepts and a skill developed uniquely in electroacoustic music. Just as film found its own cinematic techniques distinctively from filmed theater, electroacoustic music is distinct from other forms of musical composition on the basis of the "cinema for the ear." The essence of a sound collectively imagined—perhaps an erotic essence, perhaps the essence of a train—is shared among audience members, but the image, context, or narrative evoked, if indeed one is invoked, is not necessarily shared (though nothing prevents such mutual participation).

The electroacoustic game secures a performer's anonymity—unless credited there is no way to know for sure who plays. Cloaking the players' identities dehumanizes the figures and simultaneously increases the game's competitiveness. For example, Sergei Prokofiev's famous opera *The Gambler* takes place in a casino where many unidentified characters compete together in roulette. Anonymity encourages listeners to immerse in the work, to identify with the competing characters and their performed actions.[6] In *Jeu*, anonymity is secured electroacoustically, precisely in the gap between source and sound. One gambles to win, and so the game requires players to retain individual standing and fight for what's theirs. Tension is high, since, at any moment, the tables can turn. The game forces one to become hyperaware of one's neighbors on the right and left, to suspect and hence hold one's cards close to their chest; hoping never to betray an ounce of excitement, lest one disclose their hand. Substitute emotion for the playing cards and the rivalry compares to the stoicism adopted by classical or "art" music audiences.

Jeu opens with an inconspicuous sample of (an uncredited performance of) Pérotin's famous *Viderunt Omnes* (https://www.youtube.com /watch?v=9XaeYjxHglg). At a time when undergraduate university elective courses in music overwhelmingly presented surveys of the greatest hits in particular eras of Western Art Music, for many scholars of the Western classical tradition, Pérotin's famous work would have "represent[ed] the acme of development in musical technique from the time of Hucbald on."[7] That is to say, Normandeau chooses a sample that would be easily recognized by his target audience.[8] By choosing to keep intact source/cause associations, Normandeau communicates expressly with his listeners—not unlike how sampling functions in hip hop as a way of "paying dues" as well as "signifying on" a particular musical heritage and its contextual meaning.[9] Normandeau uses this reference similarly to Nas's citation of Beethoven's *Für Elise* in the opening of his song "I Can" (2002), to cite the famous sample as a product of a particular tradition and to position the new track within this history, whether to critique or reinforce the tradition as canonical. It's not that the sample needs to be recognizable (i.e., Normandeau's train) or that it should originate in a purely musical source (i.e., Normandeau's Pérotin sample), but the electroacoustic composer nevertheless employs sampling strategically to align a piece with a particular tradition just like the hip-hop producer. I sow the seeds of this comparison here but give it due diligence only in the second half of this book.

Returning to *Jeu*, after a few minutes Perotin's recognizable music dissolves into harshly panned aftereffects, flickering continuously left and right and simultaneously also ascending and descending within the constructed soundscape still occasionally littered with distant recollections of Pérotin. The familiar sustained chorus that opens the Pérotin sample is quickly defamiliarized when Normandeau superimposes the work with this flickering. The effect persists to various degrees throughout the work, functioning as a dividing boundary between the somewhat identifiable cinematic samples characterizing each section of the work. The result of this manipulation is that despite the recognizable sample of *Viderunt Omnes*, one of the earliest polyphonic works and a piece that any student of music history will immediately identify, the interference of the flickering alters Pérotin's work to such an extent as to become hardly recognizable. Normandeau says that this beginning introduces the rules of the game. The recurring audible flickering structures (i.e., fragments) the music, disturbing any sense of progression in it. Normandeau's *Jeu* thereby simulates gambling's compulsive cycle—the impulse to keep playing regardless of the stakes, to listen regardless of whether what we hear progresses sensically.

Swathed in flickering are brief scenes depicting the various idioms of Normandeau's description and supported by electronically granulated ambient samples of Pérotin joined also by Stockhausen's *Hymnen*, mechanical sounds (movement II), organ sounds and dropping bowling pins (movement III), fragmented speech (movement IV), and children's voices (movement V). Given the great disparity between samples, these sounds could not possibly occupy a single real-world environment collectively. Indeed, Normandeau deliberately fragments the various samples, isolating each by the recurring panned flickering, to ensure that the cinema that emerges is not of images, but only of sounds. In this way, similar to the "Erotica" movement of *Symphonie*, the sounds of Normandeau's *Jeu* dance around a given theme disclosed by the title of the work.

A later work is titled *Jeu de langues* (2009), "A game of tongues."[10] It is made up of women; well, fragments of women's creative practice. Three women perform the works cut up and spliced into this one, but only one woman—the composer's lover—knows of his sexual intent. With *Jeu de langues*, Normandeau claims to have stumbled upon a new field of opportunity in the electroacoustic genre: the erotic.

The piece was commissioned by the organizers of the Música Viva Festival held annually by the Lisbon-based Miso Music Portugal, an organization dedicated to the preservation and support of electroacoustic music. In 2009, electroacoustic composers and musicians involved in Miso Music Portugal gathered to discuss how apparently absent erotic depictions were from electroacoustic music as a genre. While perhaps some representative examples existed previously, those gathered observed that, compared to acoustic music and the other arts, theater, the plastic arts, or opera (which elides many arts), eroticism seemed not at all present in electroacoustic music.[11] The organizers of Miso Music Portugal hoped to resolve this gap by commissioning a concert of electroacoustic works under the heading "Cinema Dos Sons Ficções Sonoras Eróticas" (Cinema of Sounds, Erotic Sonic Fictions). Commissioned works came from five composers, Cândido de Lima, Robert Normandeau, Beatriz Ferreyra (recall she was once Schaeffer's colleague at the GRM), António de Sousa Dias, and José Luís Ferreira for a concert held on September 19, 2009, and another commission from the festival's co-organizer, Miguel Azguime's *L . . .* (2010), was later added to the archived materials. The 2009 concert was prefaced by a roundtable, "Debate Música e Erotismo," with presentations by Delfim Sardo, Vasco Tavares dos Santos, Pedro Amaral, António de Sousa Dias, and Monika Streitová.[12] Due to its popularity, the concert program was repeated under the title "Erotic Sound Stories" and broadcast on the *Arte Eletroacústica* radio program of Antena 2 on March 2, 2013, and repeated again in August 2014.[13] This is one of many recent "new music" concerts to thematize eroticism and sex, as I have learned from being contacted by organizers of such events, including the "new music & erotica" event Série Rose, staged at Warsaw Autumn in 2017 and the Darmstadt Summer Courses 2018 (more on this in the next chapter).

After the first of these concerts, I interviewed Normandeau to ask of his experiences and of how the concert came about. Though he was convinced of the importance of dealing with erotic topics in electroacoustic music, one of the things that struck him about the concert is that three of the five commissions featured only an erotic text, and did not, to his mind, capture eroticism *musically*. In Normandeau's words: "To me who doesn't speak Portuguese, the erotic part of these pieces was completely absent." Further echoing the composer's long preoccupation with composing onomatopoeically, Normandeau continued, "I thought that it was a kind of resignation from the composers to use words instead of sounds only to evoke erotic content. Words belong to literature not to music."[14] And so, Normandeau hoped to evoke eroticism in *Jeu de langues* through other, as he suggests, more universal means.

Normandeau told me that he imagines audiences responding to *Jeu de langues* as he believes they must to his many onomatopoeic works, such electroacoustic pieces as *Bédé* (1990), *Éclats de voix* (1991), *Spleen* (1993), *Le renard et la rose* (1995), *Palimpseste* (2005/2006/2009) and more recently, the acoustic *Baobabs* (2012), works in which the titles of each movement attend to a particular emotional state.[15] Below are the titles of each work's movements in what Alexa Woloshyn titles Normandeau's "onomatopoeias cycle":[16]

Éclats de voix (1991)
1. *Jeu et rythme* (Play and rhythm)
2. *Tendresse et timbre* (Tenderness and timbre)
3. *Colère et dynamique* (Anger and dynamics)
4. *Tristesse et espace* (Sadness and space)
5. *Joie et texture* (Joy and texture)

Spleen (1993)
1. *Musique et rythme* (Music and rhythm)
2. *Mélancholie et timbre* (Melancholy and timbre)
3. *Colère et dynamiques* (Anger and dynamics)
4. *Frustration et espace* (Frustration and space)
5. *Délire et texture* (Frenzy and texture)

Le renard et la rose (1995)
1. *Babillage et rythme* (Babbling and rhythm)
2. *Nostalgie et timbre* (Nostalgia and timbre)

3. *Colère et dynamique* (Anger and dynamics)
4. *Lassitude et espace* (Weariness and space)
5. *Sérénité et texture* (Serenity and texture)

Palimpseste (2005/2006/2009)
1. *Furie et rythme* (Fury and rhythm)
2. *Amertume et timbre* (Bitterness and timbre)
3. *Colère et dynamique* (Anger and dynamics)
4. *Fatigue et espace* (Tiredness and space)
5. *Sagesse et texture* (Wisdom and texture)

Looking at the titles of these pieces, we can see that Normandeau uses common musical parameters to tie the works of the cycle together. Each piece represents a state of mind characteristic to the population it is intended to depict: childhood, adolescence, adulthood, and old age, respectively.[17] Rather than use text to convey his intentions, Normandeau concentrates on certain onomatopoeic expressions he finds to be more universally recognizable. In the words of the composer: "Onomatopoeias require no translation, they may represent feelings, you have the feelings in the sounds and cannot clean them." When composing with onomatopoeias, he says, "I can make a piece directly with feelings."[18] But onomatopoeias are hardly universal. The sounds still require some interpretation and a basic familiarity within a particular linguistic idiom, as is clear to anyone who has inquired about the sound of an animal call in an unfamiliar language (a goose call, for example is "*honk honk*" in English, "*ga-ga*" in Hebrew, and "*ca car*" in French).[19] For instance, without Woloshyn's analysis, I was not able to intuit Normandeau's intended states of mind merely from listening to the onomatopoeic pieces.

Unlike the onomatopoeias cycle, *Jeu de langues* is not based in language, onomatopoeic syllables, or even in alimentations. In *Jeu de langues* we hear fragments of breath by women performers, which we may be able to determine by timbral qualities listeners attribute to the intonation and inflection of the breaths (as examined in chapter 9). *Jeu de langues* employs a common practice in electroacoustic music, sampling existing works from the composer's recorded catalog. The piece uses two previously composed pieces, for flute and saxophone respectively, and a third in-studio recording made specially for this work. Each source has its own correlation to intimacy based in an actual sexual encounter experienced by the composer.

Jeu Blanc is the first piece incorporated in *Jeu de langues*. Premiered by flutist Claire Marchand in the late 1990s, *Jeu Blanc* is an improvisation with extended techniques and, like Normandeau's earlier *Jeu*, the work receives its title from a common turn in gambling, meaning to break the bank or to lose all of one's money—a strategic failure that perhaps foretells of the work's eventual fate: *Jeu Blanc* has since been withdrawn from the composer's catalogue. The second sample incorporated in *Jeu de langues* is *Pluie Noire* (2008), a work premiered by baritone saxophone player Ida Toninato at the 2008 Música Viva Festival. For the *Jeu de langues* recording sessions held specially in 2009, Normandeau invited into the studio Terri Hron, a flutist he met at the Música Viva Festival the prior year. Hron, Normandeau's romantic partner, was the only performer who knew of the composer's intentions for her recording. Given this contextual history, the composer's intended reception of the work is conveniently situated within the compositional history of its three sources.

Jeu de langues seems to follow Ferrari's anecdotal approach in some ways, by using a collage of multiple recordings so as to construct a particular cinematic soundscape (without conjuring a definitive scene), and intimate attention to the subjects who are collected, for the most part, without their knowledge. Normandeau exerted his greatest compositional efforts in removing the pitch material (the exhalations of the wind players) from each recording post-production, leaving only the performers' inhalations, slight gasps, and hesitations, and the involuntary mechanical sounds made by the performers' bodies striking their respective instruments. The sounding result—inhalation paired with ambient electronic sustained tones—undergoes minimal if any development, which, as mentioned in my discussion of Ferrari's *Les danses organiques* in the previous chapter, can be quite taxing on a listener attending to the work for close to twelve minutes. The only change occurs toward the end of the piece (8:23–11:24), when overlapping inhalations are spliced together with instrumental *flautando* effects and occasionally reversed in playback. This arrival is climactic in comparison to the rest of the work, but, comparable to Schaeffer's "Erotica," it is hardly a notable arrival.

Outside of the throat clearing that opens the work, we recognize the voice(s) in this piece only through inhalations. And because eroticism is commonly voiced through exhalations (moaning, breath, vocalization), Normandeau jokes that *Jeu de langues* may actually be "anti-erotic," representative as it is of the antithesis of how composers might usually conceive of electrosexual music. Nevertheless, the work fits the "electrosexual" criteria as a piece in which composers and listeners come into contact by way of the sexual allusion. The sexual imaginary binds these entities in a contract of suspended disbelief, in the virtual simulation of sex through associations (heavy breathing), common tropes (pulsating and building intensity of volume and periodic repetition), and musical idioms (climax mechanism). Eroticism becomes a shared means of arriving at the *experience* of the musical work, which may be shared, though it probably is not in the end. Only the intentionality is collective, not the experience. For this reason, we might say that in the act of hearing an erotic composition, composers and listeners enter into a "plural subject-hood," a term Andrea Westlund uses for two people involved in a romantic relationship in which each person remains an *individual* rather than becoming "fused" psychologically or ontologically as a coupled unit.[20] In this way, although listeners may not actually become aroused by erotic electronic music, if the audience is told a piece is meant to be erotic, or if listeners themselves pick up on cues by way of association, each individual enters into the collective intentional state. The erotic work comes to serve as a sort of institution bearing deontic powers by way of its suggestive overtones—in the very quality of sounds employed *and not merely by facet of* "the recorded character of the recording." In short, there exist certain established or agreed-upon habits necessitated by sound, in how sound is collected, organized, and auditioned within the context of the work simply by virtue of being "music" and thus acceding to its collective experience.

Pluie Noire, sampled in *Jeu de langues*, takes its title from the play *Blasted* scripted by UK playwright Sarah Kane and with music by Normandeau.

As the composer writes in the notes, "the title is referring to pauses indicated in the play that punctuate it: spring rain, summer rain, fall rain, and winter rain. But all these rains can hardly wash the darkness of the human soul at war, especially that of Bosnia that the play makes implicit reference [*sic*]."[21] The expansive gaps between breaths, hardly supported by the wispy fluttering of the ambient background, draw out anticipation, each inhalation suggesting that an exhalation is to follow, but our expectations are never met. Thus, the pauses from *Pluie Noire*—the negative spaces between sounds—carry over also to *Jeu de langues*. The silent gaps gain additional weight, when combined in *Jeu de langues* as this previously existing piece maintains its original length running in entirety from beginning to end, burdening our ears with an absence of the sounds it once bore.

After the premiere of *Jeu de langues*, an audience member, drawn to the silences and periodic inhalations, commented on this loaded significance, exclaiming, "I don't know what your conception is of eroticism, to me the piece is about death," as if hearing the gasps of someone's suffocating last breath.[22] Such was Normandeau's deliberate allusion, a subtle connection to death and "*le petite mort*," a French idiom for achieving orgasm.

In the words of Roland Barthes, *le petite mort* is the moment in literature when language breaks down: "ce que [le plaisir] veut, c'est le lieu d'une perte, c'est la faille, la coupure, la deflation, le fading qui saisit le sujet au cœur de la jouissance [What pleasure wants is the site of a loss, the seam, the cut, the deflation, the dissolve which seizes the subject in the midst of bliss]."[23] Barthes's linguistic breakdown recalls the punctured surface of the language in Deleuze's schizoanalysis. As Elizabeth Locey elaborates, "The 'text of jouissance' (as opposed to the 'text of pleasure') is that text in which the reading breaks down. The 'coupure,' or cutting, that produces this 'inter-dit' occurs when one (or more) signifier latches onto another and carries—or cuts— the reader away from the text into that place of jouissance that is beyond language."[24] An overabundant lack is conjured in the absence of exhalations in *Jeu de langues*, the literal cutting of the tape is done in service of *jouissance*, penetrating silence with a desire to hear more. The gaping holes left in the wake of sound-that-once-was evoke curiosity, seducing us with perforations in the texture of a perceivably preconceived reality. Here, sense emerges from *non*-sense, from nothing (*rien*); absence produces presence.

This chapter has shown some confluence of composing electrosexual music between Normandeau and his predecessors, Ferrari, Schaeffer/Henry, and Lockwood. I have also shown that, within these conventions, there exist many diverse engagements with sex, including, of course, potentially diverse *responses* to the music from listeners—not least of which is the tension between those aroused by *Jeu de langues* and those repulsed by the composer's sexual manipulations of performers' recordings without their knowledge or consent. Combining several analytical methods on the basis of musical commonalities can highlight similarities across categories or genres of electronic music, such that we can redraw a disciplinary axis of inclusion that insists on sexuality as a crucial and robust paradigm at the intersection of music, society, and technology. If we bend our hearings of Schaeffer/Henry and Ferrari to similarly emphasize men's control over or violence inflicted on women's voices, we deemphasize women's roles in this significant body of music. In a way, we simply repeat and reinforce this violence.

Instead, a feminist approach acknowledges these women's works, their voices used as a foundation for a new genre of music, a genre that is sexual in content and electronic in medium—electrosexual music. Electroacoustic composers desire to assert control over how women conduct themselves in electrosexual works by cutting up, displacing, and intermingling their voices with various other sounds. And music theorists, operating in service of these composers, may very well reinscribe this misogynist dominance. But maybe, whether patriarchal modes of knowledge dissemination ultimately succeed in *taking* control is up to how we listeners interpret the works and how we correspondingly either echo, confront, critique, or contradict this desired-narrative/narrative-of-desire.

5 JULIANA HODKINSON AND NIELS RØNSHOLDT'S *FISH & FOWL* (2011–2012)

In the words of music critic Martin Nyström, Juliana Hodkinson (b. 1971) and Niels Rønsholdt's (b. 1978) *Fish & Fowl* album[1] is what happens when composers are asked to remix their own music.[2] Nyström describes *Fish & Fowl* as "a fascinating dialogue between two temperaments, between written composition for acoustic instruments and electronic sound art, where the actual breathing, muffled pulsing, groaning and sometimes excited moaning, is the cohesive link." The acoustic and electronic communicate by way of associative human sounds: breathing, groaning, and moaning. This is not to say that the vocalist is the sole focus of such a hearing, indeed, the instruments all contribute various forms of breathing and moaning. If we isolate both vocal and instrumental breathing, as I have done in figure 5.1,[3] audible breath sounds provide a constant thread throughout the work. Despite the relative artificiality of instrumental breathing—the clarinet not respiring to exist, to "live"—the significance imbued in vocal breathing is easily evoked also from the musical instruments. The clarinet's semi-tone teetering (from C to B and back again to C) recalls the ragged exhale/inhale pair, where the relaxation of the exhale causes the pitch to fall, and the tension of inhaling to rise.

Breath serves as a guiding compositional principle for *Fish & Fowl* composer Niels Rønsholdt. As he explained in our correspondence: "Musicalized breathing is a basic element of my music. I consider breathing to be a musical ground stone, a) because of the elementary musical 'phrasing structure' of breathing, and b) because of the emotional and dramatic connotations that

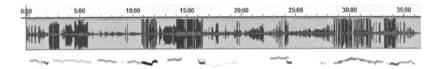

Figure 5.1

(*Top*) A time-domain representation of *Fish & Fowl* plotting amplitude over time. (*Bottom*) A map of audible breath (instrumental and human combined). The opacity of the shading indicates intensity by a change in volume and higher periodicity.

are so easily activated when you listen to human breath."[4] Though musical instruments can also embody this structural breathing in music (for instance, Salvatore Sciarrino's various works for flute),[5] Rønsholdt grants more semantic weight to human actions, which serve as the ground stone from which the semantic significance of the instruments is attained. Elsewhere, Rønsholdt describes the "protagonist" he envisions in his works as a "small, nervous girl," but one who lives inside all of his listeners, regardless of gender. His compositions *Die Wanderin* and *Hammerfall* are two works sampled in *Fish & Fowl*:

> I use "protagonist" as a word for the main character in a piece. It's not a character in the sense that they have a name or a specific history. It's not a character to identify with very pragmatically, but in a more general sense. So, the girl walking down the street in *Die Wanderin*, or the blind man by the table in *Gloomy Room*. I see the protagonist in *Hammerfall* as the same one in *Die Wanderin*. It's the same small, nervous girl. And in some way I think that we all have a small, nervous, insecure, teenage girl inside. The protagonist is representative of all of us. It is a character, but it's not a specific character. It's representative of all of us in specific situations.
>
> What I like to do is to take some psychological tendencies or things that are common to us all, at least that I feel are present, and then amplify it so that we can watch it. Like in a microscope. It's not interesting for me to tell everybody how I feel. It has nothing to do with my own story or feelings as such. I detect something in other people, in myself, and then I try to sort of dig it out or amplify it a bit. To brush the dust off. To sort of magnify these tendencies I detect in order for us to look at it. Just to look at it. Like a mirror of some sort. I have no agenda for what to get out of it. I just present it. I find it most interesting to seek out the dark sides because all of the fun stuff is out

there already. It's like digging out hidden things that we all know are there, but need to be dug out a bit to be looked at.[6]

Rønsholdt's projection of an idealized feminized image reflects a long history of such musical portrayals as we've seen in the previous chapters. But even before the possibilities of including electronic samples, composers were long concerned with constructing idealized representations of women's sexuality. One could think for example of Erwin Schulhoff's Dadaist *Sonata Erotica* (1919), "a piece in which a solo female vocalist performs a carefully notated orgasm," Richard Strauss's autobiographical allusion to marital copulation in *Symphonia Domestica* (1903), or, perhaps less overtly, of Alban Berg's opera *Lulu* (1935).[7]

Already from the opening of *Fish & Fowl*, listeners may seek to identify the music's sound sources, to assign them actual "real-world" identities. These sounds are "real" in the sense that their agglomeration sounds as music (and quite pleasing music at that), but outside of this piece, these electronic sounds have no "actual" equivalent. Well, this is not entirely true. To echo Steven Connor, as the sounds of *Fish & Fowl* are entirely electronic, they need neither be assigned to a totally encompassing persona nor should or could the sounds be heard as entirely absolute or abstract; that is, these sounds are not simply sonic but, just as any music, they are heard contextually. Hearing music somewhere in the balance between these two extremes allows listeners to intuit new meaning from what they hear each time they hear it.

It was through this kind of uninhibited self-discovery, says Hodkinson, that she gained the "fluidity" to integrate digital audio into composition for live and acoustic instruments and voice, which freed her compositional process while also affording new professional opportunities for her to express and, ultimately, to perform her music. She summarizes her notion of composition as "sonic writing," which she defines as the ubiquitous ease with which today's composers oscillate between musical composition for instruments, voices, and electronics.[8] The word "writing" refers of course to the physical notation of musical composition, but Hodkinson uses the term to also evoke its metaphorical meaning of inscription as described in Hélène Cixous's concept of *écriture féminine*.[9]

As we've seen in several previous chapters, representations of women, and particularly representations of women's sexuality, have been historically subordinated in music, art, and literature (and elsewhere) to a typical phallogocentric symbolism. This emphasis on the dominant heteronormative, masculine prerogative has subsequently led to an apparent absence of symbols representing sexuality outside of these norms. To challenge these norms, *écriture féminine* moves to make room for representations that belong uniquely to women. Though these feminist thinkers did not explicitly exclude people with intersecting minoritizing identities, given the movement's origins in France in the 1970s and 1980s, advanced primarily at the hands of white women, it was later critiqued for placing too stringent an emphasis on women's difference, subsequently implying an inherent femininity among all women and once again reinforcing women's positions as Other while also erasing the struggles of those minoritized among them, marginalized by race, economic class, and/or physical (dis)abilities.[10] In abandoning the "feminine" modifier, Hodkinson relinquishes women's inherent difference while insisting that *écriture* set off the sonic from other modes of writing. But still, in using a term associated with the feminist thinkers mentioned above, Hodkinson nevertheless references a gendered resistance to normative representations.

One of these opportunities came in the form of *Fish & Fowl*, which, after the initial release of the album, and despite its erotic eccentricities, was performed live several times throughout 2011 and 2012: at Berlin's Ultraschall Festival, the Spor Festivals in Denmark, and Huddersfield Festival. These live performances reflect Richard Maxfield's tradition of tape pieces from the 1960s, especially a series of works he deemed "a kind of 'opera for players instead of singers' in which specific 'performers, most ideally, would play themselves.'"[11] The sound "library" for pieces Maxfield dedicated to David Tudor and La Monte Young, for example, included prerecorded improvisations from the performers, and live performance would then include these samples in the mix to become a "freely improvised . . . montage" unfolding for the first time.[12] Like Maxfield's *Piano Concert for David Tudor* or *Perspectives for La Monte Young*, *Fish & Fowl*'s multiple performances include some live mixing of the existing soundtrack (the CD) with additional performers on stage, sometimes those who first recorded the catalog pieces, but also members of the Scenatet Ensemble who may never have played any of the previously recorded works. Even more intriguing, at these events Hodkinson and Rønsholdt—the composers themselves—stand facing the audience on stage, equipped with a laptop, mixer, and their respective compositional tools—Hodkinson with ping-pong balls and a mobile sampler with FX; Rønsholdt with whips and triggering the video projections. While it is not uncommon to see composers of electroacoustic works situated among the audience, controlling the prerecorded, "fixed" media at the audio desk, *Fish & Fowl* introduces an aspect of improvisation into the compositional process, enhancing the "drastic" possibility of unintended consequences.[13] Improvisation and live composition become infused in the performance situation, contributions that cannot be discounted as mere addenda to the recorded album. By sharing the stage with the musicians who first recorded the works sampled in *Fish & Fowl*, the composers accede full authorship of the work, which becomes newly defined with each subsequent and alternatively choreographed performance. Nevertheless, with the composers on stage, they remain ever-present, occupying both object and subject roles, subordinates to music's power to incite.

The music might stimulate sensations of intimacy, desire, or fear, or it might pique interest or curiosity, all while remaining in constant contact with the work. Indeed, the sounds that are most stimulating in this work, by which I mean most striking and memorable, also designate audible demarcations of the work's form. In the following analysis, I begin by presenting (how I hear) the work's formal organization, to demonstrate how digital processing facilitates but also compromises the work's structural unity through the repetition and slight variance of sound samples. The analysis shows how the source recognition of sounds is repeatedly affirmed as well as denied, inviting listeners to identify *with* the sounds of the instruments and of the human body to encourage embodied engagement with this music. The work's repetitive streams help listeners hear the work, where each repetition serves to form associations with its previous expression.[14]

Fish & Fowl opens with a clarinet playing a long sustained tone (0:00–0:08), illustrated in figure 5.2 with the initials CL. This opening utterance is almost immediately thematized when a distant electronic sound enters with a similarly sustained tone, absent the dynamic crescendo of the clarinet (0:25–0:32). The clarinet then reiterates the "sustained motive" again (1:05–1:11), joined this time by the viola, to complete the first phrase.[15]

There are three phrases in the first section of the piece, each with its own character. The first phrase is characterized by an atmospheric texture I term the "subdued rhythm" (SR), after the rhythmic propulsion given by bowed strings and blown winds. The second phrase (1:24–2:04) is characterized by more "percussive," plucked or hammered (PR), rather than bowed or "breathed," instruments. A short transition then brings a third phrase (2:11–3:07). The atmospheric texture of this phrase develops from the sustained ambiance of the electronic sound from the first phrase, this

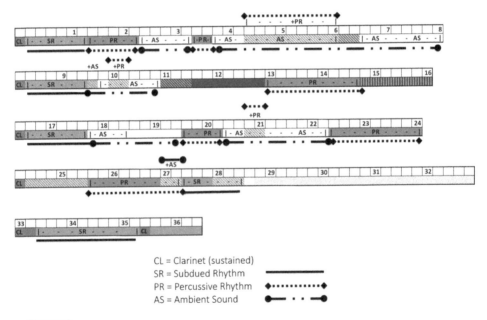

Figure 5.2
Recurring themes in *Fish & Fowl*. Numbers indicate time in minutes.

time in a more encompassing background role to parallel that of traditional harmony. In this "ambient" sounding atmosphere (AS), what sounded in the first phrase as a single melodic tone now sounds split into the simultaneity of several sustained supportive tones. Atop this sustained background, the third phrase characterized by clicking heels enters, joining the "breathing" instrumental sound of the second phrase to recall the breathing treble of the first phrase.

Already, this analysis of the first three phrases indicates salient features of the piece as a whole. First, melodic motives often shift roles to serve as harmony or support for later music, such as when the electronic tone in the first phrase becomes the background for the third phrase. Second, motives like the clarinet's opening sustained tone recur throughout the work to signify important formal events, such as the beginning and end of the first phrase. Third, two iterations of the same motive may share their duration, timbre, rhythm, or pitch, but need not present all of the same features to be identified as similar. For example, the percussive ping-pong balls of the second phrase and also the clicking stilettos of the third phrase both possess a rhythmic regularity and cracking attack; also the clarinet's sustained motive is related to that instrument's "breathing" figure.

Continuing with this first section, we see that, like the exposition of a sonata, after the initial statements of the themes (complete with transition), the themes return to close the first section, but this time slightly altered, where the second phrase (3:20–3:40) is truncated and elided with the return of the third phrase (3:40–4:15). Such elisions feature importantly throughout *Fish & Fowl*, since, as mentioned, samples that have a motivic function often recur elsewhere as accompaniment to samples that may or may not have sounded before. One such "mishmash" occurs in the conclusion of the first section (4:20–6:00), in which the familiar features of the three previous phrases are joined by what sound like a clanging chain, shattering glass, and a bass drum. The conclusion of this section builds to climax with an increased tempo (heard in the quicker pace of the clicking stilettos) and amplified dynamics (the scratching viola) only to rupture into a sudden absence of sound, a sublimation culminating in a stream of decelerating ping-pong balls.

The map in figure 5.2 shows that the first section of the piece opens with a clear statement of the "sustained motive" marked in its first statement by the clarinet (CL). The first phrase SR is characterized by bowed and breathing instruments in a particular rhythm—the rhythm of the opening of Hodkinson's *sagte er, dachte ich* shown in figure 5.3. In fact, almost every time the clarinet's sustained motive sounds, it is followed by the SR phrase, which is then succeeded by some combination of either a PR or AS phrase—again, the PR phrase recognizable by its percussive sounds, usually a piano or plucked guitar, and the AS characterized by sustained ambient tone(s). Note that each new line begins from an iteration of the CL "sustained motive," lending variably to a "mishmash" that may or may not include new material. In total, this formula returns four times in *Fish & Fowl*, with a truncated coda-like return at the fifth iteration of the CL motive.

Each return of the opening material is differentiated from our hearing in the beginning of the piece, and this differentiation—our awareness *in the present* that this is an event reminiscent from our past—is further

Figure 5.3

Excerpt from Hodkinson's *sagte er, dachte ich*, mm. 1–8.

augmented by its obvious dissonance with the interrupted climax that directly precedes the motive's iteration each time. Beginning at 6:43, the CL motive enters together with a wispily bowed viola, and development of this motive ensues. Here, the viola sounds increasingly opaque, until its final iteration at 8:11, when the AS texture enters. At 15:58, the CL returns motivically developed by recurring statements. After seven repetitions that grow in dynamics and opacity (less breathy), the SR atmosphere once again greets the figure. At this, the third refrain, the breath does not enter as "early" as it did in the beginning. Delayed by eight seconds from its proximity to the clarinet in the opening, only at 17:02 are hesitant, amplified breaths gasped, now with the accompaniment of a faint drone (AS).

THEORIES OF REPETITION

The protagonist's presence is brought to the fore at moments when the sustained motive echoes throughout the imagined ensemble, in ambient electronics (6:25), pulsating from clarinet to viola (6:57), eventually finding itself in the bowed vibraphone (7:22), and returning once more in the viola, this time gliding over several pitches in a flurry of flageolet sounds (7:51). Multiple articulations in succession, such as those that occur in this moment, recur throughout the piece in varying opacity (12:50; 19:08; 21:58; 24:08; 32:02; 33:42; 35:16), but none of these restatements is more recognizable than the refrain of the opening material at 15:58. The clarinet's initial cry is heard repeatedly throughout *Fish & Fowl*. Yet, while documenting the iterations of the sustained motive allows a structure for this particular recording to emerge, one should ask to what extent each iteration of this motive is truly repetitive.

Since the sounds of the clarinet and human breath are not so alike that we cannot distinguish between them, we recognize this relation as repetition toward an ideal, as what Deleuze might call an imminent multiplicity of "intensive differences."[16] Envisioning Bergson's "duration" (*durée*) or sustained experiential field "as a type of multiplicity," in Deleuze's later collaboration with Guattari the philosophers describe duration as "in no way indivisible, but [a]s that which cannot be divided without changing in nature at each division."[17] Pitch levels, duration, and timbre are imagined in graduated difference, as referential, relative, codependent events unfolding in time. With each supposed repetition, a newly generated difference is created, and together these elements are combined and intuited by the manner in which we listeners orient ourselves. A single sound may be discussed in terms of pitch, duration, meter, timbre, and intensity, all of which are retained, yet each is articulated anew with every iteration. Thus, the breathing voice recalls the breathing woodwind, which is, in turn, reminiscent of the bowed vibraphone or the bowed viola that join soon after. When the breath enters alone at 17:42, again the winds, vibraphone, and strings are called to mind. This is the work's "protagonist," but "she" is not human. The body is a becoming never actualized; it is a limit, a becoming-ideality that is never become. *Fish & Fowl*'s erotic *phonopoiesis* is ever productive, mobilized and resounding, always familiar and at once immanently defamiliarized.

Where Deleuze and Guattari aspire toward becoming-woman (see chapter 1), *Fish & Fowl* begins with woman to move beyond such a rigidly abstract body. And while the philosophers echo Darwin's distinction between kin and resemblance, in the digital world such divisions are nearly impossible.[18] Kin means to be of the same DNA—the same creative matter—so what is a sample if not a relation of kinship? Just as two human beings are of the same species, the clarinet and the viola are both musical instruments, though neither is from the same family. And yet, their expressions nevertheless resemble one another. Unlike Schaeffer and Henry's "Erotica" movement, which features a cooing woman atop an unrelated musical tapestry, or Normandeau's *Jeu de langues*, which simply removes any semblance of the instrumental leaving only breath, *Fish & Fowl* insists on a reciprocal relation between sounds of the "natural" human body (breath, heartbeat, footsteps) and the aesthetically shaped sounds of music.

ON BANALITY

It is no coincidence that an opera like *Lulu* would serve as fodder for sexual allusions in electroacoustic music. After all, most electroacoustic works proclaimed as erotic by their composers feature the voice. Plot and stage directions make explicit that which is implicit in "absolute" music, and lyrics make no small contribution toward semantic meaning. The coincidence of musical intonation, text, and action imbues music with added significance, but not without uncertainties. As mentioned in the last chapter, the composer's choice of language plays greatly into the audience's understanding of the plot. Yet works like Schaeffer and Henry's "Erotica" movement, Lockwood's *Tiger Balm*, Normandeau's *Jeu de langues*, or *Fish & Fowl* do not rely on language at all; these works do not even attempt to bloat vocal sounds with onomatopoeic allusions. The vocal sounds are obviously human. Maybe this is why sampled moaning and groaning has such a base reputation. Still, without words, one cannot be entirely sure what the scene betrays. After all, exclamations of pleasure and pain are not so distant; this much is evident in the frequent and parallel appeal to sex in horror films.

Opera, even when complete with lyrics, also has moments that convey meaning absent language, with musical subtext. Leon Botstein even argues that were it not for the music, Berg's *Lulu* would not have enjoyed the success it did later in the twentieth century, after sexual liberation.[19] To recall the synopsis briefly, after murdering her caretaker and betrothed, Dr. Schön, Lulu seduces his son without a second thought in the very spot the father bled out. In the third act, Lulu is brought to prostitution, while the last of her admirers—her lesbian lover Countess Geschwitz—remains faithful. In the final scene, Lulu's suitor, London's infamous Jack the Ripper, murders the two women. For Botstein, none of the developments are captivating enough to sustain the audience's interest in a regime of supposed liberated acceptance: "nonconventional sexual behavior," though still "an object of recurring fascination," became little more than "banal" in the late twentieth century. He argues that the opera maintained its popularity only on account of Berg's enthralling music.[20]

Were we to define desire within a Lacanian framework, as longing for an unattainable object, then Lulu epitomizes the antithesis to desire.[21] Interpretations that present Lulu as an embodiment of the idealized *femme fatale*, both seductive and destructive, paint the title character as easily attainable. She is possessed in every manner possible—by wedlock, sexual slavery, and lesbian obsession. As an ideal representation, it is easy to envision Lulu as an object ripe for the taking, and the same goes for the protagonist in *Fish & Fowl*. The protagonist is not real but only an illusion, a fantasy if we call again on Lacan, and as an *allusion* to woman—after all no semblance of this woman exists anywhere in reality—she is never actually attainable. *Fish & Fowl* could in this way serve as the quintessential example of the Lacanian *objet petit a*. This conclusion is Brian Kane's verdict of the acousmatic that no matter the musical content, acousmatic sound is destined for sublimation.[22]

Implicit within Botstein's assessment of *Lulu* is a belief that banality doesn't sell, that music must sustain the interest of its listeners, and even the implication that, in order to succeed, music must be continually stimulating and new.

And what about banality? Is there anything arousing about the commonplace?

As we have seen, sex scenes are common enough in music, so much so that a moaning woman and the drive to climax are generally dismissed as cliché, or worse, simply redundant of the "real" sexual act. In his rubric for the art of rhetoric, Aristotle describes banality (ταπεινεν) as one drawback of clarity; banality is merely a normative and familiar form of speech (or writing) that does little to interest the listener (or reader). Against banality, Aristotle argues for interspersing familiarity with some ornamentation.[23] Aristotle's praise of the balance between the florid and the familiar carries well into the discussion of eroticism in music: too literal, and the recorded sexual encounter becomes boring; too many theatrics, and the work moves to the realm of farce. If we take Aristotle's rhetorical definition of "banal" to mean merely "the mean," as flat (the word here is the same Aristotle uses for banal), sex scenes exemplify banality precisely because they seem inevitable, because this music's message is immediately apparent to practically anyone (remember, "flat" is how Kane describes Ferrari's "anecdotal" approach).[24] Arguably, however, sex scenes in electronic music only seem inevitable because the constructed scenarios of this music are so underwritten in human encounters with technology in the twentieth century. For the banal to appear as such requires history and tradition—banality requires banal repetition. Is the electronic medium, often framed in line with technological innovation and hence progress, enough to resist this reduction to banality?

Composer Francis Poulenc famously wrote "In Praise of Banality," where he explains that while some composers feel the constant need to innovate (implicitly pointing the finger at Schoenberg and his followers) there is something to be praised in the banal: "Being afraid of what's been heard already," writes Poulenc, "is quite often proof of impotence."[25] To explain this sentiment, Poulenc invokes the example of Schubert, whose "simple inflection of the melodic line personalize[s] an anonymous ländler at a stroke."[26] In this remark, Poulenc positions himself within a resilient twentieth-century dichotomy between the manifestos of the modernist avant-garde and the historicist's clutch on tradition. But his assessment also raises an analytical condition. Indeed, Schubert's landler is not praised for its replication of tradition, as an exercise in redundancy; rather, it is the composer's personal touch upon which Poulenc remarks. The traditional musical form thus manifests stylistic nuance, repeated variation, difference, and repetition.

Fish & Fowl draws similarly on the banality of erotica, on the clichés of the climax mechanism, but the woman's moaning and heavy breathing are only the beginning of the analysis. Instead of divorcing the banal from the innovative, *Fish & Fowl* confronts our ears with a highly complex interweaving of instrumental music and sounds of the human body. The piece is a novel expression within a tradition with a long history. *Fish & Fowl's* structure is somewhat formulaic, though what is musically formulaic is not necessarily heard as semantically recurring. Whereas Walter Benjamin described cliché in opposition to both the "serious" and the "innovative," there is something novel about repetition; such is the process of citation.

Complementing Poulenc's mid-century appeal with a postmodern twist, philosopher Giorgio Agamben writes that "To repeat something is to make it possible anew." Like memory, repetition does not restore what was but only its possibility. "Memory is, so to speak, the organ of reality's modalization; it is that which can transform the real into the possible and the possible into the real."[27] Though Agamben explores repetition within the realm of cinema, his observations easily carry over to music. He lauds creations that are "worked by repetition and stoppage," artworks that do not mask their medium in service of their ultimate expressivity.[28] Repetition is a "pure means"; it is not visible in the work, but only in the concepts that bind the work as such. Agamben's concept of "pure means" says that the perceived "image" of the protagonist in *Fish & Fowl*—the feminized voice-body responsible for producing the sounds we hear—is not a distinctive being performing the work. In fact, the protagonist is indistinct from the medium, from the sounds themselves, from "the recorded character of the recording." There is not, first, a performer and then the sounds they produce; there is both and neither, always. Agamben notes that pornography and advertising "act as though there were always something more to be seen, always more images behind the images . . ." whether or not there truly is.[29] This would be the effect of cinematic "framing," of presenting an image as if it were part of a vaster landscape. Were *Fish & Fowl*'s moaning protagonist received as pornographic, listeners may be provoked to imagine a scene unfolding within an intrinsic netherworld. In conjunction with the instrumental samples, the clarinet's motive and its performer's subsequent breathing into the instrument would serve merely as fodder for some titillating scene limited only by our imagination. But what if there is nothing more? Without any such originating scene, is it possible to be but "a voice and nothing more"? According to Mladen Dolar's book of this title, it is not.[30] The voice always signals something beyond itself. The voice "as the lever of thought" weighs down history in its habitual materialism, "as one of the paramount 'embodiments' of what Lacan called *objet petit a*."[31] In this sense, we have the voice and *something* more. Yet, rather than reflexively assigning this "something" a potentially human form, music's duplicitous musicalities speak volumes.

To ground these theories, let us turn now to "Torso" from Rønsholdt's mini opera *Triumph* (2006), another piece sampled in *Fish & Fowl*. Figure 5.4 shows the scene's diegetic opening, with a feminized voice moaning.[32] Rønsholdt introduces a scene rich with sonic inference: a whip spurring a vocal reaction from a singer, inhale/exhale. Though diligently notated, nothing of this opening suggests that what we are about to hear is music. The staged theatricality of this opening becomes apparent only after several repetitions, as if to stabilize and normalize what we hear. As the instruments

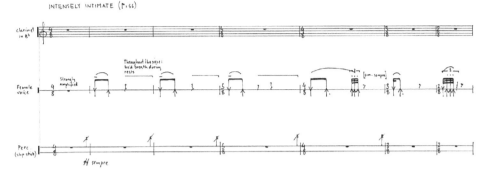

Figure 5.4
Excerpt of "Torso" from *Triumph* by Niels Rønsholdt, woman's voice inhale/exhale with whip (slap stick), mm. 1–7.

join the singer, the piece gradually succumbs to more atmospheric sounds of music. That is, the nondiegetic sounds set the stage, so to speak, for the concert hall. *Fish & Fowl*'s diegetic boundary then lies at the crossover between the so-called natural sounds of the body and the sounds made by musical instruments. When the singer is audibly in the forefront, she is the complete soloist—a character ("the protagonist")—but when her utterance recedes to and interacts with the atmospheric background, becoming essentially indistinguishable from the ensemble, we experience a blurring of the singer's surroundings and the musical canvas.

Sound artist Richard Maxfield recognized the terminality of the sound recording. In Michael Nyman's memory, "for each of his pieces he [Maxfield] composed a vast 'library' of materials out of which he could make a new realization for each performance, or each time he distributed a copy."[33] Hodkinson and Rønsholdt's *Fish & Fowl* exemplifies precisely this kind of library. The work indexes a host of compositions from its creators' back catalogs, encompassing nearly twenty years of productivity, condensed into a three-day studio collaboration to mix these recordings in Pro Tools (a popular Digital Audio Workstation). The *Fish & Fowl* project, while related to the previously recorded compositions of its respective composers, is something entirely new, existing as much in the past as in the present of both human and technology, and this is clear from the work's triumph as an independent composition, its success boosted by the CD's various accompanying materials.

The 2011 CD includes an intriguing booklet, complete with photographs and textual material in excess of what we hear on the recording (that is, these are not translations or transcriptions but original texts). The photographs by Anka Bardeleben depict the two composers in their assumed roles specific to the work.[34] And the texts that accompany the photographs were commissioned of Ursula Andkjær Olsen, who conducted interviews with the composers and the performers of the eight previously recorded works to compile several free-form fictitious communications from fish to fowl in both Danish and English (figure 5.5).[35] Presenting the metaphorical fish and fowl characters as literal subjects leaves room for interpretation of an object that is, as the saying goes, "neither fish nor fowl," neither here nor there. The animals' written exchange sets them apart from one another, while their communication remains an intimate linkage to bridge this distance.

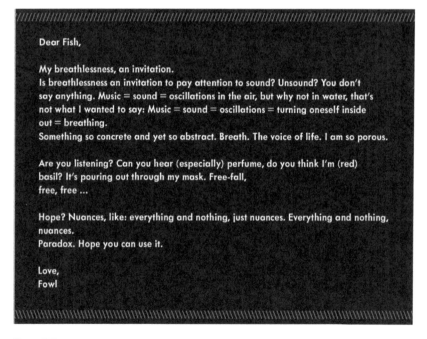

Figure 5.5
Letter from *Fish & Fowl* CD booklet.

Though there is no such indication from Olsen, the idea for the booklet letters could easily have taken inspiration from Rønsholdt's opera *Inside Your Mouth, Sucking the Sun.* The opera envisions an affair through correspondence between Napoleon and Josephine, at times written, other times portrayed through instruments—the clarinet and trombone—to represent the two figures.[36] Unlike a conventional opera, which features the voice always in the forefront atop the orchestra, Rønsholdt's *Inside Your Mouth* often interchanges the role of the voice "as extreme anima" and supportive atmosphere, to recall one review.[37] Such exchanges are common in Rønsholdt's music, and are particularly palpable in the "Torso" movement previously mentioned.

Figure 5.6 from "Torso" features the entire ensemble, where the breathing figure of the clarinet, notated in open note heads, feeds into the voice's hesitant inhalations and exhalations. Growing increasingly active as it does, the voice begs to be heard in a narrative role, where its stuttering breath in measure 32 evokes a troubled frantic gasp that spills into a reflexive pant in thirst for air. The figure is repeated (in inversion) this time at an accelerated pace, contributing to a building suspense at the *poco meno mosso.*

After some time (figure 5.7), the breathing figure becomes more pronounced in both voice and clarinet, moving from hushed gasps to fully voiced exclamations but with the familiar rhythmic inflection of belabored

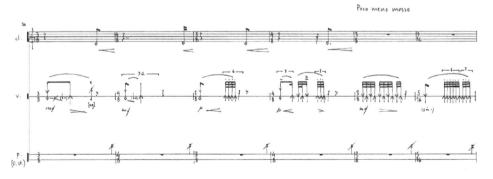

Figure 5.6

Excerpt of "Torso" from *Triumph* by Niels Rønsholdt, clarinet and woman's "breathing" with whip (slap stick), mm. 30–35.

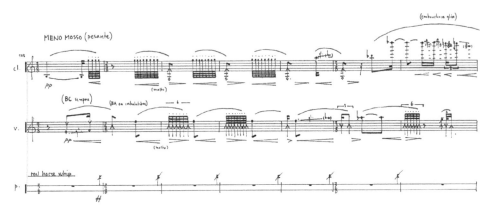

Figure 5.7

Excerpt of "Torso" from *Triumph* by Niels Rønsholdt, ensemble, mm. 108–114.

panting. This kind of development is instructive for listeners; it guides them associatively through the piece. Rønsholdt employs a common compositional technique of first assigning and then transforming meaning through repetition and a slight alteration of musical associations.

Because of the clarinet's typical association as an instrument of music, its entrance in measure 29 confirms the framing of this piece within a musical performance. The extrinsic significance of the clarinet as a musical symbol transforms the entire structure of the scene, bringing it from a pornographic moan to the concert stage. Eventually, when the voice and clarinet both become "vocal" (after the *meno mosso* at measure 108), the voice is confirmed in its typical instrumental role, at least in the role the voice has occupied since Berio's *Sequenza* "for female voice" (1965). But rather than isolate and design the voice entirely according to the music-compositional aesthetic, Rønsholdt's music is also driven by impulse. Jens Voigt-Lund ascribes Rønsholdt with "a fearlessness that is signaled through . . . sensuality. A sensuality that one doesn't find elsewhere in the genre of art music." Rønsholdt's compositions thaw "the chill of academia . . . through the sheer will of expression."[38] And once again we see a mounting tension between academic, or what was once termed "serious," music and that of the Other, the popular, the vernacular, the sexual, the banal.

Hearing sexual allure in *Fish & Fowl* is somehow telling of a particular bias on the part of listeners. Amazon reviewer Dean R. Brierly imagines *Fish & Fowl* not as a love story but as a horror scenario. Moments of austere stillness are palpable, beckoning to the inevitable tense build-ups that follow. The quickening pace of the high-heels at select spots in *Fish & Fowl* pairs well with the hesitant breathing of the feminized protagonist, and together these sounds suggest a scene that is less-than-friendly to women.[39] In our personal correspondence, Rønsholdt seemed to encourage hearings like Brierly's, writing: "Obviously, the emotional atmosphere becomes dark and sexualized when using breath i[n] this way, especially when presented together with whip sounds ... That's the underlying drama of F&F, the woman and the things that are happening to her (real or not)."[40] As if in response to Rønsholdt, Kathryn Kalinak's analysis of Otto Preminger's film *Laura* (1944) points to the "double-bind of female sexuality in *film noir*: it attracts and threatens; allures and repels."[41] Given these criteria, some might even characterize the combination of sex and violence in *Fish & Fowl* as doing a disservice to women by binding fear and thrill for viewers/listeners while reinscribing women's sexual one-dimensionality. Yet, as common an interpretation as it is, this one-dimensional impression completely flattens the nuance, breadth, and shadow afforded through music. The electrosexual as a singular concept deprives music of that which most uniquely defines this art: its frantic search for affordances lingering just beyond the scope of anthropic horizons.

For now, let us turn attention away from staid electrosexual tropes. We can leave this masculinist hearing in part I open-ended. We will return to *Fish & Fowl* and Hodkinson's impressions of the piece in the final chapter of this book, where, framed as a disturbance, it aligns with other feminist plights for sexual uncertainty.

II ELECTROSEXUAL DISTURBANCE

INTRODUCTION

Rapper CupcakKe, known for her overtly sexual performances as a teenager as well as her more recent outspoken support for LGBTQIA+ issues, bemoaned that, overwhelmingly, in reception: "The sexual music overpowers the . . . non sexual music."[1] This statement contrasts markedly with Robert Normandeau's concern about the absence of sex in electroacoustic music, examined earlier in part I. Indeed, he completely ignores popular music. Where perhaps this book threatens to replicate the problem Cupcak-Ke names, in part II I nevertheless investigate explicit sexual content in the place listeners most expect it—popular music by Black women. In the words of Ronnie Spector of The Ronettes: "Years ago we were like creations of genius men [producers]. We were little Stepford singers, interchangeable, one from column A, two from column B. We were seen as employees, not artists."[2] Part II of this book explores several such relationships between producers and performers, starting with Donna Summer in the 1970s. Spector observes the tense dynamic between authority and authenticity often has real-life consequences. Where part I examined the metaphorical cooptation of voices, part II examines physical, emotional, and mental violence and abuse Black women performers experience at the hands of their (usually male) producers.[3] By nestling this music, too, within a musical reception history of electroacoustic compositional philosophy, I hope to disrupt the common illusion that electroacoustic historiography is in any way set off or bracketed from the gendered and sexualizing rapport so frequently condemned in other contemporaneously emerging musics. Framed as a narrative of sexual agency, of choice and denial, popular music raises the same kinds of questions—even suspicions—of source/cause relations that Schaeffer, Henry, Lockwood, Ferrari, Normandeau, Hodkinson, and Rønsholdt sought to raise in their music.

The first part of this book examined tensions that arise in electrosexual music between creators and performers. I analyzed works that employed found sounds, especially voices, sampled from existing recordings to create new sexualized contexts without the performers' consent or even knowledge. I raised questions about composer integrity and the ethics of electroacoustic composition, as well as the geographic and chronologic proximity to the sources to argue that, in addition to arousing curiosity and enhancing sensory experience, the acousmatic "gap" between cause and effect can have negative consequences, particularly, as these relate to implicit gendered and radicalized hierarchies of the performers' presumed but unacknowledged and unnamed identities.

In this second part, I turn to music in which performers are acknowledged and named while still retaining aspects of electrosexual music's obfuscating veil. In addition to the previously investigated tension deliberately constructed by the creator at the expense of performers and emerging even before the music is released to the public, in part II, I extend this duality to a tripartite investigation of creators, listeners, and performers as active collaborators. Indeed, performers often control the "spacing" of source–cause–effect simply by appearing. In a world where performed hypersexuality is reflexively interpreted as an open invitation from Black women, it is a privilege to be able to perform open sexual gestures and sounds while denying the consummation of sex *acts* that could explicitly be named as acts, for example, in pornographic films (see chapter 8). As I explore in chapter 10, Ianna Hawkins Owen shows how jarring a performer's agency can be for those who seek to maintain the hypersexual status quo, for those who wish to deny Black women the right to choose whether to *act* or to *listen*. Confronting what is at once both artificial and real about the electronic medium, throughout part II we find that there exists a double standard in terms of who is seen as creating and who is imagined as listening—regardless of the kind of music. Different standards exist for those who are *allowed* to create and those who are free to *listen*. Such standards favor one constructed reality over others—namely, the white heteropatriarchal preference—and electroacoustic history is recorded according to this standard alone. As Zakiyyah Iman Jackson ascertains, "anti-black and colonialist histories have informed the very forms scientific discourse can take," and, equally, music's scientific procedural histories intersect with the libidinal discourses this book examines.[4] Jackson examines fictions that are generative and mutative of generic categories, and similarly, although the works in this book all draw on sexual themes, I consciously include music from diverging compositional philosophies, the electroacoustic, disco, R&B, and hip hop. Electroacoustic history is content to reduce performer agency to finite acts performed publicly. Rewriting this history, I show what performers are willing and unwilling to do in order to complicate and humanize them, to allow them to move from the main stage to the wings and even to join the audience as discerning listeners who have the right to say what does and does not please them—even if they, like us listeners, later change their minds. The music in this book does not capitulate any reducible structural form or representation of sex. That is, although the antiblack colonial mind prioritizes a sovereign "I," other expressions unfold simultaneously. These mutations, variations, fluctuations seek to upend any notion of one originary human and, thus, any one way—a *correct* way—of sounding sex.

Writing thirty years earlier, Patricia Hill Collins explains that though academia has not centered Black women's thought, academic scholarship values the forms Black feminist expression takes—forms like music.

> Not only does the form assumed by this thought diverge from standard academic theory—it can take the form of poetry, music, essays, and the like—but the purpose of Black women's collective thought is distinctly different. Social theories emerging from and/or on behalf of U.S. Black women and other historically oppressed groups aim to find ways to escape from, survive in, and/or oppose prevailing social and economic injustice.[5]

Collins centers historical and sociocultural context as the primary way to interpret and celebrate work(s) by Black women. Creative forms cling to humanity's socialized aspects—to gender, race, and sexuality, too.

Part I of this book presented a unified notion of electrosexual music with examples that all aspired to capture and contain erotic musical expression in the sounds of real life. Although these examples pushed the boundaries of what is achievable through human performance, they also drew on stereotypical tropes of feminized sexual expression, imitating women's orgasmic vocalizations that build climactically and subside once they've arrived. Clinging to the gendered expressions of the feminized voice also reinforces a heterosexual outlook, often with the composer sitting in the driver's seat, as the suggested object of desire. Likewise, the eroticized goes hand in hand with the exoticized, as both sex and subjecthood become coupled through their virtual estrangement.

Part II of this book features examples that retain gendered and racialized expressions of sexuality both audibly and visibly. Yet, the examples here encroach on a self-awareness absent from how those tropes arise in the works collected in part I. Electrosexual music cannot escape sexual objectification and subjugation, but the book's remaining examples seek, like Jackson's analysis, to revisit those calcified tropes of the genre and expand their limits.

UNVEILING THE ELECTROACOUSTIC

As we saw in part I, the electroacoustic medium developed as a way of disturbing the foregone link between the visible and audible. Many electroacoustic composers hoped that removing the visible performer would force listeners to imagine source–cause relations that exceed the visible world, thereby widening the inscrutable gap. Playing with source–cause associations became an "electroacoustic game." And yet, as my analyses demonstrated, time and again these efforts fell short from totalizing anonymity, since, after all, listening is a socialized behavior that relies on repetition and reinforced associations. To recall Simon Emmerson, the attempt to unveil a sound developed as an evolutionary imperative to alert and keep us safe from harm. In many instances, therefore, sound is the first warning of impending danger and, as such, it gains crucial significance in preserving human fragility.

In Europe and the United States in the nineteenth century, Black persons were the central threat to white sovereignty, citizenship, and civil liberty as defined during the antebellum and later "Post-Bellum—Pre-Harlem" period at the turn of the twentieth century.[6] Racial segregation served to protect white citizens and thereby extended to many areas of life. This white dominance became apparent also in musical participation. Listening proved a crucial privilege for enforcing racialized power structures, invisible boundaries enforced by physical separations. For instance, where the Pythagorean veil kept the master at a distance from his lesser educated disciples, white listeners for the latter half of the nineteenth century were shielded from Black performers, who were required to sing from behind actual curtains, and these separations made the performances more digestible to white audiences.[7] Following on from this practice, although the early twentieth century "race records" were actively marketed to Black audiences, these sound recordings platformed the blues—a decidedly Black musical tradition—as the most coveted genre in North America across "the color line," to borrow Frederick Douglass's phrase.[8]

African-American music publisher Harry Pace founded Black Swan Records with the intention of "meet[ing] what we believe is a legitimate and growing demand"—to "feature only coloured artists and to be produced by a company all of whose stockholders and employees would be coloured."[9] However, within a few years the wealthiest phonograph corporations, including Okeh, later Columbia Records, and Paramount had taken to distributing "race records" with an interest of securing the highest profits, which meant the medium in turn veiled the music's racialized origins.[10] And yet, this emergent practice of "audiencing" was in itself a turning point in American musical history, as Daniel Cavicchi explains, and thus is crucial in understanding electroacoustic reception—the musical practice most blatantly aligned with the gendered and racialized instance of veiling, and, as I show in the next chapters, also in popular music fandom, which is usually scrutinized on these grounds for what is visible.

Jennifer Lynn Stoever identifies commonalities between the performance practice of acousmatic veiling and overtly racist and racialized performances by tracing the etymology of the English word "listen," a word decended from the same Germanic root as *lust* (desire) and *list* (to choose). In her reading of Frederick Douglass's iconic *Narrative of the Life of Frederick Douglass, an American Slave* (1845), Stoever vividly illuminates "listening" as one expression of the "sonic color line" taut between white expectations of servitude and an enslaved people's will to exist. In such contexts, antebellum slaves were expected to enforce servitude that denied individual agency, "[slaves were] believed not to possess any of the agency associated with 'listening' in the dominant culture."[11] The enslaved should own neither choice nor desire. When slavery was abolished, and later when concert hall doors opened to Black (American) audiences, tension lingered between who would remain marked as a performer (i.e., a performing spectacle to be observed) and who would become inducted into the elite upper echelon of discerning listeners—each role distinguishable by what one wore, where one sat or stood, how one was expected by other listeners to behave, and even assumptions of how one listened. Concert etiquette was therefore born from listening practices that were not merely discerning but discriminatory.

A "respectable" listener could choose to focus sensory attention on the performer while ignoring factors that might disrupt the constructed performative framing, such distractions as noises made by other listeners or, indeed, anything that would cause them to question the illusion of performance. For example, audiences could choose to ignore whether the actor performing before them was capable of reflecting on the cultural or social significance of their performance or if they even had a safe place to sleep or eat. The ability to hear music while denying these dimensions of a performance affirmed the semblance of respectability acquired through live concert attendance and was presumed a gift bestowed unequally on members of society. Being a paying audience member afforded a sense of transcendence and superiority over the performer, which audiences expressed through listening—one semblance of the "European-style 'gentility'" Americans emulated in late-nineteenth-century concert etiquette.[12] Such superior passivity was enforced by the physical veil erected before Black performers in the "Post-Bellum—Pre-Harlem" period. Later, visibility became a key draw for concert attendees, when, as Stoever explains, white audiences craved music that paired "dialect/or black vernacular structures with bodies visibly or socially marked as 'black.'"[13] Listeners were compelled to discover the source of the sounds they enjoyed and, hence, to lift the veil, leading perhaps to the beginning of the end of enigmatic acousmatic desire.

Blackface minstrelsy—a form of "racial cross-dressing" where performers would enjoy "blacking up and then wiping off burnt cork"—was one means of securing higher footing in a Post-Bellum performing caste hierarchy, by Black performers too.[14] The more popular performers, however, were usually non-Black, often Jewish- or Irish-American women. By performing a collectively imagined and idealized "Blackness," these women embodied the acousmatic allure, and thus assumed physical ownership of the once metaphysical gap. As Pamela Brown Lavitt explains, "popular entertainers were crossing and breaking racial and gender boundaries, enacting narratives of immigration and Americanization on the Jewish female body."[15] Lavitt continues,

> Intended as comedy, coon songs ranged from jocular and dismissive to cruel and sadistic. . . . Redolent with stereotypes of blacks as cakewalking, cocksure, razor-toting, chicken-stealing, alcohol-fond idles who aspire to aristocracy but can't "fall into line," coon song sheet music and illustrated covers proliferated defamatory images of blacks in barely coded slanderous lyrics.[16]

Performing in this way, Jewish and Irish women who were not considered ethnically white could show they were in on "the joke"—on the same rhetorical plain as white audiences—by removing a black glove or wiping away their dark makeup during shows, thereby ensuring their Black counterparts remained bracketed off as performers who, on account of their skin color, could never step out of character. Thus, blackface minstrelsy ensured that both Black voice and Black body remained caricatured in the non-Black public eye, even when offstage. Jewish American comedienne Sophie Tucker explained in 1906 that she came to be a blackface performer as a way of making her performance more convincingly sexy to her audience, who had, since slavery, already been accustomed to evaluating Black women with negative sexual stereotypes.[17] Lavitt quotes a "Harlem manager who thought Tucker could not get a sexy song across. He jeered, 'This one's so big and ugly the crowd out front will razz her. Better get some cork and black her up.'"[18] Lavitt, however, argues that "coon shouting" became more than just a visible performance tactic.

There were varying degrees of asserting racial solidarity with a white dominating constituency.

> Jewish vaudeville foremothers—svelte chorus girls who obscured their ethnic roots, did not black up, and could definitively get a sexy song across. . . . They performed a doubly coded racial masquerade whereby acknowledgements of the "Jewish race" and outward signs of the "black race" (blackface) were mutually suppressed but registered nonetheless by audiences who consciously and unconsciously enjoyed the heightened subterfuge and ethnic bating.[19]

This kind of performer, who Lavitt calls, distinctly from debase "coon shouters," the "Jewish Ziegfeld Girl"

> tended to be slender elegant beauties, not the hefty, material bodies in blackface historians have documented and deemed representative of coon shouters. By virtue of their hyped glamour and assimilation efforts, the early Jewish Ziegfeld Girls generally refrained from "low," overt forms of racial disguise, namely blackface. Most concealed their Jewish roots. Race and ethnicity doubly disavowed were twin open secrets that audiences and the press playfully exhumed. Whether embraced, fervently denied or teasingly implied, the shrouded identities of Ziegfeld Girls who sang coon songs became irretrievably intertwined with familiar racialized cues, sounds and images, inevitably Jewish and black.[20]

Performers, in their performing roles, are seldom granted credentials as listeners *themselves*, hence the sense of something "broken" when the illusion is shattered (i.e., "breaking character," "breaking the fourth wall"). This is why it was crucial that audiences pick up on such racial and gendered "codes"— like a password—granting non-white/non-Black performers access to the space beyond the stage as listeners. Their ability to "contain" their racial otherness being an asset of the ability to project the quality of restraint coveted by white audiences aspiring toward the European-style "gentility" Cavicchi described above. Thus "norms" of performance come to influence expectations of how audiences receive codes, whether sexual, musical, or racial, since norms dictate whether codes are perceived as facilitating or disrupting the performative framing. This is to say that cultivation of the respectable listener involved observing and consciously or unconsciously deciphering the maneuvers of a racializing game—a game in the same ethical vein as the electroacoustic game in that the joke comes *at the expense* of the persons captured, even harmed, in the performing or recording medium.

From the perspective of Cavicchi's concertgoers, music could well arouse the senses. Hence, attending performances was in itself an act of emotional frivolity attained with the requisite status. That is, those who can pay to attend performances can afford to partake in its pleasures—paying for pleasures enacted by performers with or without their knowledge or consent; something familiar to us from the electroacoustic manipulations described earlier in this book. Quoting from the diaries of George Templeton Strong, who notably detailed events during the American Civil War (1861–1865), Cavicchi writes:

> [I]n the 1840s, [Strong] was willing to make gentle fun of the informality of his fellow elites at concerts. By the 1850s, however, he was associating lack of serious engagement with the music in terms that were more clearly, class-conscious, complaining not about chatting or sleeping among acquaintances but rather a degraded sense of taste among the anonymous "milliner's girls and their adorers, who look spoony and sentimental, and sigh out the aroma of shocking bad cigars by the hour together."
>
> Such shift in rhetoric was part of a wider social movement fostered by idealist middle-class reformers, who sought to introduce a new way to "love" music in the latter half of the century. Instead of passionate, enthusiastic attachment to a performance, they proposed what might be called a "classical" appreciation: ritualized, reverent, intellectual attention to the unfolding of a composition or work.[21]

As Cavicchi shows, the refinement of concert etiquette emerged as a way of further setting apart audiences from performers, "by removing the spontaneity and showmanship of live performance that might lead to obsession or spontaneous emotional display."[22] Sound recording was a clear extension of this logic, providing an unchanging record sure to fend off any performative spontaneity. It was also a way of secluding audience members, even from one another, so that they might express their pleasures in private—certainly a luxury—or so that listeners who could not feel comfortable expressing feelings on their own would not need to suffer through hearing others relieve themselves of their "musical ecstasy" (Cavicchi's term) in this way. In short, Cavicchi identifies the racist and misogynist plight of the power structures upholding what Searle terms "collective intentionality" from chapter 3.

Matthew D. Morrison echoes Lavitt, determining that when blackface performances became more controversial, elements of these stereotyped performances reinforced Black coded meaning also in "unmasked" performances of popular music in the United States. While visibly unmasked, performances nevertheless institute the racializing impetus behind acousmatic veiling in that they rely on what Morrison calls "Blacksound":

> Blacksound's stereotypically authentic traits are fueled by the scripts created by early blackface performers as the most consistent references to blackness—angular posture, grounded movement, broken dialect, and syncopated rhythm—were often jointly performed in blackface and subsequently viewed as ontologically black by producers and consumers.[23]

Morrison emphasizes that music enhanced staged performances of Blackness,[24] and since Blacksound supersedes the visible, it could also permeate sound recordings in which there appeared no visible performing body. The distance instituted by recorded sound was one behavior inherited from concert music etiquette, situating audiences distinctly away from performers. But, as Alexander G. Weheliye observes,

> From the onset of the mass production and distribution of recorded sound in the 1920s, black popular music functioned as the embodiment of the virtual voice. Instead of merely producing a disembodied virtuality *avant la lettre*, the phonograph harbors an always embodied virtuality, particularly in relation to black voices. Paradoxically, black voices are materially disembodied by the phonograph and other sound technologies, while black subjects are inscribed as the epitome of embodiment through a multitude of U.S. cultural discourses.[25]

To recall Mladen Dolar's insistence on the habitual materiality afforded to the voice, sound recording became one of few arenas in which Black forms of cultural, social, and personal expression remained grounded, never fully segregated, hidden, or cloaked in a veil of austerity.

Early on, the most popular instances of recorded sex music can be traced to blues singers from the 1920s, Gertrude "Ma" Rainey's euphemistic "Shave 'Em Dry Blues" (1924) or Bessie Smith's "You've Got to Give Me Some" (1928), to music by hundreds of blues women recording at that time: Alberta Hunter, Ida Cox, Ethel Waters, Lucille Hegamin, Edith Wilson, Victoria Spivey, Rosa Henderson, Clara Smith, Trixie Smith, Sippie Wallace, to name only a few such examples.[26] Women led the way in the division known then as "race records," a designation white record executives adopted from Black communities in 1923 to refer to music geared especially toward their demographic.[27] For many theorists, the entire project of the "classic blues" was an expression of sexuality, not merely as sex *sine qua non*, but as a resistance movement to the system aimed at maintaining typological hierarchies of race, class, gender, and sexual orientation. Put succinctly by Hazel Carby, "What has been called the 'Classic Blues,' the women's blues of twenties and early thirties, is a discourse that articulates a cultural and political struggle over sexual relations: a struggle that is directed against the objectification of female sexuality within a patriarchal order but which also tries to reclaim women's bodies as the sexual and sensuous objects of song."[28] Simply put, it was important and necessary for this movement's statements to be articulated by women's voices, by Black women's voices.

If when picking up this book the names Schaeffer, Henry, Ferrari, Lockwood, Truax, and Shields did not immediately jump out, perhaps, you, like many who have taken an interest in my work, thought of Donna Summer, Tina Turner, Beyoncé and Rihanna. It isn't a coincidence that the most well-known music in this book is created, indeed *performed*, by Black women. Sex and music and in music is most commonly associated with Black women. What I mean is that most people talking or writing about music today do not hesitate to mention Black women, sex, and music together. The habitual collapse of these criteria has been illuminated and resisted in Black feminist thought by Patricia Hill Collins, Hortense Spillers, bell hooks, and many more. Sound recording would never become entirely integrated—unveiled—by the sheer facet that recordings could be curated and preserved only to be encountered by certain listeners on particular occasions. One way of preventing any discomfort white audiences would feel about enjoying "race records" came by fact of these recordings being electroacoustic and therefore implicitly color-blind. Nevertheless, Black performers have crucially influenced the sounds and discourses emerging with electronic, and now, digital technologies. The chapters in this section of the book examine Black women's contributions to electrosexual music in order to illuminate the feigned and, indeed, failed acousmatic veiling of race and gender.

CHAPTER SUMMARIES

When I began this project, my initial focus was solely on electroacoustic music because, within the university, my mentors and colleagues had projected certain expectations for what my project would become. Even still, many, many, many academics asked me if I would discuss music by Donna Summer, presumably because of a reflexive equivalence between her music and her person—both wildly hypersexualized. Initially I had no interest in rehashing debates about the quality of sexual allure her music expressed and, in turn, she herself possessed. But as these provocations grew louder, in conference Q&A sessions, anonymous peer reviews, and informal conversations among colleagues, I found that there was still something that needed to be said about Donna Summer in order to point out, and possibly dispel, such habitual presumptions about the artist's sexual availability. This was the motivation for chapter 6, which, like many previous analyses of sex and music, revisits Donna Summer's first hit "Love to Love You Baby" (1975). After the song stormed radio airplay, the singer's suggestive moaning was immediately capitalized on, and everyone wanted a piece of the profits. Since the song's release, more than thirty songs have sampled it, presumably with an aim to replicate this simplistic stigma of female sexuality. Yet, samples of the song are heard in tracks ranging from hip-hop/R&B group TLC to pop diva Beyoncé, which also suggests that, beyond reductive reproductions, sampling contributes productively to progressive expressions of sexuality in music, even expressions that refuse to perform explicit sex *acts*. This chapter emphasizes the invisible electroacoustics of popular music, asserting that sound alone is no substitute for the consummation of sex *acts*. Performers might deliberately deny visible sex acts as a way of asserting a "safe" distance from listeners' conclusive sexualized fantasies.

Most importantly, however, analyses of Summer's song rarely acknowledge her as a creator in the same capacity as her producers, Giorgio Moroder and Pete Bellotte. Summer is never considered a composer with her own version of electrosexual expression. My analysis instead examines her composition of the song's chorus and verse to place her on equal footing as the composers encountered earlier in this book.

Chapter 7 examines a work composed to resist electronic music's masculinist stronghold. American composer Alice Shields was instrumental in establishing the Columbia-Princeton Electronic Music Center through the 1960s, and, though she is only recently acknowledged as influential in electroacoustic histories, she worked there for thirty years.[29] She is, therefore, well accustomed to the genre's games. Shields's electronic opera *Apocalypse* takes up this knowledge in a massive unveiling that removes any mystery delimited by the heteronormative framing or typical "concert" etiquette. By staging sex in the most obvious, visible, and audible manner, Shields confronts tacit sexual stigmas that plague both her contemporaneous American political climate and the lingering British colonial attitude toward revivals of *Bharatanatyam*, an Indian classical dance practice featured in the opera. The chapter shows how *Apocalypse* addresses sexual censorship through music, text, and choreography to envision a world in which sex is not stigmatized but instead exists as a productive and inseparable aspect of culture and music. In placing seemingly disparate traditions of dance, music, language, and mythology on equal footing—movement patterns of the *Bharatanatyam* Hindu dance dramas, gestures inspired by the *Kamasutra*, poses of Greek statues, Gilbert Austin's *Chironomia*, and text portions from Old Irish, ancient Greek, English, and Sanskrit sources, and setting these texts and directions to jazz and Americana folk rhythmic accompaniment patterns augmented with Carnatic music-aesthetics and structural elements of Western opera all synthesized together into a MIDI and analog tape realization—*Apocalypse* provides what dance theorist Janet O'Shea terms an "ongoing interculturality."

Staying the course of feminist unveiling, chapter 8 delves into the "sound score" of an actual pornographic film. Annie Sprinkle (named for the defining characteristic of her urine-drenched performances) invited experimental tape composer Pauline Oliveros to compose the soundtrack to her *Sluts & Goddesses* Video Workshop in 1992. In the vein of educational pornography, or edu-porn, the film features only women, and is intended only for an audience of women intent on exploring the limits of their orgasmic power. Although fixed as a recording, Oliveros's score was improvised in real time at her home together with Sprinkle and co-producer Maria Beatty. The analysis revisits discourses about the artistic merits of pornography to examine the congenial Oliveros–Sprinkle partnership as a mutually beneficial expansion of performance practice and audience. Where chapter 5 found correlations between sexual arousal and horror, this chapter interrogates another side of the sexual spectrum, the humor of sex play.

Barry Truax is an electroacoustic composer who has been uniquely outspoken about his own (homo)sexuality while taking many strides to amplify women's voices in the academy. Chapter 9 explores how Truax treats his concept of "homoeroticism" electroacoustically. Moving away from the gendered binary situated in the dominant heteropatriarchal retelling of music history, the chapter analyzes Truax's *Song of Songs* (1992) for oboe d'amore (English horn) and two digital soundtracks, a work setting the erotic dialogue between King Solomon and Shulamite from the biblical Song of Solomon text. *Song of Songs* gradually blurs the discrete boundaries between genders and hence also the perceived sexual orientations of the respective speakers through digital processing. Framing his composition within contemporaneous discourses in music, literature, and the vocal sciences, my analysis shows how Truax's homoerotic conceptions present novel approaches to gender, relative to the typically reductive gender presentation of the 1980s and 1990s.

Truax set out to create homoerotic music because he understood the importance of representation. He identified a lack of queer sensibilities in the history of electroacoustic music, what he recognized as the absence of "personal voice." In his words, "Art is said to mirror society, but if you look in the mirror and see no reflection, then the implicit message is that you don't exist."[30] On the one hand, Truax sees it as the composer's duty to account for this absence, to "[progress] from being an artist who happens to be a woman, gay, lesbian, transgendered, of colour, and so on, to one for whom any and all of those qualities become integral parts of their work."[31] But, on the other hand, the composer ascribes listeners with much less responsibility. Where before certain qualities were muted entirely from musical discourses, such that the universal or majority view was implicitly heterosexual, Truax, Sprinkle/Oliveros, and later TLC, THEESatisfaction, and Janelle Monáe accede to broader possibilities.

Though included as the "A" in the LGBTQIA+ acronym, asexuality is rarely invoked among the recent queer musicological literature. For example, in *Queering the Popular Pitch*, "asexual" is mentioned as "curious" in Sheila Whiteley's contribution on Patti Smith, and Sarah Kerton refers to boybands as "unthreatening[ly] asexual" in contrast to t.A.T.u.'s hypersexuality, meaning that the definition each uses opposes sexual expression of any kind to understand asexuality as lacking.[32] Other recent volumes ignore the concept altogether, including Stan Hawkins's *Queerness in Pop Music: Aesthetics, Gender Norms, and Temporality*, and Doris Leibetseder's *Queer Tracks: Subversive Strategies in Rock and Pop Music*.[33]

In chapter 10, I invoke asexuality on the basis that these performers vocally express their assignation from sex in their personal lives while performing in a way that audiences have interpreted as sexually suggestive or even advertising an artist's sexual availability. These performers are not anti-sex but only resigned from performing certain sex *acts*. The chapter looks at one instance of the "Love to Love You Baby" sample in TLC's "I'm Good at Being Bad" (1999). By incorporating Black feminist scholarship regarding sexual choice and abstinence, I demonstrate how TLC, rather than risk further objectifying Summer with gaudy samples of women moaning (like so many other instances of borrowing), choose instead to quote Summer's original compositional matter from the texted portions of the song.

Taken uncritically as evidence of *sex*, sexually suggestive performances from Bessie Smith to Donna Summer to TLC to Rihanna and Beyoncé replicate the reflexive associations of music and hypersexuality. Although hip hop is arguably the genre in which most listeners today commonly encounter sex audibly in music, this book contextualizes audible sex tropes by drawing historical connections where genres intersect on account of electronic technology. Similar to pornography's bifurcated reception, outlined in my introduction, Tricia Rose recognizes a "war" regarding female representation in hip hop, one side of which advances objectification via sexism and antifeminist agendas. The other, while appreciating the music, rejects the genre's typically demeaning portrayals of Black women. This second group, argues Rose, "encouraged a culture of black female sexual repression and propriety as a necessary component of racial uplift."[34] Rose, therefore, maintains that hip hop is not responsible for sexism, misogyny, and racism but that rap music is only one manifestation of the injustices that permeate society. Any book treating intersections of technology, music, and sex would be remiss without a discussion of R&B and hip-hop music, but these genres are no *more* explicit than their blues or electroacoustic predecessors. Given the very specific economic, social, and cultural circumstances that impact women of color—and most glaringly Black women, who are almost entirely absent in electroacoustic music[35]—the music in part II requires a dedicated analytical framing sensitive to the balance between celebrations of liberated sexuality and racialized and racist constructions of sexuality in musical contexts. As Rose attests, in order to achieve gender equity and combat sexism, scholars must consciously identify the ways in which media demeans women and the ways in which it does not.[36] Toward realizing this goal, it is not enough to include music created and performed by Black women, it is also crucial to acknowledge the scholarly impression left by music's Black intellectuals.[37]

6 DONNA SUMMER'S "LOVE TO LOVE YOU BABY" (1975)

The following quotations, one from Donna Summer (1948–2012) and one from Giorgio Moroder (b. 1940), describe the same event, the composition of the iconic "Love to Love You, Baby" (1975), one of the earliest international disco sensations and a song whose electrosexual message sent crowds rioting.

> I had this idea at home one day, and I ran into the studio, and I said Giorgio, I have this idea. Would you—do you think you could write something to it? And I sort of sang it to him, and he kept saying it over, he says "love to love you, I love to love you . . ." He kept rubbing his chin and thinking like a little mad scientist, and then he went into the studio, and Giorgio had written this track.
>
> And I began to—he asked me to go in and start singing something, and I didn't have any words other than love to love you, baby. So I was improvising on the track live. And that really became "Love to Love You, Baby," the original track. . . . And then he went from there and produced something, and then I began to sing it. And then I began to play with the, there weren't that many words. So I played with the sound of the music, you know. (Singing) "I Love to Love . . ."
>
> You know, we didn't have the same technology we have today. So I had to do everything with my own voice.
>
> —Donna Summer[1]

She started to moan, but it wasn't really, you know, the real one. So finally I said there's only one thing to do . . . not what you think [audience erupts]. So, first of all, I threw the husband out, because he was the main problem, so we dimmed the lights and I let the tape run—it was one tape, but she sang it a few times, but then the moaning she started from the beginning to the ending in one go and the lights were down. I could barely see her face, and she probably couldn't see me—although I would have been a great inspiration. [audience laughter] . . .

—Giorgio Moroder[2]

Today, "Love to Love You, Baby" is co-credited to Giorgio Moroder, Donna Summer, and assistant producer Pete Bellotte.[3] Both Summer and Moroder tell the story that Summer was the instigator behind the track's memorable sexual moaning—the quintessential feature compelling Casablanca Records executive and founder Neil Bogart to sign the singer, produce the song, and later insist on extending it from a three-minute radio reel to a seventeen-minute full-on immersive experience. Summer's mythologization as the "lady of love" and "sexy Cinderella" constructed an idealized image stemming from the popularity of "Love to Love," an ideal I equate with the electrosexual feminized voice from earlier decades. Unlike the earlier tracks from part I, the singer is credited and, indeed, she performed the song over and over again (when permitted by local authorities). Nevertheless, the unethical electroacoustic gap remains intact, since audiences collapsed any distance of the artist and her performance. Her reputation as an insatiable sexbot followed her throughout her career, even bleeding into her personal life.

Though Summer is given full credit for performing on the track (her backup singers are given credit only by first name in the liner notes accompanying the album), something about the rhetoric surrounding the myth of the song's creation discounts Summer's participation in the process. The storytellers echo one another, likely on account of having relayed this experience so many times, but each version contains a telling subtext of the work that goes into creating electrosexual music, a familiar hierarchy of labor in and out of the studio. In some respects, the stories Moroder and Summer respectively convey are unique to this song; still, similar narratives abound in singer–producer collaborations. At issue is the emphasis on Summer's role as a (mere) performer, who implicitly lacks the musical mastery required to compose such a successful track. Like so many other singers at that time (recall Ronnie Spector's remarks from the introduction to part II), Summer's performing presence is reduced to a voice and nothing more, while Moroder and Bogart claim the accolades of discovery. In this sense, Summer's story is comparable to the Robinson Crusoe tale from part I, with her in the role of Moroder's Friday. But despite Moroder trying to pillage what he could, Summer's work speaks for itself.

COMPOSING "LOVE TO LOVE"

Figure 6.1 transcribes the "Love to Love" hook Summer describes composing in her interview above with NPR's Terry Gross. Figure 6.2a transcribes the verse producer Moroder later harmonized with some melodic variation. We know from Summer herself that she composed the hook, as one of the song's studio vocalists, Lucy Neale, confirms.[4] It's uncertain who composed the verses' melody, but Neale recalls that when she and the other singers came into the studio, Moroder only had a few chords to work with, and that's what they ended up singing. These harmonies are transcribed in figure 6.2c. Neale suggests that Summer and Bellotte composed the verses together, though usually he would only compose the lyrics. In a later track, "I Feel Love," Bellotte acknowledges that they both wrote the lyrics together, while Moroder came up with the melody, but for "Love to Love," given the accounts above, it seems that Summer also wrote the melody for the verse.[5] Again, as Neale recounts, Moroder only had the harmonies, derived from the verse's descending tetrachord.

As Summer says, there weren't many words to the song, so she used her voice to emulate the kinds of technological transformations that in the early 2000s, when the interview took place, would have been commonplace in recording studios. It's not clear what exactly she sang in that session, since

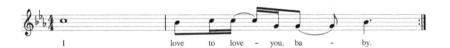

Figure 6.1
"Ahhh Love to Love You, Baby" hook.

When you're laying so close to me, there's no place I'd rath-er you be than with me –

Figure 6.2a
Melody of verses in Summer's "Love to Love You Baby" outlines a descending chromatic tetrachord every half bar, ending in a "turnaround bar" looping back to the hook (figure 6.1).

Figure 6.2b
Chromatic descending tetrachord, sustained ambient string background from "Love to Love You Baby" (11:40–12:40).

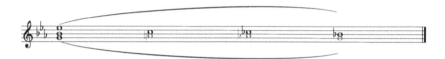

Figure 6.2c
Parallel chromatic descending tetrachords, backup singer chorus (in black) and sustained ambient string background (in gray) from "Love to Love You Baby" (11:40–12:40). E-flat and G sustained throughout.

Moroder and Summer's stories do not line up and neither were there any backup singers or instrumentalists. What becomes clear from Moroder's testimony, though, is that it is this improvisatory moaning and only the sexy moaning that he wanted attributed to her.

Isolating the strong beats of the verse melody (figure 6.2a) approximately every two bars outlines a descending chromatic tetrachord: G-flat, F, F-flat, E-flat. The motive ends with a "turnaround bar," where B-natural tonicizes the C that returns us to the hook of figure 6.1. The descending chromatic tetrachord returns in the bridge, transposed by the string orchestra (figure 6.2b)[6] and also by the backup vocals, such that together they form divergent yet harmonically complementary figures (figure 6.2c).[7] The backup singers enter in chorus to accompany the strings. Two voices sustain the outer tones—the upper E-flat and lower G—while a middle voice descends the chromatic tetrachord, this time transposed and slightly altered, B-flat, C, C-flat, B-flat. The superimposed string tetrachord appears in gray in figure 6.2c to visibly distinguish the vocal and string lines.

In the bridge, the tetrachord figure gives the illusion of descent, yet by the end of the bar we find ourselves, pitch-wise and also harmonically, back where we started with G and B-flat. Rearticulating the E-flat initiates a repeat of the bridge, as the track does numerous times. Alternatively, filling in C together with the existing G and B-flat at the end of the phrase would bring us back to the hook. In this way, the chromatic tetrachord in effect behaves like an unresolvable Shepard tone, immanently descending like an audible revolving barber's pole.

From this brief sketch, we can easily imagine how these two motives, the hook and the descending chromatic tetrachord—the very kernels Summer brought to Moroder, as recollected above—served as compositional fodder for the harmonic and melodic content of a large portion, if not all of the song. Summer came up with the melodies. Backing vocalist Lucy Neale—who sings on the bridge—explained to me in an interview quoted below that during the recording session in which she took part, Moroder only had the simple harmonic sequence (of figure 6.2c) to go on, as the improvised studio sessions with various instrumentalists thereafter had not yet been recorded. In this chapter I show that Donna Summer had an active role as a composer and an equal hand in the production process of the revolutionary "Love to Love" song, much more than the impressionable pawn her producers and record executives made her out to be.

MYTHOLOGIES OF THE DIVA

Casablanca Records founder Neil Bogart had a heavy hand in cultivating Summer's diva stature, sending her to Hollywood for a complete makeover and famously prompting radio stations to air the full-length track at midnight, with the slogan "seventeen minutes of love with Donna."[8] Radio play left open the acousmatic gap for individual sexual fantasy, but Bogart ensured that radio listeners closed the gap by transforming their electrosexual encounter into fantasies of Summer herself.

Obviously, to become as successful as she was, Summer had to be outgoing, so her perceived image was not entirely fabricated. While in the cast of *Hair* in Munich, Germany, the acting warm-ups included what she describes as "touchy-feely exercises" and she even featured in the show's nude scene, which she saw as a "celebration of the freedom of nature."[9] Given these events, "Love to Love" was, at least initially for Summer, an expression of freedom, perhaps even a form of resistance to the domesticated image of women's sexuality during the postwar years. However, Casablanca's commodification of Summer's image—her body and her voice—obscured that message of sexual protest. As Summer explains, "Part of the problem was that Neil had created a completely new persona for me, which had absolutely nothing to do with who I really was," and yet she was bound to fill the role in order to continue performing, to meet the obligations of her contract.[10]

Bellotte relays one myth of how these events unfolded: "Bogart was having an orgy at his house, there was a lot of coke going on and, to use [Bogart's] own language, they were all 'f*cking to this track' and the crowd there had him replay the song over and over again."[11] A more realistic story is one told in Summer's autobiography. In January 1975, Moroder took the track with him to the annual music business conference *Marché International du Disque et de l'Edition Musicale* (MIDEM) in the South of France, where he met Neil Bogart, who proposed that the "Love to Love" demo be extended into a seventeen-minute track to cover the whole side of an album—just enough time to "get the job done."[12] In 1975, shortly after her first divorce, Summer left her life in Germany and returned to the US to work under the Casablanca Records label, a move Summer felt was a step backward from her theatrical success abroad.[13]

So that she could easily come to represent the message of the song, Summer had to embody the typical "one-dimensionality" ascribed to this new identity of a mindless and insatiable sexbot.[14] She needed to appear both miraculous and unique, as one-of-a-kind while also seeming somewhat simple and easy (to define). In interviews, Bellotte and Moroder stress how little effort Summer exerted in the process of recording, discounting her efforts as "natural talent." Although Summer came up with the lyrics and the melody to "Love to Love," Moroder took care to emphasize that Summer's vocals on the demo were improvised and that they were even recorded in one take, contradicting Summer's own recollection of the process as iterative. In Moroder's words:

> What was remarkable about her was that she would come into the studio to record a specific song at, say, four in the afternoon, she would then talk and talk and talk for a couple of hours, and all of a sudden she would look at her watch, say, "I've got to hurry," and go to the mic, sing the track and be gone.

Moroder drives the point home, claiming, "Donna was never involved in the production in any way whatsoever, and she'd never hear any of the songs until they were totally finished and mixed." All this is to say that when interviewed about her working process, he either does not mention or significantly downplays Summer's role, diminishing the fact that the iconic melody was her own original tune and that she enhanced this melody with skills she attained through persistent improvisatory and theatrical training. Tapping into commonplace rhetoric on the laziness of Black laborers, from Moroder's perspective Summer was sitting idly by while her music simply materialized—as if in her time spent in the studio she hadn't heard or produced anything, as if her hours *outside* the studio weren't spent propping up the stage they all had set for her.

Though Summer's public image seems effortless, in reality, work was immeasurable, and likewise, unrelenting. She recalls being "emotionally exhausted," at the height of disco, plagued by insomnia and headaches, and as having undergone a mental, emotional, and physical transformation. In the same *Rolling Stone* interview, she tells the story of an Italian tour where she was transported from one plane to another in a wheelchair.[15] But Summer's physical and emotional battles offstage could be superseded by her "sexy" public persona, and certainly gender and race play no small role in this appearance.[16]

Take, for example, a feature-length interview with Summer appearing in *Ebony* magazine in 1977.[17] The article, attempting to redeem Summer's image as a strong, independent, African-American woman, challenges several circulating rumors about the singer (including the widespread belief that she was a man in drag), focusing on Summer's aspirations toward becoming a career-driven professional while balancing her domestic responsibilities. Even in this article that attempts to dispel some of the clout upholding Summer's physical, psychic, and financial exploitation, the author succumbs to raunchy, physically objectifying descriptions of his interviewee in an attempt to supposedly redeem her from some imagined gendered transgression: "As she bends to put an Edith Piaf record on the turntable and her low-cut silk dress reveals the marvelous ripeness of her breasts and clings oh so gently to the curves of her behind, it seems ridiculous that she is defending herself against the absurdity that she may not be a woman at all."[18] Subjecting Summer to such scrutiny alongside a prolonged tangent into the details of how she lost her virginity to emphasize her sexual availability, the article leaves much to be desired in the way of affirmative action.

There is no shortage of evidence that Summer was overexerted, overworked, and underpaid. It is no wonder she identified with the sex workers for whom the lyrics of her many songs advocate a shared vulnerability owing to her unabating physical and emotional labor. Her sexual availability, paired with her perceived celebrity and financial worth, appear to reinforce the stereotype of the "working girl" on a larger scale. It was easy to launch this myth of Summer's frivolity, a dual facet of both her natural talent and erotic allure, since the story mapped onto concurrently running tropes about disco music's creators and audience, as well as centuries-long racist notions of Black women's hypersexuality.[19]

EASY LISTENING THE HARD WAY

Critics of disco, namely rock and classical music fans, stressed the monotony of the disco aesthetic: it is *easy* and *fun* to dance to disco precisely because of its "four-on-the-floor" beat and the supposedly predictable structure of the songs. As Brian Ward snarks, "Just about anyone seemed capable of piecing together a passable disco-by-numbers record."[20] Commonly equated with simplicity, disco's musical and textual repetition becomes transposed onto the identity of the performer who is characterized as leisurely and unlaboring. Reinforcement of this leisurely attitude came to define disco in opposition to other forms of music, which is quite a serious accusation when the musical genre is attributed to African-American women performers with a mostly gay fan base.

Summer's "Love to Love" was on heavy rotation in dance clubs (spending four weeks at #1 on the *Billboard* Dance Club Songs chart in 1976),[21] which, in the United States, had long been populated by people at the social margins—predominantly Black and Latino gay men.[22] Like the performer, the music's audience members were pigeonholed, and mainstream musicians and social critics alike adopted negative racializing, gendered, and sexualizing stigmas commonly attached to these communities to describe the music. Such descriptions conflated musical aesthetics with stereotypes derived from idealized perceptions of actual people. These wider perceptions led to a series of hatefully motivated incidents, such as the "disco sucks" slogan of the 1979 Disco Demolition Night, when thousands of people stormed Chicago's Comiskey Park, wreaking havoc on the baseball field and burning tens of thousands of records in protestation of disco.[23] In the words of musicologist Mitchell Morris, "All the world knows by now that 'disco sucks' was a response of thinly disguised homophobia."[24]

Both rock and disco employ an electrified aesthetic, but when this music, marked by homosexual, African-American or Latino men and women, came to top the charts—like when in 1976 "Love to Love You Baby" hit number 2 on the *Billboard* Hot 100—straight, mostly white, male rock critics took to distancing themselves from the demographics, which, again, became conflated with the music's aesthetic qualities.[25] For these critics of disco, unlike the dance club focus on recorded tracks, rock thrived from the *live* evolving energy of a skilled soloist. Musicologist Judith A. Peraino writes of how "recorded repertories" served to advance virtual "homomusical communities" where individuals could entertain sexual notions that would not have been accepted in public spaces dominated by heteronormative expectations. Dance music, on account of its club-based culture, rarely featured musicians in a live performing role, since those who did "play" the music—disc jockeys—were evidently dismissed by critics and reduced to a mere playback mechanism. Here, again, we see another method of prioritizing medium ("the recorded character of the recording") above musical content and musical agents.

However, even readings sympathetic to queer culture sometimes end up replicating harmful racial stereotypes. In Peraino's words: "Recordings (on vinyl, tape, or compact disc) can be considered a stand-in for the performer, who is the real focus of attention and object of desire, even fantasy."[26] And extending this fantasy, a decade earlier, Brian Currid illuminated how queer club-goers became fueled by their identification with the sexually liberated diva to act out fantasies of performing the diva together with other dancers.[27] Though Currid's analysis importantly theorized dance clubs as sequestered queer safe havens, his hypothesis of the invisible cyber-diva reinforces the disco singer's iconic status as an immaterial sexual deviant whose digital form has no semblance in the physical world. However optimistic, and necessary as an uplifting queer (white) narrative, Currid inadvertently reprises a trope that affirms the "othered" or alienated "robo-diva"—a performer who no longer embodies the physical presence of a Black woman but rather becomes projected as an alternative and (negatively stigmatized) deviant sexuality that she does not even claim (homosexuality). That is, by way of her gender, the diva is necessarily excluded from having a perspective in this kind of analysis, enforcing divisive boundaries that do not fairly represent either Black women's sexualities, whether straight, gay, or otherwise, or listening practices of gay men, who may not identify with the music in this way and who may even be conscientious to the ways such narratives replicate the derisive tropes of "robo-diva R&B."[28]

Negative reports of electronic dance musics, including disco, appear to have been driven by the same racially motivated reception of hip-hop sampling; namely the argument that invoking technology as a means for musical production was not actually *musical*, only a kind of lazy redundancy or passively replicated sound, and inauthentic.[29] Indeed, though the stigma persists, many of disco's musicians and engineers combatted the static image marketed by disco executives, making it their job to force "dead" music to come alive by way of various production effects, including the cut 'n' mix, segue, and montage.[30] Despite being portrayed among (white) rock fans and elitist classical listeners alike as simplistic and easy to come by, in actuality, just like the electronic music of earlier decades, the cutting-edge technology behind disco was hardly "systematic."[31] Making music that sounded endlessly driving and repetitive took a lot of effort, as we see from the cumulative form arising in the brief analysis of the pitch content in the previous section.

Additionally, part of the reason disco production was so expensive was the sheer number of people it involved. Aside from the production team credited on the album—Bellotte (producer); Moroder (arranger, mix-down engineer); Michael Thatcher (string and horn arrangements); Reinhold Mack and Hans Menzel (recording engineers)—before anything could be mixed, additional instrumental and vocal tracks had to be recorded by "Molly" Moll, Nick Woodland, Pete Bellotte (guitars); Dave King (bass), Mike Thatcher, Giorgio Moroder (keyboards); Martin Harrison (drums); Franz Deuber (string section); Bernie Brocks, Giorgio Moroder (percussion); Lucy Neale, Betsy Allen, Gitta Walther (backup singers).[32]

In the seventeen-minute version, from which all subsequent singles were then edited down, even the disco groove was not easy to create. Bellotte recalls that they initially tried a generic drum machine for the basic beat, but the sound was so horrible they couldn't feature it on the album. The engineers were required to seek out prerecorded drum samples onto which a live drummer was then dubbed—and this later became common practice in many styles of music. The practice was especially widespread in the 1970s among Jamaican Dub producers.[33] In the US, such dubbing became a staple through disco and James "Tip" Wirrick, musical director and guitarist for the disco celebrity Sylvester, remembers that "In those days . . . you would literally scour records and the airwaves for any section of one to two bars of just drums . . ."—a now commonplace practice thanks to hip hop.[34]

And though Robby Wedel had figured out how to sync multiple tracks through the Moog synthesizer, there was still quite a bit of human performance involved. After recording and splicing the tracks, all the takes had to line up, and since the beat track was only so long, engineers were forced to begin recording maybe twenty or thirty times before being able to complete a single take of the entire song. Reinhold Mack recalls, "The whole thing is, in a way, a live performance. . . . Part of it is the journey of getting there."[35] And Moroder attests: "All this talk of machines and industry makes me laugh. Even if you use synthesisers and sequencers and drum machines, you have to set them up, to choose exactly what you are going to make them do. It is nonsense to say that we make all our music automatically." This description resonates with ongoing arguments still today regarding algorithms and computerized automation.

Several studio musicians at Giorgio Moroder's Musicland were brought in to play individual tracks, and these tracks were then subjected to the experimental recording techniques of engineers Reinhold Mack and Hans Menzel. Whether in the violin-heavy bridge or the bass-driven breakdown, even microphone and amp positioning became crucial to the recording process, since these determined various timbres, reverb, and other supposedly "synthetic/synthesized" effects.[36] Dismissal of dance music as leisure music not only affects creators but also stigmatizes disco's audiences, whose affinity for this music is likewise cast by critics as an aesthetic predisposition toward the simplistic and synthetic—regardless of the actual tastes of listeners who might enjoy many genres, even those that critics deride as conflicting according to aesthetics and/or demographics.

With all of the work that went into producing the "Love to Love" track and all these cooks in the kitchen, it surely took a lot of work to secure Donna Summer as the sole representative for the sexually allusive components of the track.

SEXUAL VENTRILOQUISM

Because of her training in classical music and as a musical theater belter, Summer recalls being uncomfortable singing in the soft, breathy vocal style of "Love to Love," a song she never dreamed would gain any public traction.[37] Pop scholar Jon Stratton hears Summer's breathiness as a disguise. Not only is the sexual cooing foreign to Summer's own usual performance, but the kind of timbres she uses code her performance within dominant hearings of white women's sexual performances.[38] Stratton traces Summer's rehearsed strategy through African-American gospel (though her uncharacteristic breathy singing is contrasted to gospel's typical belting) to Ray Charles's "What'd I say" (rather more overtly sexual in comparison to Summer who doesn't invoke the quintessential scream) but, surprisingly, he does not draw comparisons with familiar icons of Black women's sexuality like Josephine Baker. Rather, Stratton aligns Summer's work with a white sexual history somewhat removed from her own notion of subjecthood and self, arguing that Summer dons a white persona, to tease out something akin to whiteface.

Stratton suggests Summer took inspiration for the role from either Marilyn Monroe or Brigitte Bardot: "By thinking of herself as the most hypersexualized white woman in then-recent history," Summer unconsciously removes herself from popular imaginings of Black promiscuity (the lewd primitive Jezebel or the Mammy—subservient to the family in every way, especially the master of the house), "acting, [and] thus alleviating some of the guilt she felt as a Christian."[39] As if, for Summer, whiteness were aligned with a more forgivable sexual identity, Stratton apparently identifies white sexual expressions in music as less risqué than Black sexual expressions, reverting to a centuries-old racist dichotomy between the hypersexualization of Black performers and white supremacist rhetoric that valorizes sexual deprivation in white women. In Stratton's words:

> [Donna] Summer did not imagine herself as Monroe experiencing sexual pleasure but as Monroe acting the experience of sexual pleasure. At no point in this chain is there any claim to there being an actual sexual experience. What we have is Summer's idea of what Monroe would have sounded like performing sexual pleasure; that is, the representation of sexual pleasure on the long version of "Love To Love You Baby" is Summer's idea of what the woman who was thought by many, perhaps we should add men, to be the most sensual woman in America would have sounded like experiencing sexual pleasure.[40]

Certainly, Stratton's account resonates, in part, with Summer's experiences. But rather than attribute sole responsibility to Summer, as if *she wanted it*, in Summer's account, it was her producers who insisted she emulate these actresses, and, she adds, "I even dared, as a black woman, to portray the

famous *Seven Year Itch* Marilyn Monroe pose, which continued to perpetuate my sexual image, to Neil's [Bogart's] delight."[41] In many interviews,
and her autobiography, Summer repeatedly stresses coercion as the fuel for
her sexually available reputation.

Like most of Summer's reception owing largely to "Love to Love," Stratton's analysis focusses on the extended passage of moaning Summer performs
at the song's unusual "break."[42] For hip-hop scholar Mark Katz the break is
"often called the 'get-down part,' in other words, the most danceable part of
a song."[43] Stratton points out in his analysis that Summer's moaning enters
as new material and therefore serves as a formal "break" from what happened
earlier in the song, and yet Summer's subdued, breathy vocal quality couched
in percussion-less instrumentals departs from the anticipated climactic arrival
typical of most breaks. Rather than a singular arrival, the song taps into new-
found public awareness of women's ability to experience multiple orgasms,
music, too, adapting at that time to a "new possibility of end-less female
sexual pleasure."[44] Robert Fink similarly alludes to the anticlimactic nature
of Summer's endless moaning, comparing this moment in the song to minimalist art music from Philip Glass and Steve Reich, as freeing in its mutation
and metastasizing of the classical tension-release via cyclical repetition.[45] An
interesting reversal here is that rather than note Summer's performance as an
expression of abnormal sexuality, Fink pathologizes a teleology that is decidedly associated by critics of disco and minimalism alike with "straight white
masculinity."[46]

Fink identifies the singular building drive to climax as the historically
valorized goal of teleologically directed "desire" in phenomenologies of both
music and erotics. Whereas electronic dance music has been derided by fans
of both rock and Western Art Music alike for its anti-teleological stasis, Fink
shows how disco and minimalism do not deny the "tension-release arc," but
that such unabating repetition "reconfigure(s) a fundamental phenomenological aspect of the Western listening experience: the sense that the music has
a coherent *teleology*. . . . The feeling that the music is 'going somewhere.'"[47]
In Fink's definition: "a complete tension-release arc might be much smaller
than the piece, perhaps as small as the four-bar rhythmic cycle (three bars of
groove plus a final 'turnaround' bar that leads back to the beginning again)

that is the primary building block of disco, house, and techno, or the four-to-six-fold repetitions of measure-long modules in a typical minimalist process piece."[48] One such example of this reconfigured tension-release arc occurs in the successive combination of figures 6.1 and 6.2a. Together, these two motives, the verse and the hook, relay a complete tension-release cycle and a quintessential teleological drive on a diminutive scale, but one that does not really reach any determinable goal or "climax." Indeed, Fink's analysis revolves around and builds to what he identifies as a momentous key change much later in the song:

> After eight minutes at one pitch level, the music lurches up a whole step from C to D. . . . This is not a modulation in the traditional tonal sense—the bass moves right back down again. But it does represent the first time in the entire song that the harmonies have not been either the simple presentation of a chord based on C, or an obvious circle of fifths turnaround leading right back to C. Over the course of the next "variation" (8:30–10:45), this C-D bass oscillation recurs as the music for the first time explores other sonorities than Cm_7 and its functional preparation.[49]

In other words, outside of this moment, the melodic, harmonic, and in essence structural framing of the entire seventeen-minute track rests on the two opening figures, the hook and the verse that Summer composed prior to meeting up with Moroder.

Most notably, in this break we hear the only major harmonic departure from Cm7 and its functional preparation—and even this departure lasts only moments.[50] Fink frames his analysis around this moment and its structural surroundings in the song's break, which he characterizes as "a remarkable instance of controlled musical improvisation"—this we know from Summer and Moroder's accounts—as well as "a stretch of careful, gradual process leading to a clear teleological release that sounds, in this case, like exactly what it is supposed to be—a sexual orgasm."[51] And, this, my friends, is key. "Sounds . . . like," but to whom?

Fink's justification for the analysis is to appeal to listeners who might prioritize certain kinds of music over disco and minimalism on account of their perceived repetitiveness, which these critics dismiss as somehow absent the teleologically driven desire such listeners value as a facet of "good" music. Fink, therefore, seeks to demonstrate how "It is possible for music to sound entirely different than Beethoven [i.e., Western classical] or Little Richard [i.e., rock] and still do cultural work by affirming the construction of desire."[52] This may very well be true, but it's not immediately clear—and maybe this is my cultural distance both as a woman of color (who also admittedly has sex with women) and as someone approaching this music a decade-and-a-half later—why we should expect to hear desire in music. Is desire a quintessential feature of *all* music? Such an assumption leans desperately close to the presumption that equates music with pleasure, beneficial health properties, or even music as (mere) entertainment with no further purpose or function.[53]

Absent this requirement to arouse, an alternative hearing would be that Summer's "soft-core" virtual elation does not necessarily move people to dance or even entice any physical movement in its listeners. Though sexually allusive, Summer's sexy digression sounds more like an "anti-break," a moment that distracts and detracts from dancing both in its musical qualities—including downplayed rhythmic impetus—as well as the overt and perhaps exaggerated sexual urging, its over-the-top performance arguably more of a put off than a turn on, especially for listeners clued in to the forced performativity of it all. Club dancers may feel awkward hearing the overt sexualization in the dark or find themselves in competition with her unending vocalized ecstasy and thus alienated from their own terminating bodily experiences. Stratton identifies Summer's orgasmic moaning as near musical extension to her broad melodic range, thereby idealizing the "lack" of visible sexual evidence (which we might construe in a Lacanian framing) as an open-ended invitation to explore the imaginative extremes of listeners' own sexual fantasies. His hearing in effect enables a stereotype of hypersexual availability to be transposed from this singular performance onto multiple performers, who are perceived to embody similar musical and performative characteristics. Such hearings, including Currid's and even Fink's, frame desire as implicitly goal-directed, even if the goal is never achieved because they assume a misogynist cycle of sexual performance → sexual availability → sexual consummation, as a teleological narrative that *ought* to end in the consummation of an act, even when no such act occurs. This Lacanian lack—this "gap" between what we perceive to hear and what may have actually happened or what is happening as we listen—returns us once more to the constraints of the electroacoustic game. And, again, the joke is on her.

SUCCEEDING "LOVE TO LOVE"

Donna Summer has expressed concern about her influential role as a diva, the roots of her iconic image mostly having to do with clothing (or lack thereof)—the white dress designed by Bill Gibb in which Summer appeared on the "Love to Love You Baby" album cover, and, of course, her iconic sound: her undeniably sexual moaning in the song. The dress became a symbol of her sexual prowess, clearly inspiring Rihanna's "good girl" image in the video for her single "Umbrella" from the album *Good Girl Gone Bad*. Robin James observes that the "good girl" in this video dons a "white dress, fresh face, warm strings, major mode," an image juxtaposed with the "bad girl," evinced by a "minor mode, heavy synths, black fetish-gear-like clothes, heavy makeup, Æon Flux hair, and showers of sparks," an image reminiscent of Summer's *film noir* inspired track "Bad Girls" (1979), which, like "Love to Love You Baby," was also produced by Moroder and Bellotte with Casablanca Records.

When Summer approached Bogart in 1979 to release the single "Bad Girls"—the first of many songs in which she personified a sex worker—she was flat out refused by the executive who insisted the track was "too rock . . . and not dance enough" for her. Given the typical rock/disco opposition, Bogart's decision was clearly motivated by racially coded genre policing, an aspect punctuated by the fact that he attempted to secure Cher for the recording.[54] In Summer's words: "Although I grew up on and loved R&B music, I was much more of a pop-rock, folk-oriented artist. But my skin was brown, so I was automatically packaged as an R&B act."[55] Despite all this, Summer aspired to evolve as a musician and perform styles of music beyond disco. She eventually went directly to Moroder and the track was recorded. But, in 1980, after several run-ins of this sort, including financial misrepresentation, Summer broke contract with Casablanca. Her next album *The Wanderer* (perhaps fodder for Rønsholdt's "die Wanderin" from chapter 5)—the inaugural album launching Geffen Records—cemented her expanded range as it was subsequently recognized by a Grammy nomination for Best Female Rock Vocal Performance for the song "Cold Love."

Even with her diverse talents, Summer was not fairly compensated for her music. Breaking contract with Casablanca Records caused her to lose out on royalties. She was already embroiled in a ten-million-dollar lawsuit accusing Neil Bogart and his wife Joyce (Summer's manager under the label) of financial fraud and misrepresentation, arguing that "the Bogarts jointly made managerial decisions primarily to benefit Casablanca rather than herself."[56] But this sort of fiscal undercutting was par for the course in Moroder's studio. Backing vocalist Lucy Neale referred to several such incidents:

> Dave King, the bass player, came up with that iconic bass line, and after the song was a hit, he tried to sue Giorgio for some of the composer rights because he considered his inventing that bass line as part of the composition. He did not prevail, but I understand where he was coming from. . . . Once the instrumental tracks were laid down, we laid down our layers of backing vocals. . . . We got our 100 [Deutsche Marks] for that title [Silver Convention's "Fly, Robin, Fly"] and for "Love to Love You" and every other title we sang back up on. Never mind that they sold millions!![57]

Though King may not have had a case against Moroder, R&B band The O'Jays maybe would have, considering how closely the baseline from "Love to Love" resembles their 1973 hit "For the Love of Money," which achieved #3 on the U.S. *Billboard* R&B chart.

As we will see when I return to "Love to Love" in chapter 10, for listeners in the 1970s, the 1990s, and today, the song carries a dose of sex, yes, but sex is not timeless, and the song necessarily encapsulates other musical and cultural representations at various registers. Most importantly for our purposes, the form of desire "Love to Love You Baby" captures does not implicitly collapse into gendered and racialized stereotypes confirmed by the cycle: sexual performance → sexual availability → sexual consummation; rather, this happens only in reception. Of course, that so many listeners hear "Love to Love You Baby" as sexually suggestive indicates a consensus of sorts. But whose hearings are being privileged in these analyses? In other words, who hears the music as arousing? And, most importantly, what kind of sexuality is being performed? Despite the critical emphasis on moaning, the moaning part of "Love to Love You Baby" is not the part that other Black artists, including TLC and Beyoncé, have chosen to sample. Indeed, as pervasive as women's sexualized moaning has been throughout the hundred-year history of recorded music, moaning on its own is hardly *representative* of the cultural weight of any particular song. In the next chapters I turn to examples of electrosexual music conceived deliberately in contrast, perhaps antagonistically, to what I termed in part I electroacoustics of "the feminized voice."

7 ALICE SHIELDS'S *APOCALYPSE* (1990–1994)

On a warm summer's eve in 1990, American composer Alice Shields (b. 1943) visited her friend, Columbia University colleague and fellow composer, Daria Semegen (b. 1946). Under the stars outside Semegen's home in Stony Brook, Long Island, the conversation veered—as it often did—to contemporary US politics.[1] Enraged by "bigoted puritanism," Shields and Semegen criticized the recent influx of conservatism, specifically, the increasing volume of antiabortion groups. The composers began improvising playfully in call-and-response on the risqué behavior they supposed had triggered these groups, and Shields eventually settled on the chant "Your hot lips, Apocalypse," what would later become a line from "Apocalypse Song," the title aria to Shields's electronic opera *Apocalypse*, written in 1993 and released a year later on CD.[2]

> Your hot lips, Apocalypse,
> Your words divine made flesh in mine,
> Turn my blood back into wine.[3]

Obvious Christian themes in the text include the divine flesh paired with the reference to blood transformed into wine.[4] In the "Apocalypse Song," Shields reverses the meaning of the Eucharist accompanied in the Catholic tradition by chant and ritual movements by exchanging the symbolic "wine" of disembodied spirituality for the embodied, sensory life. However, the composer imparts wider significance beyond Christian themes to the transformation described in this passage. Alongside Christianity, Shields explains, miraculous transformations of this sort are also variously described in Tibetan, Japanese, and Indian dance and theater, Native American shaman rituals, and Egyptian burial rites.[5] Exchanges of religious and cultural signification across different systems of belief are important in the opera, and, more specifically, it was important to Shields that although each tradition realizes the theme differently, many shared the concept of transformation. Shields cites particular influence from the Indian *Bharatanatyam* dance drama, a practice the composer studied and performed throughout the 1990s. As is characteristic of *Bharatanatyam*'s *devadāsī* dancer, the sole performer takes on various sacred personae and easily transforms from one character to another. In *Apocalypse*, Shields's multiple roles as composer and sole performer allow her, like a *devadāsī* dancer, to make connections through movement, text, and music between several traditions and histories.

The plot of *Apocalypse* is relatively simple. WOMAN embarks on a spiritual journey of self-discovery. After a setless act of independent searching, in the second act WOMAN encounters SEAWEED the sea goddess, who teaches her the strength to pursue a path of enlightenment with the support of an accompanying chorus. The opera culminates in the third act when WOMAN meets the Hindu God SHIVA, and together they engage in a choreographed sexual union. The opera parades typical phallic imagery—a "phallus . . . two feet high, with balls the size of grapefruit"—in its culminating scene to combat the post-Reagan conservatism to which Shields and Semegen were reacting.[6] But instead of a climax driven by the stereotypical male sexual drive, WOMAN's encounter with SHIVA is mediated on her own terms: the performance centers around her experience of the act through sound, timbre, voice, text, lights, and physical response. Staging sex in the most obvious, visible, and audible manner, Shields avoids replicating the played-out tropes of male sexual fantasy. Rather, Shields confronts tacit sexual stigmas that plague both her contemporary American political climate and the lingering British colonial attitude toward *Bharatanatyam* revivals. In this chapter, I show how *Apocalypse* addresses sexual censorship through music, text, and choreography to envision a world for which sex is not stigmatized but instead exists as a productive and inseparable aspect of culture and music.

ALICE SHIELDS, SOME BACKGROUND

Alice Shields's early success at the Columbia-Princeton Electronic Music Center (CPEMC) is evident. She worked at the center for over three decades as an instructor and in various important administrative roles, but despite years of service, she hardly received the recognition of her predominantly male colleagues.[7] As she recalls, when joining Columbia in 1961, "supposedly no one, not even the musicologists specializing in medieval or renaissance history, had ever heard of a woman composer, including Hildegard von Bingen or Élisabeth Jacquet de La Guerre, who were famous in their time."[8] During her early days at the CPEMC, in the 1960s and 1970s, Shields says she rarely encountered other women, and given the figures most commonly associated now with the center (Vladimir Ussachevsky, Otto Luening, Milton Babbitt, Bülent Arel, Mario Davidovsky, Charles Dodge, and Charles Wuorinen), it remains apparent still that women were scarcely represented there. Pril Smiley and Daria Semegen, who themselves went on to become successful composers of electronic music and, like her, were active for several years in the CPEMC, though they too, until recently, were omitted from printed histories about the center.[9] In her book *Women Composers and Music Technology in the United States*, Elizabeth Hinkle-Turner suggests that failure to acknowledge Shields among her contemporaries occurred possibly on account of academia's typical "cost-cutting and exploitation," but Shields's gender, as one of few women working among a predominantly male staff, was likely also a factor.[10] In conversation, Shields recalled to me a confrontation with a fellow composer in the Columbia music library:

> I remember, still an undergraduate, walking into the music library looking for a score and [seeing] a guy who was in one of my classes—I was the only girl typically in classes in composition. We were talking about composing, and he said, "No woman could compose rhythms like Beethoven, if she were *normal*." I realized afterwards he meant if I really was a straight woman I wouldn't be able to write exciting music with rhythms like Beethoven's. It was a curious, twisted insult, I think ultimately reflecting a belief that only what he defined as male sexuality can express itself through powerful rhythms. But it of course was a defensive statement: he was feeling threatened by me and my music and wanted to make himself believe I couldn't write music as good as his, or better. That was very typical.[11]

Whatever the reason for the neglect of her influence in the history of electronic and electroacoustic music, it seems that Shields's official status at the center, along with her unusual compositional choices, certainly set her apart from the studio's more recognizable names.[12]

Shields turned to folklore and mythology looking for poetry and drama and music that had meaning and transformational power. She immersed herself in writing that represented women as powerful, knowledgeable, and liberated, as in the classic Greek, Hindu, and Gaelic myths, to name a few recurring sources. What especially piqued Shields's interests was the freedom with which non-European mythology and religion portrayed women's sexuality, praising every part of the body, from her facial expressions to her feet. In Indian mythology, says Shields, "each part of the body of the 'Great Goddess' is described with love and reverence. I thought it magnificent that such a personage could be celebrated for her femaleness and at the same time feared for her power. Such goddesses are not just to be addressed for compassion in moments of grief, like the Virgin Mary can be addressed by Christians, but are dangerous as well as intensely sexual."[13]

Shields has devoted a large portion of her music to literary and religious icons. The goddess Devi is a character in *Apocalypse*, and also appears in the mini opera *Shivatanz* (1993) and is invoked in *Sahityam* (2000) for solo marimba or bassoon, based on the intonation pattern of a Sanskrit poem to Devi. The Virgin Mary, a Catholic spirit not present in Shields's Protestant childhood, appears in *Ave Maris Stella* (2003) for SATB chorus and *Kyrielle* (2005) for violin and tape, based on Gregorian chants associated with the Virgin Mary. According to the composer's note: "Since it has the French 'elle' ('she') embedded in it, the word Kyrielle can also evoke a female deity, not just the Christian Mary, but all compassionate female spirits such as the Chinese Kwan-Yin, the Japanese Kannon, and the Tibetan Tara."[14] In a reconstruction of Chaucer's *Troilus and Criseyde*, the composer returned to Boccaccio's Italian text to retell the story from Criseyde's perspective in an opera named after the title role, *Criseyde* (2010), and, not restricting herself only to religious and mythological texts, Shields has also drawn inspiration from living (or once living) women such as the Japanese poet Komachi (*Komachi at Sekidera* [1987/1999]) and Rachel Corrie, a woman crushed to death by an Israeli forces bulldozer as she stood before it trying to prevent the demolition of a Palestinian home in the Gaza Strip (*Mioritza—Requiem for Rachel Corrie* [2004] for trombone and fixed audio).

The composer's initial attraction to these women may have been motivated by a need to represent in art the idols she longed for in life. Yet her compositions betray greater ambitions than mere fantasy. Art is a reflection of lived experience, but the context of art also affords certain liberties that are not normally tolerated in "real" public spaces, especially when it comes to sexual situations. In an interview conducted in June 1982, Michel Foucault remarked that contemporary art was one context in which individuals could express sexual tendencies that are ordinarily ignored or suppressed by society:

> When you look at the different ways people have experienced their own sexual freedoms—the way they have created their works of art—you would have to say that sexuality, as we now know it, has become one of the most creative sources of our society and our being. My view is that we should understand it in the reverse way: the world regards sexuality as the secret of the creative cultural life; it is, rather, a process of our having to create a new cultural life underneath the ground of our sexual choices.[15]

Foucault confronts the artistic autonomy on which the modernist European avant-garde prided itself, arguing instead that postmodern (and postwar) art enriches real life, such that life should be modeled after art and not the other way around, a model such as one finds in Shields's *Apocalypse*.

On the recording, the composer, who trained and performed as an opera singer, casts herself in the opera's three roles: WOMAN, DEVI the SEAWEED sea goddess, and SHIVA.[16] Since the performer is expected to sing along with the recording as the "tape" accompaniment, any live performance will inevitably rely on the composer's own interpretation in terms of timing and delivery. This limitation is perhaps one reason the opera has yet to be staged.[17] As is possible only in the electronic medium, Shields herself also sings the variously sampled voices that make up the opera's chorus. Beyond her immediate compositional aspirations in the opera, the possibilities afforded by electronic processing situate Shields conceptually with other women electroacoustic composers who, to paraphrase Andra McCartney, create worlds in which their music can exist.[18]

CLOSING THE GAP

In the 1980s, most composers who allied their music with electroacoustic practice were men.[19] And even still more recently, when voices sound in the electroacoustic world, however manipulated—whether or not they remain recognizable—they are most often the voices of women.[20] Such seem to be the criteria for women's participation in electroacoustic sound, that their presence be at once both implied and denied. This contrasts with men's controlling hands at the mixing desk or their roles as representatives on prestigious institutional boards and committees.[21]

Sexually speaking, men were also more present than women. Men's pleasure has long belonged to the visible—at least since the Dionysian phallic processions and, more recently, in pornographic films. Though women's bodies silently masquerade on the covers of magazines, women's sexual pleasure is generally not represented visually; the visible realm is reserved for the phallus. According to philosopher Luce Irigaray, women's sexual pleasure lingers in folds of the skin. It is *ec-static* (Gr. ἔκστασις), situated "elsewhere," out of place and removed from the site (sight) of intercourse.[22] Women's pleasure is not usually seen, though it is often heard; their satisfaction most often equated with the "quality and volume of the female vocalizations," and, in *Apocalypse*, Shields capitalizes on this trope.[23] In the staging of Apocalypse WOMAN's sexual ecstasy is seen, as well, in her choreographed movements.

In the third part of the opera, typical phallic symbolism is taken to its literal extreme. The chorus is instructed to rip SHIVA to shreds, but failing to reconstruct him, the chorus erects a comically large phallus and testicles instead of a man's body, thereby reducing the god's status to the extent of his biological apparatus. When SHIVA is finally restored to his original form through a ritual performed by the WOMAN, he commences the sexual union with her as if that is his sole purpose. A torch "held behind his erect phallus silhouette," ensures that SHIVA maintains a figurative and literal spotlight on his manhood.[24] Breaking with tradition, in this scene WOMAN does not resolve herself to the typical moaning and cooing à la Donna Summer's "Love to Love You Baby"; instead, as the libretto instructs, "The God and the Woman sing together in a voice that is neither male nor female, but both at once."[25] In this way, the Woman's sexual pleasure retains a link with the sonic, not with a reflex inferior to the man's control but by way of shared song. Such a vision of unity, the title of "Woman" notwithstanding, leans away from binaristic notions of gender and aligns more closely with Shiva's dual apparition (as detailed below).

After three ritualized orgasms, "a chaos of light and sound takes over," and the human forms vanish into a choreographed light show. Here, Shields calls on the electroacoustic tradition, with its history of severing the link between voice and body. Instead of dwelling on this "spacing," Shields closes this gap. She uses the electronic operatic medium to address sexual censorship, to explore the ways in which the medium grants her greater flexibility and freedom than she would have in her everyday life. In addressing sexual taboos through performance, Shields reinterprets and hence moves to alter her society's attitude toward sex through its "cultural life," to recall Foucault. In particular, the analysis will compare the right-wing conservatism of the post-Reagan political climate in which the opera was composed to the stigma against sexual expression in early twentieth-century colonialist revivals of the Indian *Bharatanatyam* dance drama to show how *Apocalypse* confronts the silencing of sex in both by borrowing symbolism common to both. Before exploring the *Bharatanatyam* revival and its role in *Apocalypse*, first a summary of the ancient theory of emotions in the Indian drama.

Bharatanatyam is a form of South Indian classical dance with origins in the *Natyashastra*, an aesthetic theory and dramaturgical text written by Bharata Muni ca. 300 BCE.[26] The dance form, known originally as *Sadir*, was performed in temples by *devadāsī* dancers who would personify and embody the spiritual presence of the deities to which they were devoted. The *devadāsī* dancer could transition between several personas without a moment's notice, each character possessing its own attributable gestures. For example, the *abhaya* or "fearless" *mudra* invokes the god Shiva (figure 7.1), the raised right hand protecting from evil and the lowered left hand, outstretched or giving the sign of the elephant, leading through the jungle of ignorance.

Figure 7.1

Shiva as Lord of Dance (Nataraja), Chola period (880–1279), ca. eleventh century. New York, Metropolitan Museum of Art. Copper alloy, height 26 7/8 in. (68.3 cm); diameter 22 1/4 in. (56.5 cm). Gift of R. H. Ellsworth Ltd., in honor of Susan Dillon, 1987. Acc.n.: 1987.80.1. Photo: Bruce White. © 2017. Image copyright The Metropolitan Museum of Art / Art Resource / Scala, Florence.

Shiva's embodied appellation, Shakti, appears as an expression of his feminine energy, her hair flowing like fire. In characteristic statues, Shiva and Shakti are frequently depicted with the bodily form of the man and woman in ecstatic entwinement, the transition from Shiva to Shakti achieved through erotic union, or *maithuna* (figure 7.2). But where sculpture captures only an instance of movement, dance brings gestures to life. Dancers can transform from Shiva to Shakti instantaneously.

NAVARASA THEORY AND BHARATANATYAM

In *Bharatanatyam*, not only does the performer imitate the poses of the characters she invokes, but her gestures are also said to imbue emotional inflection. The *devadāsī's mudra*s (signifying gestures) are expressive of *bhavas* (moods) that draw on a long history of associations. The *mudra*s convey the dancer's experience but are also intended to arouse a specific response, or *rasa*, meaning "flavor or essence," from spectators.[27] *Navarasa* theory, as expounded in Bharata Muni's transcription of the oral tradition in the *Natyashastra*, serves as a guide for Indian drama performers on how to evoke the emotional complex in spectators. The eight *rasa*s arise in four pairings, where the first of each pair is dominant.[28]

love/humor (*sringara/hasya*)
valor/wonder (*vira/adbhuta*)
anger/sorrow (*raudra/karuna*)
disgust/fear (*bibhasta/bhayanaka*)

The treatise likens the experience of *rasa*s to flavors enjoyed in a delicious meal, where a diner does not relish each taste distinctively but rather takes joy in the entire experience.[29] A drama arouses many *rasa*s, the tones of one *rasa* lingering behind another, though Bharata advises, in a successful drama "only one *rasa* must be predominant and others subordinate to it."[30] According to feminist scholar of Indian art and philosophy Ranjana Thapalyal, of the eight *rasa*s, *sringara*—love, either romantic separation (*viyoga*) or sexual union (*samyoga*)—is the "king of rasas." This is because "love is the one emotion that, by definition, encompasses all the others."[31]

Figure 7.2

Loving couple (*Mithuna*), Eastern Ganga dynasty, thirteenth century. New York, Metropolitan Museum of Art. Ferruginous stone, height 72 in. (182.9 cm). Purchase, Florance Waterbury Bequest, 1970. Acc.n.: 1970.44. Photo: Bruce White. © 2017. Image copyright The Metropolitan Museum of Art / Art Resource / Scala, Florence.

Whereas in European classical traditions of dance, music, and performance art the audience may observe and even identify with the action onstage, there is no expectation that the audience take part by entering the mind of the character to embody that character's feelings as their own. However, *Sadir*'s movements draw on a convention of complex physicality and emotion, and the *devadāsī*'s gestures are expected to arouse a web of associations between narrative, meaning, and emotion: "[Bharatha's] ideal spectator (*sahrydaya*), absorbed in the religious stories evoked in the conventionalized mode of representation, experiences *rasa*, or aesthetic delight—a state of joy characterized by emotional plentitude. Endowed with superior artistic and intellectual capabilities, Bharatha's sympathetic spectator harmonizes differences into unities by the power of his own mind. He, like the performer, perceives the sublime in the erotic, the divine in the human."[32] This depiction has as much to do with the dancer as it does with the ideal spectator, who was expected to possess certain knowledge and be invested in the performance along *with* the dancer.

Dancer Avanthi Meduri writes of the shifting ideology surrounding Indian dance aesthetics from the height of the *devadāsī*'s importance in the ninth through twelfth centuries, under the rulership of Chola kings, and until the dance revival of the early twentieth century. In former times, the dancer was wed to the deity of a particular temple in a sacred thread-ceremony (*tāli kettu*, colloquially "tying the knot"), and she resolved never to marry a mortal man, yet it was expected that she would conduct "discreet sexual relations with priest or king."[33] The realities of the *devadāsī*'s position, therefore, extended beyond her religious devoutness and Bharata's idealized image of her.[34] Such duties became problematic in the nineteenth century, when, under the British occupation of India, the dancer's reputation was sullied by her extracurricular activities. By the 1920s, *devadāsī*s were deemed "a seedy symbol of a perverse and backward Indian culture," such conservatism seeping into the rhetoric of those attempting to revive classical dance practices at this time.[35] A movement to preserve Indian classical culture gained momentum, and in 1932, important societal figures were attempting to redeem the practice from its association with *devadāsī*s by renaming the practice *Bharatanatyam* and therefore solidifying its association with Bharata's more idealized theory than with the realities of the practice.[36] In 1947, the Madras Devadasis (Prevention of Dedication) Act was implemented, which "officially outlawed the social, ritual, and aesthetic practices" of *devadāsī*s.[37]

REINVENTING BHARATANATYAM

In the revival, *Bharatanatyam* was taken outside the temple and brought to the secular stage. New interpretations of the tradition, as described by practicing artists like Rukmini Devi (1904–1986), an upper-class Brahmin and wife of Lord Arundale, the head of the Indian Theosophical Society, leaned on Bharata's ideal of the practice as interpreted by Abhinavagupta, an Indian aesthetician who shifted the significance of dramatic love from a physical to a transcendental spiritual form. Meduri argues that the secularization of the dance removed its carnality and hence compromised the role of dancers, who were now charged with recreating a tradition from its idealized traces. "Today's *bharatha natyam*," writes Meduri, "with its danced stories of God evoked in a secular world, is analogous to a human being walking forward with his face turned backwards."[38] Dancers who aimed to cleanse their association with prostitutes while also restoring their "respectable" position reinterpreted *sringara rasa* by removing the dance's explicit "amorous mood [conveyed] through posture, gait, gestures, glances, adornment, perfume, and accompanying song" and instead substituting inoffensive hand waving.[39] According to Rukmini Devi, "love was not sensuality . . . but rather devotion (*bhakti*), and she therefore began to exalt devotion in the presentations."[40] In order to convey this more exalted version of the practice, practitioners deliberately excluded some of the more blatantly erotic gestures from the repertoire to give the semblance of respectability and modesty.[41] By removing actual *devadāsī*s from the picture, (male) spectators could shirk their shared responsibility in shaping the practice and hence relieve themselves from the guilt of association with individuals of questionable character, "the concubines, mistresses, or 'second wives' of South Indian elites . . . implicated in a larger world of servitude focused on the fulfillment of male desire."[42] Members of the right-wing Indian high caste invented a nationalist narrative that was simultaneously anti-British colonialism, anti-Muslim, anti-Dalit (the lower-caste groups to which most *devadāsī*s belonged), and, in many ways, antiwoman, as caste, class, and religious discrimination also diminished the role women played in redefining Indian nationalism.[43] Right-wing high-caste Hindu patriarchies eventually renounced *sringara rasa* altogether, and as the *Bharatanatyam* revival grew in popularity throughout the twentieth century both within India and internationally, the movement retained a stigma against sex. Severing the link between music and dance, revivalists redrew the lineage of the practice to emphasize its ancient link to Bharata Muni and thereby lessen the importance of practicing *devadāsī*s in favor of the perceived "purity" of some disembodied (desexualized) classical ideal.[44]

American composer Alice Shields obviously has no hereditary ties to the practice so, although she attempts to reinstate some aspects of *Bharatanatyam*'s lost traditions, to some extent she also reinforces aspects of its new interpretation. After all, *Bharatanatyam* was traditionally a *hereditary* performance practice of music *and* dance, the revival period being the first time in history the dance became divorced from its paired music, a division that was wholly a result of the colonial dismantling of the practice. But, as with many dance and music practices, Shields learned *Bharatanatyam* from another form of inheritance: by apprenticing with her teacher, Swati Bhise. In reparative fashion, Shields worked to reunite the dance gestures with the music, albeit outside of the *Sadir* tradition, which had been fragmented and fractured many times over at the hands of both revivalists, who appropriated hereditary musicianship and dance from many parts of India and from non-Indian practitioners. Instead, as the next section demonstrates, Shields uniquely incorporates features of Indian music and dance together with many other traditions so as to rekindle the *Bharatanatyam* spirit of creating music and dance side by side.

Furthermore, *Apocalypse* confronts sexual stigmatization by bringing the classical dance moves back into contact with explicit sexuality. Though *sringara rasa* is more commonly depicted by the deity Krishna and his compatriot Radha, Shields casts Shiva because of his dual appellation as Shakti. As she explains, "Instead of Krishna's emphasis . . . on duty and emotional detachment . . . Shiva teaches identification with all life, the breaking of the illusion of separateness, through the vivid metaphor of sexuality."[45] In our conversations, Shields repeatedly emphasized a distinction between classical Indian dance and classical European ballet on the basis of liberties *Bharatanatyam* extends to the female(-presenting) body. Shields recalled her excitement at first seeing her teacher Swati Bhise's performance of a traditional dance in which the solo dancer changes alternately from Shiva to Shakti: "Shakti's gestures wind around her curling hair, her round breasts, her broad hips, immediately followed by a complete turn in which she becomes Shiva in his bold posture as Lord of the Dance, his hair the fire that circles the world. They alternate, fast turning, the dancer becoming both Shiva and Shakti alternating in time. In the maithuna sculptures, the two are seen simultaneously as one being whirling in suspended time. In dance or in plastic art the transformation from Shiva to Shakti is so well defined that the two are one, and this is where the *maithuna* in *Apocalypse* comes from."[46]

Shields's incorporation of *sringara rasa* is an example of how the composer contributes to *Bharatanatyam*'s continued development, providing, as Deleuze would say, theater as a "condition of movement under which the 'actors' or the 'heroes' produce something effectively new in history."[47] She does not attempt to evoke the original setting of the language, music, or movements she chooses; rather, her setting rereads and newly envisions these texts. For example, the opera's performer is expected to have experience as a dancer in *Bharatanatyam* style and singing in a European operatic style. The gestures and postures in the libretto draw from the classical Indian tradition but are not so advanced that a novice could not perform them. The hand gestures (figure 7.3), for instance, recall Shiva's *abhaya mudra* (figure 7.1), while the postures (figure 7.4) are inspired by poses from the *Kamasutra*, Greek statues, and Irish educator and clergyman Gilbert Austin's *Chironomia*, an eighteenth-century treatise on rhetorical delivery.

In fusing Indian *mudra*s with the gestures preserved in Greek statues and the postures of Austin's *Chironomia*, Shields contributes to the ongoing choreographical reformulation typical for centuries of *Bharatanatyam* practice, thereby placing these traditions in dialogue. The gestures carry deliberate and meaningful significance beyond the boundaries of the opera but fused

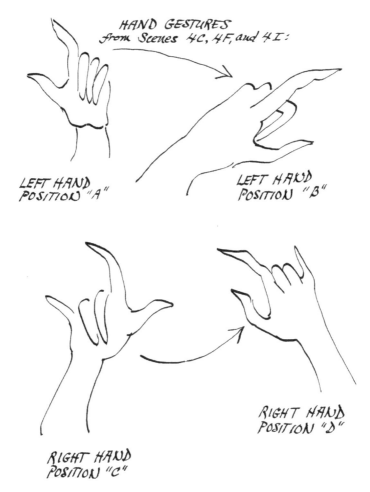

HAND GESTURES
from Scenes 4C, 4F, and 4I:

LEFT HAND
POSITION "A"

LEFT HAND
POSITION "B"

RIGHT HAND
POSITION "D"

RIGHT HAND
POSITION "C"

Figure 7.3

Hand gestures illustrated by Alice Shields in the *Apocalypse* libretto, page 82.

together they act as no simple citation of an already existing practice. Finding gestures common in two or more traditions, Shields elaborates in one direction or the other, thereby taking existing movements as inspiration toward new intercultural realizations.

Beyond the simple overlapping of gestures, Shields uses the same practice of commonality to address the sexual stigma both in *Bharatanatyam*'s revival and her contemporaneous political climate in the United States. The

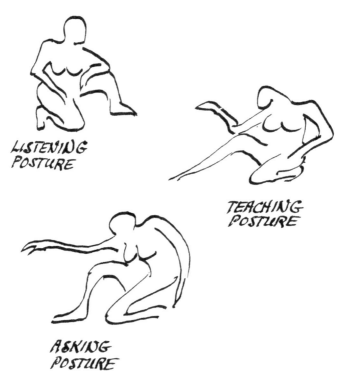

SELECTED POSTURES
from Scenes 4D, 4G, and 4J:

LISTENING POSTURE

TEACHING POSTURE

ASKING POSTURE

Figure 7.4
Postures illustrated by Alice Shields in the *Apocalypse* libretto, page 81.

resistance to *sringara*'s carnal message parallels the censorship and abstinence campaigns that later emerged under Ronald Reagan following the outbreak of AIDS, at the same time when Shields was composing her opera. Shields confronts both traditions by way of a staged insurrection of sexual norms. Like the great Greek tragedian Aeschylus or the *devadāsī* dancer, Shields casts herself in the principal roles of *Apocalypse*, performing every character, WOMAN, SEAWEED the sea goddess, the divine SHIVA, and even the voices of the Aeschylean chorus.[48] When asked about WOMAN's role—her role—the composer replied definitively, "I consider art transformative at best, so for me, writing a piece, I'm looking for not just healing, but transformation. And that was one of those moments, when I created that piece. The woman is the hero—the true hero and no one is dying, as in so many operas by men."[49] By performing and magnifying the "malestream" symbolism of the sexual act, Shields instigates a social transformation—a reevaluation of sexuality—by placing the act front and center, in plain sight.[50] In reclaiming female sexuality from its relegated "elsewhere," to recall Irigaray, Shields's *Apocalypse* makes sex audible and visible, both current and present.

In the scene just prior to the culminating *maithuna*, WOMAN's overprotective chorus stands fearfully before the god SHIVA. When he arrives, they attack him and shred him to pieces. Shortly thereafter, however, they feel remorse, and the group attempts to reassemble the god. They fail, being able only to repeatedly reconstruct a giant two-foot phallus with testicles the size of grapefruits. The WOMAN arrives thereafter and, to everyone's relief, recites a chant to revive the god in his original form. In the *maithuna* scene, the Woman lies atop a conveniently placed sofa shrieking ecstatically in "three choreographed, ritualized orgasms, one at the end of each verse of music."[51] Unlike the chorus, for whom the phallus is the male's sole representative purpose, the woman appeals to higher powers. Carefully choreographed movements combine with the composer's electronically modified voice to confirm her multiplicitous presence, which flows like a *devadāsī* effortlessly between the characters—WOMAN and SHIVA, as well as the remorseful chorus. More than phallic imagery, sex for her is a powerful spiritual union. The *maithuna* scene celebrates "the ecstasy of sensuality" to confront the "bigoted puritanism" Shields recognized together with her friend Daria Semegen.

Despite common parlance today, the premodern etymology of "apocalypse" is not associated with a literal end to the given world. With roots in Christian eschatology, the concept is more closely attuned with the process of revelation, an end to life *as we now know it*. After the chorus attacks and devours SHIVA, following his subsequent reincarnation, the god and the chorus chant together "ahpohkahléoh" (to call back); the woman then responds "to give up the ghost again"; and the god and chorus echo her in Greek, "ahpohkahpüeyen tsükáyn" (to give up the ghost once again).[52] The metaphoric and sonic repetition apparent in this slightly transformed exchange of the *ahpohkah* prefix (from the same root as "redeem" and "reclaim") echoes throughout the lyrics of the opera, and Shields likewise embeds this repetition in the opera's structure. For example, the dialogues between the WOMAN and SEAWEED and between SHIVA and the WOMAN turned DEVI are repeated three times. In our conversation, Shields acknowledged the common three-part structures in *Bharatanatyam* dance, connecting these to her training as a psychoanalyst. "In music and maybe other things too," says Shields, "you shouldn't repeat beyond three times. The pattern seems to be a cross-cultural, human thing—three times, and that's it." Shields likens the therapy session to a rondo, where the patient and/or therapist "[return] to a certain concern, feeling, or thought, alternating with new material."[53] In this way, she explains, both music and therapy are cyclic and hence temporally relational activities, always in motion. Writing on the relationship between music and therapy, Shields cites Melanie Klein's definition of repression, which results from the patient's projection of a "bad object." The "bad object" or Jungian "shadow" dwells in the patient's subconscious. It is the therapist's duty, writes Shields, "to help the patient become aware of their internal Shadow, the complement of verbal expressions and nonverbal, sound expressions."[54] Where sound casts no shadow, electronic production techniques can work to distort or veil certain sonic elements. Like therapists, dancers are uniquely capable of facilitating such revelations: "Dancers are trained to kinesthetically feel, see, and improvise with spatial relationships, weight shifts, repetition, and mirroring of movement themes, boundaries, and rhythm. Like visual artists, dancers are trained in an aesthetic mode of perception that has elements in common with other arts, but which also has its unique kinesthetic dimension. This language, which articulates forms of process, can be helpful in describing the process of the therapeutic dance."[55] Unlike sculpture, dance animates transformations from one state to another to do away with any flickering gap of uncertainly.

Alice Shields's *Apocalypse* helps spectators to identify, recognize, and acknowledge sex, to retrieve sex from the shadow, and also to shed light on the erotic currents of (electroacoustic) music. Her manner of making sex sound (literally and metaphorically) is to stage sex in the most obvious, visible, and audible manner so as to avoid any ambiguity in the matter. For Shields, more than sexual intercourse, *maithuna* expresses "the universal joy of being alive" and presents an opportunity for continually developing artistic expression. Music, in addition to gesture and narrative, is yet another way of forming this artistic intercultural dialogue.

"COCK ROCK" REVISITED

In their seminal article "Rock and Sexuality," Simon Frith and Angela McRobbie recognize rock as a contributing force to constructions of male sexuality. Frith and McRobbie dub the genre "cock rock" because the music exhibits a hypermasculine aesthetic in which "mikes and guitars are phallic symbols; the music is loud, rhythmically insistent, built around techniques of arousal and climax; the lyrics are assertive and arrogant, though the exact words are less significant than the vocal styles involved, the shouting and screaming."[56] It seems that these are common stereotypes with which Alice Shields was familiar. As one reviewer put it, *Apocalypse* is "a full-blown electronic opera, based in Indian classical music and . . . 'heavy metal rock,'" and it is the strong image of "cock rock" to which I believe *Apocalypse* responds musically.[57] Pairing certain characteristic traits of rock music—including the occasional electric guitar riff—with ornamentation, instrumentation, and modes typical of Indian Carnatic music, as Shields does, opens an exchange between musical epistemologies and also between the gendered and sexual epistemologies constructed within such musical expressions.

In a similar manner to Shields's pairings between gestures common to *Bharatanatyam* and Greek chironomy, she uses musical mode and ornamentation to forge bridges across musical traditions, and these traditions all meet under a synthesized electroacoustic canopy. Instead of recording a real sitar or piano, Shields chooses electronically synthetized timbres to limit her peculiar sound solely to the electronic world—the only medium in which these particular timbres are available. The sounds of electroacoustic music are not mere imitations of some real, preexisting world; these sounds belong entirely to virtual simulation, an electronic setting for Alice Shields's utopic vision of a world in which sex is freely expressed.

The opera's central number, "Apocalypse Song," discussed in the opening of this chapter, is more readily identifiable with an album "single" than an aria. The song features extensive vocal acrobatics, jam-band guitar flights, and a synthesizer accompaniment to rival any psychedelic improvisation, not to mention the jocular humor that stands so starkly in opposition to how Frith and McRobbie imagine female sexuality (in music). Alice Shields's voice soars multiple octaves, at times tinted with a rock guttural rage, and the song's lyrics boast typical rock buzzwords ("blood," "flesh," and "hot lips"—think Foreigner's "Hot Blooded" for comparison), though, as mentioned, the context points as much to Christian themes. Additionally, the violent imagery alludes to the scene in the opera in which SHIVA's flesh is torn apart, a form of punishment recalling the Hindu *Vishnu Purana*, in which a thief's body is to be torn apart by scorching iron balls and tongs. That several traditions share this imagery allows Shields to easily depart from and again return to the rock tradition. Combining throaty "shouting and screaming" with heavy vibrato and an operatic "head voice," as Shields does, brings 1970s rock into dialogue with theatrical vocal techniques, thus inviting members from each audience to engage jointly in the experience.

The campy synthesizer accompaniment in "Apocalypse Song" centers on G, unfolding through F to E and back again. Here the vocal motive D E-flat sounds together with the G in the left hand of the piano, harmonized by an open-position blues accompaniment but with straight eighth notes that are occasionally ornamented by diminutive rhythms—sixteenth notes, quintuplets, and triplets. In the A section of the song (figure 7.5, mm. 39–46) the voice lingers on the half-step D E-flat, while the B section's melody (figure 7.6, after m. 47) breaks away into a scalar ascent D E-flat F, returning to D and then repeating the inversion of this fragment D C B-flat C, as in the first section.

Figure 7.5
"Apocalypse Song," A section, composer's transcription.

As mentioned, "Apocalypse Song" was the compositional seed for the opera, and many of its fragments are elaborated elsewhere. The repeating half-step motive D E-flat from the A section is heard much earlier, in the First Greeting (Part II, 4C), when WOMAN first encounters SEAWEED. WOMAN chants to DEVI "An tu? An tu?" (Gaelic, "Is it you? Is it you?") to the pitches B C B C, and DEVI responds, likewise adding a descending Phrygian tetrachord. The quintessentially rock Phrygian modality is used by Iron Maiden, Rush, Metallica (think "Wherever I May Roam"), and many others to achieve a "dark sound," but similar variations also exist in Indian music, the influence of which is apparent in the music of this scene. Whereas

Figure 7.6
"Apocalypse Song," B section, composer's transcription.

"Apocalypse Song" features an electric guitar played by Jim Matus, the Greeting features electronically synthesized bells and a chorus, and whereas "Apocalypse Song" retained more or less straight repetition, in the Greetings (six in total), repetition becomes ornamented by turns from above the note or *zamzama*, both in the voice and in the synthesizer accompaniment. As the scene progresses, the voice becomes more acrobatic, with more twists and turns and heavy, exaggerated vibrato reminiscent of the Hindustani *kampan* (quivering or undulation between two pitches).

Again, like the pairings of gestures common to multiple dance practices, Shields uses instruments, modes, and ornamentation to forge bridges across musical traditions. She imitates the rock instruments and the traditional instruments of Indian classical music while also creating sounds that exist wholly within the electronic soundscape by employing synthesizers, a half ring modulator called the Klangumwandler, MIDI, and the Csound computer programming language used for audio programming.[58] Today it is easy to dismiss the synthesized (i.e., *synthetic*) "dulcimer" or "lyre" sound patches, since they fail to elicit the sounds of the *real* piano, lyre, sitar, or whatever. We might say that Shields's playful music has an audible, low-tech artificiality, given the capacity of synthesizers in the early 1990s, when the opera was composed. And one may attribute this failure to remain faithful to "high fidelity" to the hardware limitations of the digital synthesizers, keyboards, and software or describe it generously as a characteristic sound of the times. But I don't think this is a fair assessment. Pairing the lyre and piano already provides a strange, pan-stylistic dissonance, one that I believe Shields institutes deliberately. She purposefully couples the instruments of disparate traditions, both classical and modern, to create a unique soundscape, not unlike Led Zeppelin's "The Battle of Evermore" or the Indo-jazz fusions of Coltrane's "India."[59] However, instead of recording a real sitar and piano, Shields chooses electronically synthetized timbres over "concrete" referential associations, and in doing so limits her peculiar sound to the electronic world.

If we envision the 1990s and the context in which Shields was composing, we can understand that what the composer saw in her everyday life, both in and outside the music profession, was a hypermasculine attitude toward music and sex, but what she imagined in *Apocalypse* is a musically rich universe in which sex is no unique act but one of life's many pleasures. *Apocalypse* employs typical phallic symbolism—vulgar lyrics, electric guitars, and even a two-foot papier-mâché penis—but it does so in order to confront the distantiation of women's sexuality from rock music, opera, *Bharatanatyam*, and the everyday politics in which music is ensconced. Introducing the visible phallus alongside the electronic guitars, mics, and amps embellishes their significance to the point of banality and farce. In pairing familiarity with unfamiliar language, novel electronic sounds produced by invisible instruments, and choreographed gestures more subtle than the typical sexual gyrations, viewers are made aware of sedimented and accepted norms so as to examine and question their own preexisting cultural investments.

In the early 1990s when *Apocalypse* was composed, *Bharatanatyam* performances were becoming more common in America, but skeptical cultural theorists characterized these displays at first as inflexible idealizations and appropriations of an authentic Indian culture. Yet with the growing international popularity of *Bharatanatyam*, perceptions of its cultural heritage also changed. By the early 2000s, practitioners of the dance pointed out its part in constructing, defining, and delimiting the roles of women of Indian heritage in India and abroad. Subsequently, postcolonial studies began to question the need for its practitioners to be of Indian heritage.[60] As mentioned, Shields brings her contemporaneous regional politics into dialogue with other cultures through *Bharatanatyam*. She confronts her audience with particular sexual symbolism both visually and aurally so viewers and listeners will "examine their own cultural investments" and stigmas alongside the subtleties of traditions with which they may have little to no familiarity.[61] In thus challenging the audience to engage with unfamiliar texts, choreography, and music, *Apocalypse* is an example of what dance theorist Janet O'Shea has termed "ongoing interculturality," in which "methods of exchange between epistemologies . . . circumvent or reverse an orientalist problematic" that might emerge in non-Indian performances of *Bharatanatyam*.[62]

It is easy to fall into the Orientalist model of co-option when attempting to invoke some expressive music or dance movements foreign to one's time and/or place. For example, Antonin Artaud's famous Theater of Cruelty was inspired by a Balinese dance performance he witnessed in 1931 at the Paris Colonial Exhibition. His review of this performance admires the "mystery" of Balinese gestures and movements, describing them as "hieroglyphs" that "Occidental" intellectuals had long since forgotten.[63] Artaud's ambition to return to a "repressed" reality, as he calls it, presumed that Balinese dance somehow preserves some primitive and preexisting practice from which "the West" has since moved on. Whereas Artaud demanded that Western theater adapt and adopt new cultural practices, he did not afford the Eastern tradition the same opportunities to evolve. O'Shea describes this problem as "the orientalist model of translation," which rests on the assumption that an interlocutor with "specialist knowledge" "could unlock [the] mysteries [of Eastern practices] *for* 'the West.'"[64] Shields, in contrast, does not presume to uncover any hidden meaning in *Bharatanatyam* or any other tradition from which she draws. Rather, her works draw on the epistemologies of disparate traditions without attempting to reconcile their differences for an uncomprehending audience.

O'Shea's "ongoing interculturality" refers specifically to intercultural exchanges of the language, dance, and music associated with *Bharatanatyam*. While *Apocalypse* certainly participates in intercultural exchanges within these dimensions, I expand O'Shea's term to include less overt concepts, such as gender and sexuality, concepts conveyed through performance but not explicitly in any one aspect. Prior to incorporating *Bharatanatyam* movements in *Apocalypse* and other works, Alice Shields took years to study the practice. She employs choreography in a productive way without attempting to convey supposedly foreign symbols through explanation or translation. Those already familiar with its particular movements will benefit from the added significance without any further explication; however, the text assists the audience in making more overt connections. In addition to movements, the libretto is comprised of text portions from Sanskrit sources, fragments of ancient Greek, and Old Irish, *alongside* the composer's own English text, which does not serve as a primary language or translation but as an equal counterpart to the other languages. Furthermore, the music's text setting, ornamentation, and instrumentation also participate in this "ongoing interculturality."

Shields's role as *devadāsī* functions as what Janet O'Shea would call a "transnational interpreter," placing the history of the stigma against sex from the *Bharatanatyam* tradition into dialogue with the opera's contemporary American political atmosphere. Reagan's Mexico City Policy (initiating the "global gag rule" on abortion for institutions receiving federal funding), Nancy Reagan's "Just Say No" campaign, and the feminist antipornography agenda of the 1980s and 1990s all opposed forms of explicit sexual representation, even those that presented sex in a positive light. By using the phallic symbolism typically representative of sexual desire, Shields cites a history long dominated by masculine sexual imagery, but she does not reinscribe this history. The comedic size of the two-foot papier-mâché phallus forces the audience to confront the typical imagery in order to question its function as a symbol of sexual arousal or pleasure. Whereas the visible symbol serves as a literal confrontation with masculine sexual imagery, the music more subtly converges on a metaphorical reappraisal of such symbols.

SEXUAL STIGMA TODAY

In our interview, Shields told me that *Apocalypse* was a reaction to the suddenly more pervasive resistance to abortion and women's reproductive rights in the United States in the 1980s. On the coattails of the AIDS scare, when the US government finally began to acknowledge that the disease was spreading indiscriminately—without heed for the host's sexual proclivities—Ronald Reagan brought into law the 1984 Mexico City Policy, a "global gag rule" that effectively prohibited family planning centers receiving government funding from performing abortions or even from providing counseling and referral services regarding abortions.[65] As Shields recalled in our interview, "I remember feeling outraged with Reagan. When he took over in the 1980s the country radically changed. . . . I consider women's rights to be one of the primary problems in the world. Unless women take over or become at least 50 percent of the rulers everywhere, we're done for. Though many women have fallen into the male practice of dominance too, so that isn't the only answer."

Political scientists Barbara B. Crane and Jennifer Dusenberry, who specialize in reproductive rights issues, assert that in the 1980s and 1990s "opposition to the government-supported family planning services, like abortion, grew from a set of beliefs about the role of modern contraception in promoting promiscuity, moral breakdown and the weakening of the traditional male-dominated family structure."[66] Reagan's policy proliferated widely through his wife Nancy's "Just Say No" campaign, which in one fell swoop waged a "war on drugs" while simultaneously condemning (implicitly heterosexual) premarital sex and sex acts understood as homosexual or queer by linking these two issues through self-control and abstinence.[67] In addition to an obvious resistance from the conservative Right, the stigma against promiscuity was also advanced by vocal antipornography feminists, who masked their biases as women's liberation.

Dissident feminist Camille Paglia explains that the US of the 1950s and 1960s was dominated by Protestantism, which "systematically repressed both sex and emotion as part of the Puritan bequest." Writing in the 1990s, Paglia insisted that "repression continues in current American liberalism, which is simply Protestantism in disguise." She recognized sexuality as a deeply complicated expression of humanity: "Above all, to understand sex and emotion, you must study the world history of art, music, and literature, which is the precious record of the strange, kaleidoscopic human imagination."[68]

When articulated as an attitude about sex rather than any quantifying measure of actions, the "promiscuity" buzzword deceptively amplifies the supposed risks associated with sex. Under the canopy of "promiscuity," all sexual activities are magnified and scrutinized to the point of censure, where only certain sexual behaviors are permitted. According to sex activist and academic Gayle Rubin, those who publicly lampoon pornography "have condemned virtually every variant of sexual expression as anti-feminist," Rubin herself having been a firsthand victim of these accusations. Anti-pornography activists, which most feminists were in the 1980s and 1990s, claimed to be performing a social service, while, in Rubin's words, the discourse presented "most sexual behavior in the worst possible light. Its descriptions of erotic conduct always use the worst available example as if it were representative."[69] To punctuate Rubin's theory, we can recall Theodor Adorno's description of the supposed sexual liberation of the 1960s: "Everywhere prostitutes are being persecuted, whereas they were more or less left in peace during the era when sexual oppression was allegedly harsher."[70] Ultimately, both conservatives and anti-porn feminists pushed similar agendas: both aimed to remove sex from the public eye altogether, thus condemning sexual intercourse (an evolutionary trait designed to promote population growth and to prolong the human condition), ironically, as an antihuman act. Not only does such censure delimit which sexually charged images, ideas, and acts are permissible in the public domain, but, given Reagan's proposed restrictions on family planning services, these prohibitions inevitably invade private domestic spaces, the typical venues for viewing pornography and engaging in sexual activities.

Recently, Reagan's "global gag rule" was in the news again. As has each Republican administration since Reagan, Donald Trump's administration once again instituted the rule, barring institutions receiving federal funding from mentioning abortion as a viable option to patients and expanding Reagan's original policy by prohibiting not only foreign NGOs but all global organizations receiving funding from mentioning abortion. This amendment most harshly impacts centers working to prevent the spread of HIV/AIDS, as abortion is hardly the sole or even primary treatment at many centers. Just as in 1984, when Reagan first instituted the policy, today's supporters of the ban have a vested interest in maintaining silence, which is not so much abstinence as secrecy. Secrecy is another way of limiting sexual freedoms—freedom to perform sexually deviant behaviors—only to those who have the privilege of privacy. While the main motivation for securing the norms of (hetero-/homosexual) monogamous sex has remained, rather than shy away from acknowledging such consequences, today's constituents have enthusiastically expressed the benefits of such policies to the "silent majority"—Trump's largely white constituency. Many of Trump's adversaries have made the connection between antiabortion and pro-war rhetoric, citing racial and ethnic discrimination as a major link. In forming an alliance across geographic and chronologic borders, Alice Shields's work anticipated criticism of these ongoing developments by aiming to illuminate and sever the false separation between sexual policy and cultural influence.

Partha Chatterjee points out that the Indian nationalist movement and its enduring history have only recognized women insofar as they *contribute* to the national reconstruction, as if the movement were advancing independently of women, regardless of their roles.[71] Similarly, the history of electroacoustic music, as told until now, is often presented as being merely enriched by select compositions by women without acknowledging how innovations by women like Alice Shields actually molded the tradition into its current form. I understand Shields's *Apocalypse* against this social backdrop as a response to and commentary on the community in which its composer was living. The work serves as an important example of how composers contribute not only to musical culture but more to issues of social relevance beyond music's sonic profile alone. Shields's utopian enterprise truly reflects a world absent sexual stigmas, a world very different from the one in which she lived but not so distant from the world envisioned by many people today.

INTERTEXTUAL RUINS

Alice Shields's methods of continual variation are not only musically apparent, but she also exerts similar treatment on her textual sources. Woven throughout *Apocalypse*'s libretto are varied texts from Sappho, Archilochos, Aeschylus, the Bhagavad Gita, the ancient Gnostic Gospels, and unidentified Gaelic folklore. Quoting none at length, Shields says she combines phrases or adjectives "crudely together, creating ruins of language analogous to the ruins of temples."[72] In building and reshaping the traditional imagery, music, and text, Shields expands traditional practices through new interpretations rather than enshrining these practices into idealized representations of a permanently fixed culture, nation, or time. Shields is sensitive to the traditions and history of her sources, where her reinterpretation aims to amplify connections *across* cultures, geographical borders, and historical periods. As a white American woman of Presbyterian background, Shields says her attraction to Indian, Greek, and Old Irish mythology, art, and religion was first motivated by her need to move away from her own sheltered heritage. By incorporating languages and mythology from various origins, her music does more than merely replicate predisposed idealizations. *Apocalypse* offers a new universe, a constructed but inspired collage.

Though Shields cites some sources, many more arise only through allusion. Shields's fictional SEAWEED speaks in both English and Gaelic (which Shields specified as Old Irish in our interview). As with the other non-English texts in *Apocalypse*, SEAWEED's text is an agglomeration that includes the composer's English adaptations of Krishna's words from the Bhagavad Gita.[73] Given the prominence of Gaelic text, I was driven in my analysis to identify some of these mythic sources. By the time we conducted our interviews—over twenty years after the opera was composed—the composer was no longer able to recall precisely which texts were used in *Apocalypse*. And so I set out to hunt down probable sources on my own.

Inspiration for the "Dismemberment" (III:7D) and "Organ Screaming" (III:11B) scenes likely came from an adaptation of a popular Old Irish folktale transcribed in the lay *Laoidh Chab an Dosáin*. This is one of many tales featuring Conán Maolis, a frequent target of fantastic misfortunes in Celtic Folklore. The story told in *Laoidh Chab an Dosáin* is as follows.

LAOIDH CHAB AN DOSÁIN

Conán enters *Bruidhean*, a fairy dwelling where one can always expect a mishap. Conán is generously hosted, and so, sated with food—five lines of the lay detail how stuffed he is—he prepares for bed.[74] Nearly satisfied, Conán voices one last request before sleep: "It's well I would sleep now till day, / If I would get one foray on a woman."[75] His companion Diarmaid Ó Duibhne, cautious of the fairies, shushes him. Unfazed, Conán insists he needs a woman that night. Just as Conán's head hits the pillow, a woman appears in the bed beside him.

On the woman's first appearance, only her legs are visible, and Conán takes the opportunity to pounce. He launches toward her, but in a magical sleight the attacker finds himself suddenly dangling from a plank suspended high above a moving body of water—the image recalling the seaside of *Apocalypse*'s part II. The plank splits and Conán is sent plummeting. He hits the water and is awoken, screaming in terror from inside a cauldron. The commotion wakes the residents of the home, who have rushed into the kitchen just then to chastise Conán for waking them.

Conán returns to bed.

He begins to drift back to sleep and again is taunted by the appellation of the beautiful woman. He launches at her for the second time. Twice deceived, he finds himself transported to a dark enchanted forest surrounded by a band of clawing and snapping cats—recalling the ravenous chorus of *Apocalypse*'s "Dismemberment Dance," which, as mentioned, has resonances also in Indian mythology. The cats attack and viciously gnaw at Conán, driving him up a tree. He rouses once more from slumber with the house's inhabitants again surrounding him perched on a table with a kitten licking his lips. Again, they reprimand him.

Conán returns to bed for the third and last time. When the apparition reappears, he approaches her. This time he catches her. Repulsed by Conán's unwelcome advances, the fairy "punishes" her attacker by turning him into a woman—recalling the Woman's transformation to Shiva in *Apocalypse*—and if that weren't demeaning enough, *she* then proceeds to sexually assault *him*.[76] Throughout his violent rape, as his bones shatter beneath her mighty force, Conán begs the relentless spirit for mercy. In the end, the fay impregnates Conán, and through a mystical acceleration of the gestation period, he is instantly in labor. In his confused state Conán implores his companion Diarmuid to help him give birth. Shouting in pain and terror, he asks Diarmuid to tie a rope to the baby and pull it out of him. The commotion again rouses the residents of the dwelling who rush into the visitors' room for a third time to discover Conán, untransformed, and Diarmuid pulling with all his might at a rope tied to Conán's testicles!

Though the lay resembles some aspects of the opera, *Apocalypse* does not incorporate the violence of the third episode. This omission should not, however, be taken as evidence exempting the story as a possible candidate for *Apocalypse*'s narrative. According to historian of Celtic folklore Seosamh Watson, though *Laoidh Chab an Dosáin* was preserved through a long oral history, the woman's third appearance was later omitted from many of the lay's transcriptions, of which there exist nearly twenty, the earliest of which survives from the 1730s.[77] "The obvious reason" for the omission, writes Watson, was "that the literary author was too prudish to include these details."[78] Though fit for oral communication, writes Gerard Murphy, "[the author(s)] deliberately omitted the third episode on account of its being too uncivilised for the literati."[79] And, when I asked Shields about the story's similarities to *Apocalypse*, she acceded that Conán's first two encounters with the fairy were familiar to her.

"Dismemberment Dance" (III:7D, for the scene's accompanying physical movements see figure 7.4) engages the lay's imagery of "devouring" hunger, where the chorus, like the cats, represent a protective entity to guard the woman from the looming threat of sexual assault. Complete with chewing, belching, bird sounds, and a light show, this is the most elaborate of the nonvocal musical scenes in the opera.[80] Guitarist Jim Matus, in whose studio many of the effects were recorded, is also credited in the "Dismemberment" scene as "a gifted belcher who can perform upon command." Additionally, Shields's liner notes read, "Jim also soloed in this scene on broccoli; I accompanied him with yogurt. The various 'bird' and 'sea-lion' sounds are also my own voice; I'm sure my operatic voice teachers would be proud."[81] After devouring Shiva, sated and placated the chorus grows suddenly remorseful and makes efforts to reassemble the God's body from its strewn remnants (III:8A).

Scrambling around on their hands and knees, they frantically squash together gobs of flesh with unexpected results. The following stage commands are enacted silently:

> They find sections of his phallus, and put it together again. The newly assembled phallus is two feet high, with balls the size of grapefruit.

> They carry the phallus lovingly to a pedestal mid-center stage, and place it thereon.

> Sitting in a circle around the Phallus-Monument, they pray to it to come alive again.[82]

The chorus then begins to chant in Greek along with the reverberating and gradually crescendoing tape part, desperately trying to awaken the God.

> Stroking the Phallus, they cry out.

> They try to lift the balls of the Phallus.

> They press their lips to the Phallus, kissing it, stroking it, trying to turn it on.

> They step back and watch expectantly for it to come alive:

> But it doesn't move.

> They go up to the Phallus again and move it from side to side as if it were walking, making it jump up and down, stroking it, rubbing it, etc.

> But nothing works: it doesn't move.[83]

Only the WOMAN-turned-DEVI is able to awaken SHIVA. After chanting an invocation in both English and Greek (III:9A), the fog parts and the God is revealed in SHIVA's typical manifestation, sitting in silent meditation, "his skin covered with white ash, his long hair matted, his chest bare" (III:9B).[84]

Where the ignorant and bigoted chorus attempt to rid themselves of the God's sexual proclivities, the Great Goddess, who is more powerful than all of them combined, easily invokes SHIVA, explaining, "Only by knowing thee, / do I pass over death."[85] Obviously, one could interpret this metaphorically that the DEVI and SHIVA's union will lead to procreation and thus yield a continuation of her line, but Shields also hints at a less literal consequence—love's extemporal, intangible, unlocatable universality.

In part IV of the opera, SHIVA returns in a scene entitled "The Beginning of the End" (IV:10B). Commencing the courting scene in "Heat Drum" (IV:11A), the woman lays upon "a convenient couch which the Chorus push under her. The Chorus holds torches between the Woman's legs as she opens and closes her thighs in time to the music." Their job is to illuminate the shaded regions of WOMAN's body, to locate the "elsewhere" of female sexuality.[86] SHIVA moves to approach the WOMAN and arriving at long last he holds his phallus against the WOMAN's vulva. She "throws her thighs wide apart, her mouth wide open, and head thrown back, in ritualized ecstasy." Here the two commence in "Organ Screaming" (IV:11B), "three choreographed, ritualized orgasms, one at the end of each verse of music."[87]

The phallus, in many cultures, is the symbol of fertility and a vital sign of sexual conquest. But, returning to Irigaray's critique in the opening of this chapter, this universal symbolism has nothing to do with female sexuality. Shields cites this common symbolism but deliberately exaggerates the proportions of SHIVA's penis to allude farcically to the size and virility of male genitals. It was rare to encounter a penis in opera in the cloistered conservatism of 1990s American society (or since?) and would have been a spectacle to present one of such magnitude. Paired with Dionysian gestures and Greek language, the phallus becomes quite another apparatus, a tool with which to critique the history of its symbolic illusiveness.

The Chorus's devotion to the phallus in *Apocalypse* is no doubt inspired by the traditional ancient Greek "Phallic Processions." As famously described in

Aristotle's *Poetics*, the processions celebrated Dionysus—an alternate identity also bestowed on SHIVA in the opera's "First Naming" scene (III:6C).[88] Like Aeschylus's tragedies, Shields's opera features no set design, and only minimal instructions for lighting. As the composer explains, "Geography plays no place in the opera. *Apocalypse* is a work for theater, and is meant to be universal; thus, it is deliberately not situated in any one culture or geographic locale."[89] The absence of scenery and set grants *Apocalypse* a placial universality, which cuts across geographic locations and historical periods—at least this is what the composer says. I would take Shields's compositional vision one step further and argue that the over-the-top phallic procession references the "cock rock" manifesto by seizing its most potent imagery while reappropriating the phallic symbolism toward feminist empowerment. Thus, one of the most recognizable signs of male sexuality (but sometimes a sign of dominance and violence) becomes a tool for illuminating—making present—women's sexual expression.

CONCLUSION

Frederic Jameson characterized the 1960s in literature, in art, and in music, as a "reinvention of the question of Utopia."[90] In 1969, science fiction author Ursula K. Le Guin made space for an "ambisexual" population in the "*other* world of *The Left Hand of Darkness*." Like the *science* of fiction, electronic equipment affords music a technical credibility, which for some composers in the 1960s meant a gradual phasing out of what was "human" about the work, its context, or the work's performance. The characters of science fiction, like the sounds of electronic music, do not exist in the world of the experiencing subject, only in some other, fabricated world. The utopic visage allows readers to approach the world at a distance, to admit its "irresistible reminiscences" without fully believing it exists.[91] In this way, the "new sciences" of the 1960s awakened a healthy fact-based speculation, while sustaining a suspension of disbelief in practice (in both literature and music). Unlike mythology, which is wholly entrenched in a devotion to the unknown or *never*-known, *science* fiction has the potential to be proven. In music, though the exacting execution of electronic equipment phased out the faults of human performance, it did not compromise the spectator's perception. Shields's colleague composer-theorist Milton Babbitt praised the incorporation of electronic equipment into musical composition, because for him these machines opened new musical possibilities. "The precise placement of time points and their associated durations, though easily and exactly specifiable, takes one into the area of rhythm, which is not only of central concern in contemporary compositional thought, nonelectronic as well as electronic, but the most refractory and mysterious perceptually."[92] For Babbitt, the electronic medium was a way to break the glass ceiling suspended above the post-war avant-garde. On this limitation, Herbert Eimert, who mentored the European electronic music faction,[93] similarly recounted: "In the history of the 'Music of our Time,' electronic music might be regarded as a final chapter or even as a postlude." Electronic music was the music of the future.[94]

Alice Shields takes the electronic tools and the concept of a fictional patriarchal "elsewhere," to move toward a yet unrealized feminist utopia. Her *Apocalypse* confronts opera's typically gendered and sexual distanciation, where sex is relegated as the most extreme of human actions. It shows that intercourse can be beautiful, artistic, and enjoyable without all the negative repercussions of being a woman in opera. *Maithuna*, says Shields, is the culmination of the spiritual and physical journey. What Alice Shields sees in her everyday life, both in and outside the music profession, is a hypermasculine attitude toward sex, but what she wants to see are fully formed women for whom sex is no special issue but one of life's many pleasures. In *Apocalypse*, WOMAN gains powers, which endow her with the ability to change the destiny of both earthly and divine characters. It is ultimately her performance that causes the narrative to unfold as it does. Where Irigaray critiques the invisibility of women's sexuality in phallogocentric discourses, Shields puts sex in the forefront. Shields views WOMAN as an ideal or hollow representation. As a hull filled with remnants of DEVI the GREAT GODDESS, the WOMAN is the woman of an Irigarian "elsewhere" brought back before us.

Unfortunately, the opera's wild and diverse performance requirements make this piece difficult to stage. At the time of its composition, the opera's electronic components, though now widely in use, were viewed as an undue burden for performance. Few, if any, opera halls had the necessary equipment to realize such features, and operatic singers may have been uncomfortable with the amplification required to properly balance the voice and electronics. A second impairment to its staging is the opera's combination of classical dancing and operatic singing. Ideally, Shields would like the opera to be fully choreographed, where the WOMAN is a trained singer experienced also in *Bharatanatyam*. Realistically, it may be difficult to find a performer who is skilled as both a singer and a dancer, but Shields nonetheless includes what she calls "intermittent choreography" in the libretto, the most essential movements that are basic enough that anyone can perform. As she quips, "anyone can learn *muhdras*."[95] Shields seeks to reformulate ideals of gender and sexuality beyond those represented in art, aiming in the process to break the generic mold of what constitutes as "electronic music." Ultimately, however, Shields succeeds only partially, since, first of all, the medium itself is restricting: *Apocalypse* exists only as a CD with Shields's voice. Shields's operas atypically require classically trained singers to sing along with the electronic part. Singers are then also expected to dance in the highly specialized *Bharatanatyam* style as choreographed according to Shields's reinterpretation of this style. In short, there is almost no foreseeable way to stage *Apocalypse*, as the composer's fixed approach leaves little flexibility for the live performers. Using a recording in performance would prove difficult, as shown in Shields's experience with *Mass for the Dead* where not only were the singers required to sing along with the restricting electronic accompaniment, but where the entire performance relied on the inflexibility of Shields's fixed voice while also requiring singers to perform *Bharatanatyam*—although these challenges did not prevent the *Mass* from being performed.[96] She explains,

> What I found with a couple of other electronic operas, including *Mass for the Dead*, even if you're dealing with opera singers with fine technique the separation coming out of the loudspeakers and the live performers on stage, even with the loudspeakers on stage not in the audience, they're coming from

different acoustical spaces. So I found it necessary to convince the singers, what I did was tiny amounts of amplification of the singers and put it into the loudspeakers and then it mixes just enough. It has to be extremely subtle, the singer must not hear it, nor must the audience hear it—[the sound] has to come from the singer on stage, but it has to be enough to overcome the separation from the electronic sound. This isn't just spatial, it seems to be psychological too.

Such a performance stipulates either that the composer herself be present for a realization that is true to her vision or that we invent/train a new kind of performer for such an opera. New artistic media provide new opportunities, but in the meantime this opera's conception lingers only in abstraction.

Hindering the opera's performance is another unspoken point of resistance: Shields's unabashed confrontation with sexual politics, the explicit physicality and uninhibited display of sexual imagery uncommonly (if ever) broached in European classical music. In the words of musicologist Paul Attinello, "The whole point of shame is to *prevent* performance."[97] Even in art one is always under obligation to perform (in Judith Butler's sense) in a manner that is consistent with the site of the performance.[98] Through Attinello's eyes, "The primary link between shame and performance is that performance is in its essence *exposure*, self-transformation into a sign for public examination and judgment."[99] Failing to replicate the correct gestures in the appropriate context, argues Attinello, is the source of shame. If made aware that one does not meet society's expectations of appropriate behavior, a performer will grow overly self-aware and self-conscious. In real life individuals may attempt to conceal obtrusive traits or enhance—that is, perform—alternately, but what is customary in musical and theatrical performances is far more exaggerated than the typical behaviors of polite society.[100]

Though SHIVA propositions WOMAN first, it is she who physically instigates their lovemaking. She is not afraid to seize her desires, and though she appears in the insular realm of the diegetic stage, her creator, Alice Shields, does not. Where WOMAN and SHIVA perhaps consummate privately, Shields is all too aware that wide-eyed spectators are watching her unconventional ideas come to life. In a sense Shields is simply performing *in the guise* of these characters. As their creator, she foresees their motivations and desires, and her controlling hand is ultimately responsible for realizing their actions. Turning the spotlight on the most sacred, and hence veiled of music's elements, Shields faces her examining and judging public in a full-fledged performance of that of which they are most ashamed. As in life, music also has its norms and ideals, whether in style or in content, and composers must adhere to these by meeting the expectations of a discriminate and discriminating public. If a composition does not meet what the audience believes a musical work to be, then either the work has failed to replicate existing norms, or perhaps the work has boldly superseded them.

8 ANNIE SPRINKLE'S *SLUTS & GODDESSES* VIDEO WORKSHOP, SOUND SCORE BY PAULINE OLIVEROS (1992)

As far as I know, Pauline Oliveros (1932–2016) is the only "art" composer to write music to a pornographic film. Pornographic actress, stripper, and former prostitute Annie Sprinkle's (b. 1954) *Sluts & Goddesses* video workshop boasts a soundtrack by Oliveros and draws on a pairing of sight and sound common to much pornography. Novelly, the film resituates this dynamic anew within a feminist paradigm of women's empowerment.[1] Coupled as it is with Oliveros's soundtrack, I wondered if *Sluts & Goddesses* (1992) invites electroacoustic philosophical introspection into the pornographic—a space electroacousticians do not usually occupy publicly or in scholarship—or, as the case may be, absent visuals, what Darshana Sreedhar Mini terms the pornosonic:[2]

> pornosonic media are not opposed to pornographic media, but both are held in a supplementary relationship in which the work of sound is affective (working at the level of the body), social (working to create arrangements of public order around sonic artifacts), and cultural (filtered through cultural norms).[3]

The sonic is never completely isolated from the visual, even when sex is not immediately visible but only imagined.

In the reverse of cinematic composition, which reassigns sounds from elsewhere to a (new) visual field, Oliveros turns to sounds in their natural state to listen deeply in context. Her Sonic Meditations act as organically unfolding improvisations that, while also experimental, in comparison to electroacoustic practice tend to be more unwieldy, uncontrollable, unfixed. Such attention to sensory experience, as it unfolds, aligns with Sprinkle's more recent vision of work as an "ecosexual" "sexecologist," from her attention to the environment and earth as a lover.[4] We can find roots of Sprinkle's current "Sexecology" project already in this earlier work.

SEX PLAY

As an electronic and experimental composer, Pauline Oliveros became known through her work as one of the founders of the San Francisco Tape Music Center 1961–1965, serving as co-director with Ramon Sender and Morton Subotnick and then as director 1966–1976, when the Center moved to Oakland and was renamed the Mills Tape Music Center. She then became affiliated with the music department at the University of California San Diego, where she directed the Center for Music Experiment 1976–1979. In 1981, she left the West Coast of the United States for Upstate New York to expand her work as an independent musician, scholar, and composer. Martha Mockus's book *Sounding Out: Pauline Oliveros and Lesbian Musicality* examines select compositions from 1960 to 1985, including, perhaps most significantly for this book, Oliveros's collaborations with Annea Lockwood and her partner Ruth Anderson.[5] Elizabeth Hinkle-Turner's examination of *Women Composers and Music Technology in the United States* associates Oliveros, Lockwood, and Anderson "aesthetically and philosophically," remarking that, "Additionally all three of these women have been relatively open about their lesbianism and have helped to foster a comfortable creative atmosphere for women composers in an academic setting regardless of their sexuality."[6] Indeed, Lockwood first moved from London (following her collaboration with Alston, explored in chapter 2) to New York at Oliveros's invitation. The two had corresponded from 1970–1977, exchanging letters about Oliveros's Sonic Meditations, about feminist theory, and connections between sounds and mediation with and through the body.[7] In Mockus's assessment, the correspondence is "an invaluable record of shared ideas about music, creativity, the body, and feminism during an extremely productive decade of feminist activism in many spheres—social, political, cultural, artistic."[8] And, certainly, we see the effects of this feminist exchange in Oliveros's later collaboration with Sprinkle and feminist fetishist Venezuelan filmmaker Maria Beatty on *Sluts & Goddesses*.

Sprinkle and Oliveros met through performance artist Linda Montano, but before this encounter, Sprinkle first met Montano after responding to a flier. Montano had placed an advertisement for her Summer Saint Camp in 1987 at the ART/LIFE Institute outside Kingston, New York, one of the mid-Hudson Valley's prime locations for experimental art by and for women-loving women in the 1980s. Each of the camp's visitors attended a one-week residency to practice a "self-imposed exile" and engage in a self-exploratory "hermetic research project" together with Montano. Residents at the camp would keep a "monastic schedule, eat rice and beans, live frugally (a little toilet paper, lights off, cold water baths, etc.)." In terms of arts practice, Montano provides the following description: "In 7 days we visualize and experience the 7 chakras, one a day, and then 'perform' from that chakra each night, after having spent the day coming from that particular energy."[9] Montano published about these residencies, placing particular emphasis on Sprinkle's visit with her then professional partner Veronica Vera, who was similarly engaged in feminist activities as well as acting in porn films. The article opens with general details about the camp and then transcribes an interview between the three.

In the interview, Sprinkle explains why she got involved in porn and prostitution: "Getting into porn was perfect for me. I love sex. I also love creativity—making movies, video and photos, acting, making costumes, etc. And it involves a lot of play—the world puts down sex and also play."[10] And, although Sprinkle enjoyed that life, she explains, she's pivoted towards art to seek out a new audience. Sprinkle and Vera agree that contemporary art is more open in terms of the kinds of performances its audiences will accept.[11] Playing, teasing, pleasing, alluring, Sprinkle is drawn to porn for the same reasons electroacoustic composers are driven to make electrosexual music: the game of it all.

Whereas scholarly investigations of Sprinkle's cinematic catalog could easily justify her performances as art over pornography (on this distinction, see my introduction), Linda Williams, a scholar of musicals, film, and founder of the academic discipline porn studies, insists that Sprinkle successfully blurs any clear-cut distinctions between the two, celebrating the performer with her dual capacity as "artist *and* whore."[12] For Williams, Sprinkle expands feminism as a concept because of how she plays with definitions of key feminist concepts like "woman," "whore," "sex," "desire." Williams links the prepositions pre- and post- in their correlation to the terms "woman" and "feminism," both of which have a reciprocal relationship to the word "whore." When Sprinkle began working as a "whore," liberatory discourses surrounding sex workers' agency did not yet exist.[13] Because the word "whore" exceeds and postdates the concept of woman, its definition necessarily excludes women's agency, and in particular, women's own "desire."[14] In Williams's understanding, pornography has the potential to unite woman *and/as* whore and in doing so grants women space to examine and claim desire. Pornography upholds the subjugated and objectified positionality women occupy historically, as summarized in previous chapters, but also potentially grants performance opportunities that cause firm boundaries to shift between concepts or across performance practices.

Where the traditional whore–client relationship is based on a contract of the customer's satisfaction, Williams speaks to the gap introduced by pornographic media:

> The intimacy with a flesh-and-blood client is the one thing that is not possible within the mediated form of the porno film and video; the whore-client relation of proximity is necessarily replaced, and in a sense compensated for, by the ideal visibility of sexual performers who are not physically there with the spectators viewing the film.[15]

In Williams's observation, in order to project such an ideal, pornographic films usually abandon the "sexual performer's address to the client." As we've seen throughout this book, distance as idealization is often presumed as an electrosexual requirement for arousal. But not for Sprinkle. Like Shields's *Apocalypse* or Lockwood's *Tiger Balm*, Sprinkle maintains a "paradoxical, quasi-parodic rhetoric of intimate address," speaking to viewers while toeing the line between formal address and dirty talk.[16] Like Shields's performance of Woman in the previous chapter, Williams writes of Sprinkle: "By performing sex differently, though still within the conventional rhetoric and form of the genre, Annie Sprinkle demonstrated a provocative feminist agency that would fruitfully contribute to her later feminist performance work."[17] This is to say that expressly feminist electrosexual works necessitate breaking the fourth wall; they require full frontal address.

Central to Sprinkle's art of pornography are pedagogical aspirations she dubs edu-porn. Williams determines that Sprinkle therefore commits to the Greek etymology of the word "'pornography,' literally, whore-writing: the *graphos* (writing or representation) by *pornei* (whores)."[18] In its historical meaning, pornography refers to biography written in the first person, not lewd posturing. Much like today, in ancient rhetoric there is no set form or genre for "obscene" sexual content.[19] Classicist Holt N. Parker explains that "obscene content did not therefore in itself determine a specific genre," appearing in "Old Comedy, satyr plays, mime, iambic verse, hendecasyllabic verse, and satire." In contrast, Parker shows that pornographic sex manuals and handbooks were commonly encountered as a particular genre of sexual adoration and biography.[20] Sex's didactic function figured importantly in early scriptures, and Williams revives this lineage in her analysis of Annie Sprinkle's performances. Sprinkle's films retain continuity with and thus belong among the pornographic. Her films rescript the contemporary whore/woman proxy by reinvoking pre-feminist notions of porn's didactic function as well as maintaining the whore–client rapport. I would argue that her films push the limits of genre and gender in sound too. By incorporating Oliveros's artistic practice, Sprinkle opens her pornography to new audiences, musicologists, music theorists, electroacoustic enthusiasts, and so on. When paired together, Oliveros and Sprinkle's cumulative pedagogy in the *Sluts & Goddesses* video workshop builds on a collaborative, improvisatory, and humorous practice that consciously accounts for the inevitable overlap between audiences of both "art music" and porn, and any parodic dissonances arising therein.

In our interview, Annie Sprinkle explained to me that it was Beatty's idea to approach Oliveros for the video's music.[21] They arrived at Oliveros's house after sending her the soundless fifty-two-minute tape in advance. According to Sprinkle, Oliveros sat in front of a screen with her accordion and recorded the entire soundtrack in a single take. She then added some additional sound effects in the same way, leaving her perch to pick up the occasional rain stick or other instrument out of arm's reach. That is, like the six women in the film entranced in dancing, vocalizing, masturbating, and so on, Oliveros's process of creating the music was spur-of-the-moment and it involved her whole body.

The narrative arc of *Sluts & Goddesses* is driven by its pedagogical aim, where the first twenty minutes—almost half of the film—are dedicated to instructions on how women can explore and play with their sexuality without any of the typical features of a porno. Sprinkle first appears to viewers (presumably women) dressed in office attire and an updo. She explains that in contemporary "Western" culture of the time, the sex life of the "ideal woman" is straightforward, boring, and inaccurate: she is born asexual → loses her virginity on her wedding night → remains in a monogamous heterosexual marriage → loses all interest in sex after menopause. The film proceeds to destigmatize sexuality that departs from this ideal, where the titles "sluts" and "goddesses" describe two such departures. These initial lessons provide examples from "the tantrics," "Taoists," and "ancient sacred prostitutes," which Sprinkle says are all more progressive in their relation to sexual pleasure than Western culture. She then advises viewers to experiment with makeup, body paint, names, and physical movements, like dancing and yoga, to find what pleases them, all the while providing examples of either a "slut" persona or a "goddess."

Visual theorist and cultural critic Johanna Drucker's review of the video describes it as "funky," "minimal," and even "hokey, kitsch and a little silly." In her summary, "While not overly artsy, it has clearly been crafted." As a visual expert, Drucker's review pays little attention to the audible, except in one line: "The sound values are clear, and the Oliveros sound track moves between unobtrusive emphasis and musical interest in a way which complements, with equally light touch, the interplay of humor and serious intention in the visuals."[22] Indeed, film and media scholar Constance Penley has examined how humor in pornographic films works to enhance sexual arousal, naming dirty jokes a core feature of the genre.[23] Notable examples in *Sluts & Goddesses* are when a butch-appearing woman conveys exaggerated facial expressions of pleasure when she applies her makeup raising a question about whether she really enjoys it, or when one of the women of an intense sexual threesome steps back from the action to mow down an ear of corn.

Still *Sluts & Goddesses* remains faithful to the porn genre in a number of ways, featuring the "essential" sex acts or "numbers," as Williams refers to them after Stephen Ziplow's *Film Maker's Guide to Pornography* (1977). Ziplow includes a checklist of essential sexual acts performed in porn films: (1) masturbation; (2) straight sex ("male-female, penis-to-vagina penetration"); (3) lesbianism; (4) oral sex; (5) ménage à trois; (6) orgies; and (7) anal sex. Surveying films only from 1972, Linda Williams observed that the list should include (8) "Sadie-Max," or "a scene depicting sadomasochistic relations such as whipping, staking, or bondage, performed with or without paraphernalia."[24] While comprehensive, Williams points out that Ziplow's list limits itself to "heterosexual porno" and for this reason leaves out a few common subjects (presumably because of their taboo nature and not for their lack of popularity): "there are no male-to-male relations of any kind, nor is there any bestiality or 'kiddie porn.'"[25] *Sluts & Goddesses* stays true to these main numbers with two major differences.

First, because only women appear in the film, the number excludes "straight sex," which it substitutes with penetration using dildos and strap-ons. Secondly, the numbers take an explicit pedagogical aim with more transparent and descriptive explanations of how women can enrich arousal and pleasure—however tongue-in-cheek. The film also makes a point of demonstrating safe sex practices, for example, by introducing vaginal penetration with gloved hands (25:27), and explaining and demonstrating that a dental dam is a thin, flexible piece of latex that can be used in oral sex (26:18). In a later scene, a woman secures a condom with the camera zoomed in closely to a strap-on (32:50). In this aforementioned ménage à trois scene—the climax of the film—a third woman uses a dental dam, and several hands penetrating Sprinkle's various orifices don gloves.

In terms of narrative and visible numbers, the film could be divided into three parts:

A. Edu-porn warm-up (0:00–30:50): features short numbers inviting viewer participation and experimentation with makeup, clothes, "sexercise," etc.

B. Slut sex (30:56–42:30): what Sprinkle describes as the "wild, frenetic side of sexuality," one long number ending with the "megagasm."

C. Goddess sex (42:33–52:00): what Sprinkle calls the "soft side" of sex; personal masturbation together as a group.[26]

Turning to Oliveros's "sound score," the sexual numbers more or less line up with the musical numbers, though there is less variance in the musical themes. Since Oliveros performed the track in one take, combining real-time acoustic and electronic improvisation with live mixing, few themes recur and never in exact repetition. The instrumentation includes an accordion, various percussion instruments (cymbals, drums with or without sticks/mallets, bells), and electronics including the recorded and processed voice.

The film opens with rhythmic, percussive tapping. Gradually, different percussive sounds enter, first a Tibetan cymbal, and soon after high-pitched bells (0:06). The tapping instruments phase in and out, keeping a firm beat ornamented with an occasional roll or bifurcating subdivision. Next, low bells enter (0:45), then slide whistles (1:25), which arouse anticipation as several whistles slide alternately, incrementally gaining speed throughout Sprinkle's pedagogical introduction until around the two-minute mark. Here, the introductory music ends with an abrupt cut scene. Sprinkle has finished telling us about the "ideal woman."

In a new scene, at 2:12, a wooden flute enters playing a one-and-two rhythm, randomly alternating pitches of a major triad. Soon the fluttering flute is joined by another flute-like sound (possibly synthesized) playing sustained tones—a sound reminiscent of the sustained motive in *Fish & Fowl*, recalling the suspenseful "waver" drones in horror films and film noir, which for Roger Hickman are "quite effective in generating suspense and for supporting scenes of nightmares, drug states, and madness."[27] Oliveros's accordion then joins in fluttering improvisation with both flute sounds. Because these fluttering sounds recur numerous times throughout the sound score, I call them Motive X (shorter than XXX!). This motive recurs four times in Section A, and the fifth time it coincides with the beginning of the feature number, the ménage-à-trois scene opening Section B. Motive X is sometimes prefaced by sustained flute-like tones, which, because of the prominence of

Table 8.1

Simplified formal scheme of Sprinkle/Beatty's *Sluts & Goddesses* Video Workshop

Form	INTRO	Section A				Section B	Section C
Narrative Description/ Sexual Numbers	Ideal Woman	Lessons	–	–	–	ménage à trois (climax)	group masturbation
Musical Numbers	I N T R O	"soft" motive + Motive X	Motive X	"soft" + motive Motive X	Motive X	Motive X	"soft" motive
Timing		2:12	24:50	26:50	28:55	30:56	42:33–52:00

these tones in Section C, I will call the "soft" motive. Table 8.1 presents a summary of this formal scheme to show how visible numbers align with the recurring musical themes. Concatenating the introduction, Motive X, and the "soft" motive provides a loose structure to the film, and given the salience of these features and their placement in the background's audible soundscape, I would venture a guess that these are the product of Oliveros's single-take recording. Atop this background, we also hear more prominent sounds, the occasional electronic bleeping and blooping (14:00), feminized breathing and moaning as we would expect of a pornographic film, and, at times, this voice is enhanced with reverb (15:00; 19:20).

On two occasions, a voice transgresses the spatial boundary between background and foreground, as if mocking such cut-and-dried divisions. The voice embodies the playful electrosexual character of sound and cinema by modulating to a frequency well below or well above the spoken range of any of the film's visible actors (35:40; 38:00; 47:16). In these instances, we might wonder if Oliveros herself is vocalizing as if alongside the film's various sexually active instructors. Oliveros's audible signature also sounds in other ways, for example, when we hear insect noises (10:47—like in the "Who Said What?" movement of *Cicada Dream Band* [2014]), bird song (48:16—like in the "birds" movement of her *Springs* [1999]), and sampled, processed and distorted sliding accordion (throughout).[28]

REAL SEX

In the same way in which I have until now ascribed instability to electro-sexual music as a genre, Williams describes pornography as a medium that "catches up its viewers in the impossible question of the ontological real of pleasure."[29] That is, even when seeing and hearing it with our own eyes and ears, no spectator can confirm what's actually going on, whether inside the woman's body or to the extent that the performance is either rehearsed or spontaneous. Certainly, part of the allure arises precisely from the "variety of different truths" operating simultaneously for viewers and performers alike.[30]

When it comes to sound, porn films as a whole tend to augment the visual experience with dubbed-over voices such that the above-described disruption between background and foreground is all part and parcel. Williams compares this "dubbed-over 'disembodied' female voice'" to the "nonsynchronization of female body and voice in avant-garde practices that deconstruct the dominance of the image, especially the patriarchal fetishized image of women."[31] As I have shown in previous chapters through sound alone, experimental practice, while often performed improvisationally, usu-ally still holds firm to pre-given philosophical beliefs held by its creators and practitioners. Schaeffer and Henry spent years refining *Symphonie* even if, in the beginning, the sounds originated from "happy accidents." And Ferrari, Normandeau, and Lockwood recorded prolonged scenes from real life, only to edit these later to conform to specific notions of sex or musi-cal form. Electrosexual music is fixed in some sense and therefore could be understood to represent what Carolyn Abbate recognizes as music's "gnos-tic," or textual, predictable, and intellectualized form. And yet, the impulse to peek behind the curtain, the sonic allure of its erotic envelope casts doubts on this music's reality—its origins as well as its projected narrative context. Sprinkle/Beatty/Oliveros composed and prepared their respective works ahead of their performance, rehearsing like jazz musicians many possible scenarios that would ensure the appropriate execution in the moment. Such is the habit of pornographic films, where "sex as a spontaneous *event* enacted for its own sake stands in perpetual opposition to sex as an elaborately engineered and choreographed *show* enacted by professional performers for a camera."[32] With its many bodily shapes and sexual tastes expressed, *Sluts & Goddesses* departs from the stereotypical cover girl to such an extent as to appear unstaged, lifelike. Yet, in the same hand, Oliveros's sonic backdrop disturbs and confuses this reality, where the instrumental over-dubbing provides a closeness and clarity well beyond the usual cinematic frame by reducing the distance between performer and viewer, or rather, enhancing their intimacy.[33]

Williams argues that in porn and musicals, alike, the most successful numbers "have been the numbers that most gave the illusion of being created spontaneously out of the materials at hand."[34] Sprinkle as a sexpert, compared with the presumably inexpert workshop participants, further augments this uncanny reality, wherein no "ideal" woman can exist because the film does not present just one distinctive model. The film doubles down on this disruption of the ideal, presenting performers across the vaginal spectrum. Williams acknowledges that one of hard-core pornography's biggest conflicts lies in "figuring the visual 'knowledge' of women's pleasure. Although the genre as a whole seems to be engaged in a quest for incontrovertible 'moving' visual evidence of sexual pleasure in general, and women's pleasure in particular, this is precisely what hard-core could never guarantee."[35] In this sense, Sprinkle compares to performers like Dolly Parton ("It costs a lot of money to look this cheap"),[36] singer and record producer Dusty Springfield,[37] or TV personality and stripper Anna Nicole Smith,[38] amplifying the essential characteristics of the idealized sexually available woman—ready-to-go at any moment. Sprinkle's performed persona acts toward an ideal that ends up succeeding because it works in excess of any one woman but also of any one man.

John Corbett and Terri Kapsalis have observed that equal to men's propensity for the money shot is an arousing *soundscape* dominated for the most part by women's pleasure.[39] But Sprinkle dominates both stereotypes, performing in excess and therefore lapsing over any either/or boundary. In several films, from *Deep inside Annie Sprinkle* (1981) to *Sluts & Goddesses* (1992), she emphasizes the orgasmic power of her climax, showing her pleasure both in the visible ejaculation she projects and in how she *sounds*, as if taking on stereotypical roles from opposite poles of the binary. *Sluts & Goddesses* does away with penises entirely, though Sprinkle encourages all her viewers to experiment with gender the way she does and to try drag. One performer is introduced as a specialist for the threesome, sporting an elaborately drawn goatee. In this sense, drag does not diverge from the ideal, rather it rescripts the limits of gender by relying on excess, exaggeration, and, in this way, redefining gender as limitless (I take this up again in the next chapter). Sprinkle has fun with sex and gender in a way that "continuously plays with the terms of *norm* and *perversion*," as Williams explains,[40] and because she exaggerates the sexual tropes that characterize men and women alike, her performances resist reduction to any mold of the stereotypically authentic porn actress, which itself is a performance of an identity.

A crucial scene in the film presents a graph detailing the qualities of single and multiple orgasms, with orgasmic energy charted on the y-axis and time on the x-axis. Sprinkle proceeds to demonstrate what she terms the "megagasm," a type of orgasm that, when charted, is riddled with "peaks, valleys, rocky tops, and plateaus, like the high Sierras" to defy and outperform the previously set ideals. Williams compares this five-plus-minute orgasm to the man's "money shot," evaluated for its force and projectile of the ejaculate[41] and often the feature number of many pornos. While this scene does superimpose the narrative climax onto Sprinkle's physical climax, *Sluts & Goddesses* is not your typical film. Sprinkle "exhibits not only a female money shot but also the performance of a six-minute orgasm, [which] would seem to so spectacularly imitate the male standard of the pornographic evidence of pleasure as to destabalize and denaturalize its 'normal' meaning."[42] Williams believes Sprinkle deliberately invokes comparisons to men in order to surpass and outperform them. Likewise, Oliveros greets this moment with howling and screeching electronically synthesized tones as well as a synthesized voice. The manufactured voice begins low, chanting repeatedly in disbelief a word that sounds like "what?" Gradually the voice shifts upward in anticipation to sound more like "yeah, yeah, yeah" with each passing moment. Sounding higher and higher, it also grows progressively faster until ululating, as if shaking. A timer on screen shows Sprinkle has been climaxing for a full five minutes. The scene and sound fade out.

Sprinkle's performance in *Sluts & Goddesses* takes on a didactic role similar to sexual education pamphlets that aim to convey information to a possibly undereducated or unsure audience, at the same time giving a campy wink to knowing feminists. Rather than adolescents, who stereotypically occupy the role of uninformed audiences when it comes to sex, Sprinkle targets an audience of women who, presumably on account of patriarchal suppression and a longtime favoring of men's sexual pleasure over women's, are ignorant to their own desires and ways of achieving sexual pleasure. Not unlike traditional porn, which entices audiences with the promise of forbidden knowledge, this educational bent introduces an element of mystery. Oliveros's music enhances this mystery and viewers are left wondering not only what they just watched, but also "What did I just hear?" Where the typical porn soundtrack enhances the atmosphere by dubbing over feminized moaning to immerse the viewer, Oliveros's whimsical sound score plays into the random quirks that mark Sprinkle as unique. Of course, hard-core Sprinkle fans view the film with firm knowledge that women's ignorance of their own pleasure is only a myth, a joke that falls at women's expense but hardly acknowledges any actual experiences.

In her chapter "Crackers and Whackers: The White Trashing of Porn," Constance Penley famously raises concerns about stratifications of class and race in the United States by comparing these to a parallel stratification in pornographic content. Drawing a boundary between content worthy of defense and content ignored by staunch cultural defenders, Penley argues that porn scholars tend to downplay pornographic oddities—the cheesy, the trashy—perhaps to augment ways in which the genre might intersect with experimental art.

> But where are the defenders of contemporary work like *Tanya Hardon, The Sperminator*, or *John Wayne Bobbitt: Uncut?* The more mass-cultural the genre becomes, and, it seems, the more militantly "tasteless," the more difficult it is to see pornography's historical continuity with avant-garde revolutionary art, populist struggles, or any kind of countercultural impulses.[43]

Penley's argument often serves as justification for including Sprinkle's oeuvre in scholarly analysis as an instance of crossover between art and smut.

And it's clear that Oliveros's appearance serves as further justification. Conversely, by appearing alongside Sprinkle, Oliveros is positioned in an even more hospitable light for her eclectic followers, who might interpret such a collaboration as further resisting the inculcated cultural stasis of the "contemporary music" scene as I've shown in earlier chapters. That is, Oliveros earns cool points for slumming it, and Sprinkle's otherwise trashy whoring is elevated from (mere) porn. Might we say the same about the excesses of "Western" culture adopted by both Sprinkle and Oliveros? What of the Orientalist co-option of loosely defined "tantric" knowledge and exoticized sounds? A white woman in yellow face doing yoga, belly dancing classes, accompanied by Native American flutes, drumming, moaning . . . How much clearer the electrosexual genre appears from the opening of this book.

9 BARRY TRUAX'S *SONG OF SONGS* (1992)

Writing almost thirty years ago from the Canadian West Coast, composer Barry Truax (b. 1947) explained how significantly "granular synthesis" informed his approach to musical composition. Despite concerning itself with "seemingly trivial grain(s)" of sound, as explained below, the digital process of granular synthesis not only changed how he conceived of the greater forms of musical composition but also changed the way he thought of music and its larger social and historical function.[1] The practice concerns the musically global as much as the minute, "[it] clearly juxtaposes the micro and macro levels, as the richness of the latter lies in stark contrast to the insignificance of the former."[2] Musical relationships forming the "inner complexity" of the work, he says, are necessarily shaped by "their possible relationships to the environment or society at large"; what he terms the music's "outer complexity."[3]

As a researcher working with R. Murray Schafer, Truax groups himself within the "soundscape composition" movement, a compositional approach that combines "artistic creativity with social concerns."[4] But he has gradually moved away from a broadly conceived soundscape philosophy, which sometimes risks overly "aestheticising the sounds of the environment," gravitating toward what he calls "context-based composition" to describe how his "homoerotic" compositions fit into their situated sociohistorical context. Whereas soundscape composition often refers to the real world by maintaining connections with sound's recognizable sources, many of Truax's context-based compositions begin by introducing referential sounds like the voice or the environmental sounds of animals and insects, only to modify these sounds gradually with digital processing. In this way, the composer encourages his listeners to relate recognizable sources to heavily processed and often unrecognizable results; in short, inexperienced listeners learn to listen acousmatically.[5] Truax explained this compositional orientation in a call for submissions for a special issue of *Organised Sound* journal on the themes of "context-based composition": "A key distinguishing feature of context-based composition appears to be that real-world contexts inform the design and composition of aurally based work at every level, that is, in the materials, their organization, and ultimately the work's placement within cultural contexts."[6] Thus sound itself becomes entrenched, and, indeed, drenched with context at every register, engaging not just with contexts extrinsic to the work, but insisting and relying on transcontextual knowledge. "Perhaps most significantly," reads Truax's call for submissions, "listeners are encouraged to bring their knowledge of real world contexts into their participation with these works." In context-based composition, musical structure is defined by parameters that extend beyond the "purely" formal to such structuring parameters as gender and sexuality, for instance, if one looks to Truax's own compositions. Disappointing for Truax, my article, from which excerpts for this chapter are taken, was the only submission to his call to address "gender concerns in composition," a subcategory mentioned explicitly in the call for the issue.[7] However, this may be unsurprising to those of us invested in diversifying representation in electronic music.

Truax's *Song of Songs* (1992) for oboe d'amore, English horn, and two digital soundtracks is one example of context-based composition.[8] More specifically, it is a piece that references gender and sexuality at several registers—text, music, meaning, as well as socially and historically constructed knowledge—moving beyond such binary categories as the male and female[9] genders or homo- and heterosexual orientations to a more fluid and dynamic understanding of gender and sexuality. The voices of two speakers (a woman and a man), a singing monk and environmental sounds from birds, bells, crickets, cicadas, flowing water, and a burning fire are all subjected in varying degrees to the technique known as time-stretching to shift a sound's time scale without varying its frequency. In *Song of Songs*, time-stretching is used to subtly modify the rhythm of the spoken text to make it more "song-like," and, as I show later, when accompanied by harmonization, the technique prolongs sounds into sustained timbral textures.[10]

In the opening material of the piece, Truax interferes with the samples minimally in order to retain the sonic identities of the voices. The voices of the two speakers, the man, recited by Norbert Ruebsaat, and the woman, recited by Thecla Schiphorst, are therefore arguably already identifiable from the beginning with their representative genders. By having both speakers recite the same portions of the biblical text of the *Song of Songs*, without altering the gendered pronouns of the original text, the composer gives the speakers the opportunity in the beginning of the piece to seemingly address either opposite- or same-gender lovers.[11] While the speakers in *Song of Songs* are introduced within the recognizable pitch range, timbre, and inflection attributed to their respective genders (more on this later), as the piece progresses the voices are often time-stretched or harmonized, such that the role of the voice is changed. The voices of the speakers sometimes sound ambiguously gendered, as if spoken through a vocoder, and, in more extreme cases, the voices are stretched and filtered so far out of proportion that they are hardly recognizable as voices. At times, in addition to the voice's more conventional solo performing role, they may even be heard as sustained environmental ambiences setting the contextual scenery for the music. The following analysis explores the normative gender categories listeners are likely to assign to voices they first hear in the opening of the piece, at least as the piece would have been heard in the early 1990s. The analysis then provides examples to demonstrate instances in the piece in which gender "blurring" occurs as identified by the composer.[12] The final section of the chapter situates *Song of Songs* within its contemporaneous musical climate through a comparison to Susan McClary's well-known analysis of Laurie Anderson's "O Superman," published in 1991, which also provides some context for how theories of gender and sexuality may have informed Truax's compositional approach.[13]

GENDER AND THE VOICE

Vocal music clings to gender, whether or not we attend to it as meaningful for a given analytical orientation. In their introduction to *Embodied Voices*, editors Leslie C. Dunn and Nancy A. Jones compare feminist uses of the term "voice," as a metaphor for women's historical and political struggle to be heard with non-verbal "vocality" as the literal and sonorous expression referred to in Roland Barthes's notion "the grain of the voice" (see chapter 3). Dunn and Jones's term "embodiment" adjoins these two meanings to emphasize "the performative dimension of vocal expression . . . [its] dynamic, contingent quality."[14] The embodied voice thus functions as the "material link between 'inside' and 'outside,' self and other. . . . [And] since both language and society are structured by codes of sexual difference, both the body and its voice are inescapably gendered."[15] Women's voices have often been used in aural reference to women's bodies—think, for example, of the enticing sirens in Homer's *Odyssey*.[16] Barbara Bradby has addressed links between "gender, technology, and the body" in electronic (dance) music to show how a woman's voice can be used to represent a distorted female perspective compromised by the auspices of "male fantasy"—by the gaze of the (male) producer or composer (more on this in chapters 10 and 11).[17] In this regard, a monumental study by Hannah Bosma expanded Bradby's findings into the electroacoustic realm, showing the prevalence in "electrovocal" music of partnerships between male composers and female vocalists. In these ubiquitous partnerships, as Bosma observed, women's voices are often attached to a conjured image of women's bodies to argue therefore that "gendered voices" have a "symbolic significance" on account of their reflexive embodiment.[18]

Gender not only shapes how we intuit bodies individually, but, according to critical theorist Eve Kosofsky Sedgwick, it is also the primary means by which we explore relations between and among bodies. Sedgwick perceives gender (and not age, physical appearance, ability, sensation, frequency of behavior, species, etc.) as the most dominant dimension by which sexual orientation is determined in contemporary Western society.[19] Electroacoustic music, however, does not easily resolve itself to reductive gender determinations, since, without visually accompanying evidence of a body or the recognition of the performer's name or identity, it is not always clear if the represented figure is indeed a woman, or otherwise.[20] The disembodied voice presents ambiguousness, particularly in the slippery categories of identity and orientation. As Judith Butler notes, "[gender] identity [i]s a compelling illusion, an object of *belief*," and in the case of electroacoustic music, the voice is at the heart of this illusion.[21] When the voice is recognizable as such, a question as to whether the voice references its originating speaking body still lingers, and we might also wonder about the speaker's originating gender, whether before or after transition.[22] If I recognize the voice at all, it is likely that I will attempt to understand, categorize, and parse it, whether I do this consciously or unconsciously.[23] Given any number of transformations, including digital or physical operations, I may never successfully identify its source or cause.

When it comes to the acousmatic voice there seem to be two competing, though not absolutely exclusionary, theories: the first being that, once recorded, the voice no longer serves as a referent to its original source and cause, to its originating body; and the second that the voice can never be disembodied. Advancing this second theory, Steven Connor declares: "dissociated voices always [seem] to summon in their wake a phantasm of some originating body, effect convening cause."[24] In the first case, it is tempting to explain the recorded voice as a sound "object," whose identity is open and unrestricted by the delimiting categories that organize actual bodies. And yet, a position that does not account for such identifying factors as gender, however optimistic, potentially relies on a myth of music's absolutism, perpetuating the long-standing resistance by some to acknowledge that context might not arise entirely from "music's inner relationships."[25] The second case, the opinion that the voice is never disembodied, implicitly accepts that we retain gender and sexual markers as structuring determinants even when no body is visible. But with this view there is still a risk in *insisting* on the "phantasm of some originating body," to recall Connor's description. Maintaining that the voice is always embodied presumes that gender categories are always-already inherited into the acousmatic context, thereby precluding voices from potentially sliding from one gender to another or anywhere in between, a fluidity of identity not easily attained with actual bodies (either in the real world or in performance) but easily and commonly encountered electroacoustically. Truax's solution is to constantly negotiate the territory between voice recognition and denial of its referential significance. In this way, relationships develop between the voices and the personae they are perceived to embody as well as between the other musical and environmental sounds common to Truax's pieces.

Despite the frequency with which electroacoustic composers employ the gendered voice towards erotic ends, relatively few aside from Truax explore erotic significance outside the confines of normative heterosexuality. In "Homoeroticism and Electroacoustic Music: Absence and Personal Voice," Truax writes, "Art is said to mirror society, but if you look in the mirror and see no reflection, then the implicit message is that you don't exist."[26] The composer therefore tries to find representative ways of including minority identities—primarily gender and sexual identities—to amend their habitual absence from the electroacoustic music tradition. Many of Truax's works exhibit the composer's acknowledged "homoerotic" representations. These include *Androgyny* (1978), *Androgyne, Mon Amour* (1996–1997) after Tennessee Williams's poems, *Twin Souls* (1997), several pieces on the theme *Powers of Two* (1995–1999) including an operatic version (2004), and *Skin & Metal* (2004).

In many pieces, Truax's samples undergo digital "transformations," a term composer and electroacoustic theorist Denis Smalley discusses at length in the context of electroacoustic music. "Transformations," says Smalley, "concern changes in the state of sonic identity. . . . A transformation may be regarded as travelling a certain distance from its base, and the type of change may be defined in terms of its direction—whether the source-cause implications are specific, implied or free."[27] Given these transformative possibilities, I have chosen to focus on *Song of Songs* as an example that explores a wider spectrum of gender and sexual representations than the homoerotic perspective Truax explicitly mentions in his article. In this chapter, I take listener perceptions of gender and sexuality in electroacoustic music as context-dependent criteria, meaning that notions of gender and sexual orientation depend as much on the processes by which a piece is composed as on a contemporary understanding of gender—both contemporaneously to the work's composition and, separately, to its hearing. Since composers attempt to wholly construct the electroacoustic music context, many overlook how situated and embodied knowledge regarding gender, sexuality, race, and other socially constructed bodily markers can, nevertheless, stick to sounds referentially. With supporting evidence from psychoacoustics and empirical studies, I consider the inheritance of gender and sexuality within a genre of music often represented historically as devoid of bodily markers.

TEXTUAL CONSIDERATIONS

Of the many sensuous moments recorded in the Bible, King Solomon's Song of Songs is probably the longest, most detailed, and most celebrated portion. Remarkably, two thousand years after the scripture was first recorded in the second century CE, Solomon and Shulamite's mutual desires continue to be recognizable to us today. Their love is described in vague enough terms to still be relevant, while the specifics of the story—the lovers' mutual descriptions of one another's physical bodies, their movements, and even details of their environment—remain alluring for new interpretations. The Song of Songs is commonly interpreted in one of two ways. It can be read literally as a dialogue between King Solomon and his lover Shulamite, the two named characters in the poems, or in a metaphorical reading (the typical religious interpretation), as a dialogue between God and the people of Israel. Identifying the characters one way or another of course changes what relationship we perceive between them, whether romantic love or mere companionship, and this is without mentioning alternative levels of meaning yielded by retaining both the metaphorical and the literal narratives as viable interpretive registers.[28]

There is much to suggest that these songs depict erotic desire beyond mere companionship, as, for example, in many physically descriptive passages, "thy stature is like to a palm tree, and thy breasts to clusters of grapes" or "I sat down, under his shadow, and his fruit was sweet to my taste." Song of Songs expert Michael V. Fox convincingly disputes a singularly metaphorical interpretation, arguing that metaphorical poetry from this time typically speaks of "love in generalities," whereas the first-person accounts in the Songs reads more like personal testimony.[29] Although the characters sometimes refer to one another in the third person out of respect, they often also refer directly to each other, something uniquely intimate for texts from this period. But Fox warns against interpretations that distort the text's palpable allure by overemphasizing its anatomical symbolism.[30] Love in the Songs is an orientation towards a lover. "It is mainly the presence or absence of the beloved, or the expectation of that presence, that determines what effects love will work."[31] And so, although only two characters are named, Fox suggests that one can read multiple personae in this text in addition to the named Solomon and Shulamite.[32]

Likewise, attempts to unveil the voices in Truax's piece—to reconcile the voices with real-world sources and causes—oversimplify the complex identities evoked by the work. Truax's setting exploits the text's multitude of possible voices, employing various "blurring" effects to expand the meaning of love through metaphor. As Truax sees it, electroacoustic composition is an ideal setting for sexually explicit engagement with sound, precisely because of the ambiguity of attempting to recognize sound sources. This ambiguity allows for flexible relations to form among and between elements, people, and things.

The titles of the four movements are derived from times of day suggested in the text, and each movement features sounds characteristic to the time of day indicated by that movement's title. Monastery bells and an Italian monk's song sound in the "Morning," and the hazy "Afternoon" is adorned with the seasonal charms of singing cicadas. As "Evening" descends and the heat disperses, crackling fires create a perfect setting for amorous activities. In "Night and Daybreak," a canopy of chirping crickets first beckons sleep, and another day dawns at the familiar tolling of the bell from the work's opening.

The daily cycle is apparent not only in explicit references to morning in the first song or in lines like "until the day break and the shadows flee away" to signify evening and daybreak in the third and fourth movements, but also in the speakers' more subtle allusions to their environment—references to shadows, light, sun, flowers and insects—indicating times of day or the season. The composer's organization of the text emphasizes these environmental references. By structuring the sonic environment according to textual descriptions, Truax ensures the sounds "themselves" are inextricable from the overlapping intimacy conveyed linguistically by the adoring lovers. Like the warm memories invoked in the unaccompanied *saetas* of Federico García Lorca's (homo)erotic poem "Before the Dawn" or Langston Hughes's "Songs to the Dark Virgin" (also after Song of Songs), Truax's environmental sounds convey several levels of familiarity.[33] Truax's "Evening" recalls Lorca's forbidden "green hour," the time when out-of-sight lovers cast all rules and social expectations to the wind to unite briefly in hidden bliss. "But like love / the archers are blind," writes Lorca, and in stride, Truax arouses erotic connotations but grants his listeners their own particular associations.

VOCAL QUALITY AND IDENTITY

Truax is an electroacoustic composer who, since the 1980s, has worked extensively with the granulation of sampled sound.[34] Granular synthesis, as it is more commonly called, affords a process of time-stretching. The grains are not merely extracted from the source material but are manipulated according to the flexibility afforded by the time scale. Notably, this temporal flux also contributes to perceived differences in timbre. Truax notes that the time-shifting he introduced in *The Wings of Nike* (1987) and developed further in *Song of Songs* "[prolongs] the sounds into sustained timbral textures," which allow listeners to engage more closely with sound. Granular time-stretching is one manner by which Truax "magnifies" sound.[35] Using a short, fixed sample, he introduces grains in the same order as they sound in the originating sample but staggers overlapping grains to deliberately extend their offsets (the point from which the sample begins to play) and smooth out their envelopes so as to give the impression of the sound in slow motion. Even if, for example, two adjacent grains are of the same aural quality, source, and envelope, one's perception of how a sound unfolds in time depends on how the reintroduced grains are staggered in relation to one another. The dragging effect is not merely acoustic but extends also to any meaning listeners may attribute to a sound. More specifically, when applied to a recognizable voice, the technique affects the quality of that voice and hence the listener's perception of it. All the recorded sounds on the two soundtracks—every sound in *Song of Songs* except the live oboe d'amore and English horn—undergo time-stretching.[36]

When Truax composed this piece in the early 1990s, it was presumed in psychoacoustics and psychology that pitch was the greatest determinant of gender perception in the voice. Although the composer himself does not refer to any supporting literature from the sciences, his compositional intuitions in terms of gender and psychoacoustic identity very much support such informed hearings. A study conducted by Wolfe et al. demonstrated that without a visually corresponding figure, for example when speaking on the telephone, a speaker is generally identified as "female" when speaking at a fundamental frequency *above* 160 Hz, roughly E3 or E below middle C, and listeners identify a "male" speaker when speaking at a frequency *below* 150 Hz, or D below middle C (figure 9.1).[37] Although the composer has not specifically referred to it, as we will see later, Truax's processed voices in *Song of Songs* often intersect at the ambiguous range between 150 and 160 Hz, particularly when the text takes on an erotically suggestive tone.

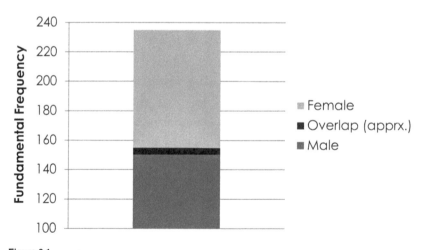

Figure 9.1
Fundamental frequency and gender perception.[38]

Building on the findings of Wolfe et al., Weston et al. found that, when given a choice between the male and female gender, listeners correctly identified the speaker as either a man or a woman even when the fundamental speaking pitch was altered, so long as the spectral envelope remained unaffected by pitch-shifting.[39] Weston et al. asked listeners to gauge a speaker's gender in three distinct sources of auditory stimuli. Listeners were introduced to unadjusted recordings of the speakers and also to two examples that had been digitally pitch-shifted to the ambiguous range around 160 Hz: one using "pitch-synchronous overlap" that *retained* the spectral envelope, and another that changed the pitch and *distorted* the spectral envelope using a "scalar factor" or asynchronous granular synthesis.[40] They found that when the spectral envelope remained intact, listeners could correctly identify the speaker's gender regardless of the spoken pitch level. These studies show that pitch could determine gender perception in the voice even when the spectral envelope is altered, but if unchanged, spectral flux (the rate of spectral change over time)—which is statistically greater in women[41]—remains the primary determinant of gender perception in the voice. Hence, using a time-stretching process to alter the speaker's inflection disproportionately to the original may impact perception of the speaker's gender. While these findings are useful for determining a basis for the normative gender categories that listeners are likely to attribute to the voices they hear in *Song of Songs*, they do not, in my opinion, denote inherited biological traits pertaining to the genders of either speakers or listeners.

Similar to the techniques employed in the above studies, in *Song of Songs*, Truax plays with the fundamental speaking pitch of the two speakers through comb filtering and harmonization to elicit various sonic effects as a deliberate commentary on the relationship between the speakers. He does not employ any pitch-shifting in this piece; instead, he preserves the spectral envelope of his samples and variously extends sample duration without altering pitch content. However, his setting often layers numerous samples of the two speakers on top of one another, sometimes using harmonization to emphasize other pitches or overtones more strongly than the fundamental frequency. Hearing each speaker recite the text in its entirety with the original gendered pronouns and then layered with several more voices may cause listeners to question the number of perceived characters in the work, their respective genders, and, given the erotic tone of the text, also the characters' various orientations towards one another.

REDEFINING CATEGORIES

In my hearing, Truax's setting first establishes binary categories, such as the normative male and female genders or homo- and heterosexual orientations so as to break away from these tropes as the piece progresses. The composer's measured departure from normative constraints, I believe, functions to question and confront *a priori* categories—including the orientation each speaker takes towards their speaking partner(s) when reciting the erotic text. This blurring of categories fragments and hence reorients the respective identities listeners might attribute to each voice.

For example, in the opening of *Song of Songs*, Truax establishes the respective identities of the two speakers by introducing the voices with minimal processing. Alexa Woloshyn suggests the opening line, "Return, return O Shulamite, return, return that we may look upon thee" (0:37 [i, 0:44 score])[42] is recited by both speakers so as to evoke a polyamorous relationship with Shulamite, their female lover.[43] This hearing is supported by literary criticism that Shulamite "gives the appearance of being a loose woman . . . , but she upsets all conventions. Her love . . . rather than being a source of shame, it is gloriously proclaimed."[44] In system 5[45] of the first movement (1:31 [i, 1:28 score]), a man speaks "Who is she that looketh forth," and we hear a corresponding interjection from a woman (Schiphorst) on the word "fair" (figure 9.2). The prominent pitch of this word sounds around A below middle C, or 220 Hz, with a characteristically feminine upward inflection, as determined by Wolfe et al. and in subsequent studies.[46] The upward inflection of the "fair" voice brings its fundamental frequency to D or 294 Hz, the pitch at which this voice is met by another voice (Ruebsaat), speaking the word "terrible," absent the "t" attack. Starting here, from the supposed female range (as determined by Wolfe et al.), and then gradually dropping in pitch by exploiting the original downward inflection on the word "terrible." Now made evident by a hundred-fold time-stretching

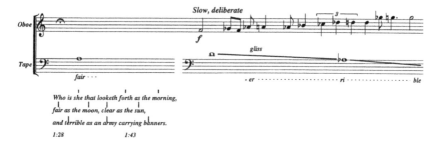

Figure 9.2

Song of Songs, i: "Morning," line 5. Transformation of perceived gender identity and sexual orientation in "fair" and "(t)errible" voices. Oboe d'amore in A.

and downward harmonization, the "terrible" voice descends through A and G-flat to join the oboe for a moment before sliding progressively lower and out of the normative female range.

The mixture of the "fair" and "(t)errible" voices descend from the female range through what could be inferred as the male range but quickly depart from the human-speaking range altogether. This gender ambiguity leads a minute later to the first exclamation of the refrain, "I am my beloved's and my beloved is mine." When the line repeats, the two voices return to sound in their respective male and female ranges. After additional declamations, the repeating voices progressively begin to offset one another until eventually a third speaker seems to join the duet, another male voice—the same speaker doubled. This additional male partner completes the sentence of the first two speakers: "I am my beloved[']s and my beloved is mine. He feedeth among the lilies." Whereas at first the man and woman may be heard as lovers, soon one voice is transformed to sound ambiguously gendered, mixed with the female voice harmonized both upwards and downwards. Then, after the contraction of the voices back to their respective genders, the doubling of male voices introduces the possibility of a homoerotic relationship among two of the speakers if we take this "he" as the man's lover. Alternatively, by the end of this phrase, three speakers are engaged in a perceived polyamorous relationship.[47]

While sexual identity is presumed to betray an individual persuasion, it is, in actuality, an individual's relation to others that defines one's sexual *orientation.* Sedgwick theorizes the concurrent stability and fluidity of sexual definitions in American society of the 1980s and 1990s in an investigation of the intersection between the fixed homosexual minority (or minoritizing view) and the commonalities shared by individuals across a spectrum of possible sexualities on account of gender or other social factors (the universalizing view).[48] The fixity of the minoritizing view, on the one hand, is important as an identity from which to relate and refer, but, on the other hand, the presumed stasis of sexuality—that individuals should be grouped together only on account of what arouses them sexually—limits the political potential of identification across many more articulations of sexuality. In distinction from Woloshyn, whose analysis remains faithful to Truax's categorical separation of male/female and homo-/heterosexual identities, my analysis views such identities not as inverse categories but as relationships determined by momentary orientations.

SHARED TERRITORY

In *Song of Songs*, Truax disguises the process of transformation, such that the original and processed sources sound simultaneously, thereby bridging associations between recognizable sounds and their transformed counterparts. Such pairings can easily bring about false cognates; meaning, similar sounding sounds may originate from different sources. In the second movement, "Afternoon" (figure 9.3), we see an example of time-stretching and harmonization that introduces a change of identity and an instance in which the fragmented voice alters the perceived number of speaking personae. The male speaker enters as a recognizable or determinately male voice reciting the text: "A garden enclosed is my sister, my spouse, a spring shut up, a fountain sealed" (4:52 [ii, 0:52 score]). Here the sample is only minimally time-stretched and retains its spectral envelope. At the oboe d'amore's entrance, the word "fountain" is repeated, this time by the female speaker, and subjected to heavy distortion, granulated and harmonized (both upwards and downwards) at a time-stretched ratio of 100:1.[49] Interestingly, the pitches of the time-stretched "fountain," which emerge through a harmonization of

Figure 9.3

Song of Songs, ii: "Afternoon," lines 3–5. This transcription shows the strongest sounding fundamentals in the bass clef, the time-stretched "fountain" sounding in the ambiguous register, in a range around E-flat₃, or 155.5Hz. X noteheads on lines 3 and 5 indicate the aliasing caused by the consonants "t" and "n." Oboe d'amore notated at concert pitch.

Schiphorst's voice, are the same frequency as Ruebsaat's initial recitation of the word "fountain" directly preceding the stretched passage—both sounding in the ambiguous register at a range around E-flat3 or 155.5 Hz, thereby blurring any contrast of the two speakers as well as their perceived gender identities. In this example, Truax introduces timbre and pitch ambiguity precisely when the speaker's identity might determine one's orientation towards a partner. In reciting the same text, a bond is formed between background and foreground voices, and because the gender identity of the background voice is ambiguous, their relationship becomes uncertain. Neither an opposite-sex nor a same-sex partner can be fully ruled out.

The two-part texture here occasionally fragments into a chorus of many more voices, for example in system 4, when F-sharp in the lower voice is met also with C-sharp, a fifth plus an octave higher.[50] Truax employs this technique of harmonization frequently in the piece, in places where the text and voice are recognizable to provide an opportunity for attentive listeners to associate these more ambiguous background streams with time-stretched voices. Listeners acquainted with granular synthesis and time-stretching will likely also recognize the voice given its timbre and the distortion caused by harmonizing the unvoiced consonants "t" and "n" (represented by X note heads in my transcription)—not coincidentally recalling the sound of the flowing fountain water. The same techniques contributing ambiguity to our perception of the speaker's gender identity also cause some blurring of musical categories. For instance, the initial overlap and subsequent trans-formation of the two voices (at the word "fountain") is an occasion when the prominent speaking voice drops into the background to take on a sup-porting role. The composer likens this "merging of sonic elements . . . [to] the extended metaphor of the original text which compares the Beloved to the richness of the landscape and its fruits."[51] According to the composer, the blurring between genders occurs primarily on account of the gendered text recited by each speaker, "sometimes reflecting an opposite sex form of address, and sometimes a same sex form."[52] Yet he believes real-time granular synthesis enhances the sexual allusions of the text:

> Granular stretching of a voice, by adding a great deal of aural volume to the sound with the multiple layers of grain streams . . . often seems to create a sensuousness, if not an erotic quality in the vocal sound. A word becomes a prolonged gesture, often with smooth contours and enriched timbre. Its emotional impact is intensified and the listener has more time to savour its levels of meaning.[53]

The transformation brought on by granular stretching recalls a similar ana-lytical interest invoked in gender and sexuality studies by Sedgwick's term "transitivity," which she recognizes as the "grounds [from which to find] alliance and cross-identification among various groups."[54] Sedgwick notes here a distinction between "transitive" and "separatist" tropes of sexual-ity on account of gender. Whereas members of the latter group insist on separating men and women, advocates of the former acknowledge alliances between groups and across genders on the grounds of shared sexual inter-ests. By deliberately instituting disjunctions or transformations of textual, musical, and gender and sexual identities, more than merely "blurring" the distinctions between the male and female voices within the music, as the composer acknowledges doing, I find Truax's method of composition exemplifies the transitivity model by simultaneously evoking both man and woman simultaneously in the same vocal range. Granularity also causes the woman's voice to sound huskier, thereby departing from the breathy quality typically attributed to women,[55] and therefore further cementing a connection in both pitch and timbre to the supported (male) speaking voice as well as to the more heavily granulated Monk's song and cicadas in the background. By the end of the *A tempo* (5:28 [ii, ~1:16 score]), the stretched "fountain" meets the oboe's pitch to fully solidify the music's many contextual registers.

Notably, Truax's text omits the common question of Shulamite's ethnicity, exclaimed in the lines:

I am black and beautiful,
O daughters of Jerusalem, like the tents of Kedar, like the curtains of
   Solomon,
Do not gaze at me because I am dark.[56]

However, his technique of emphasizing certain qualities through granularity, both in darkening and deepening Shulamite's voice, suggest Truax's music retains aspects that mark speakers ethnically. As I mentioned with respect to Donna Summer (chapter 6), a "breathy" vocal quality is not only gendered but also racialized white, while the full-bodied woman's voice, even when disembodied via electroacoustic practices, reliably references Black women's habitually hypersexualized bodies.

OVERCOMING MUSIC'S "NEUTER" ENTERPRISE

Truax identifies the romantic tryst in the third movement, "Evening," as the height of homoerotic sensuality in *Song of Songs* on account of its sultry text, extreme stretching of the word "desire" (12:27 [iii, 3:38 score]), and a greater frequency of same-sex pairings.[57] Yet, in my opinion, the fourth movement, "Night and Daybreak," is more exemplary of the music-historical context in

Figure 9.4

Song of Songs, iv: "Night and Daybreak," lines 7–9. Ambiguously gendered voices, as if spoken through a vocoder, notated in the bass clef. English horn in F.

which the work was written. Here (figure 9.4), a gender ambiguous vocoder-type voice awakens their lover calling: "My beloved spoke and said to me: Rise my love, my fair one and come away" (15:40 [iv, 1:58 score]). Truax labels this line a "duet" in the production score, as both speakers repeatedly recite the line in counterpoint.[58] Chanting alternately therefore draws timbral similarities between the two voices on account of their similar processing through comb filters, while still maintaining the identifying timbral qualities of each gender. In this way, the passage insists simultaneously on difference and sameness.

The lower vocoder voice sounds alternately on the pitches E-flat and B-flat, where the impulse provides the music's rhythmic propulsion. Through alternation, redundancy forces the suspension of functional harmony, not unlike the accompaniment in Laurie Anderson's "O Superman" (1981).[59]

Truax's *Song of Songs* shares more than a few features with "O Superman." Both reinterpret biblical texts, but more than this Truax's observations on personal voice and absence compare easily to Anderson's text, in which she states in the opening line, "This is not a story my people tell. It's something I know myself."[60] Sharing a desire to push beyond tradition, both Truax and Anderson seem to rewrite history in their respective retellings of biblical stories. Anderson's recitation paraphrases the story of Adam and Eve from the woman's perspective, while Truax's *Song of Songs* preserves gendered pronouns to subvert heteronormative assumptions common in the text's historical reception.

Beyond textual considerations, both works elicit bodily concerns, as each uses electronic means to challenge the performers' respective gender identities. Attending to Anderson's voice, McClary's analysis of "O Superman" recognizes the importance of the perceived "physical source of sound" in this piece.[61] As the composer herself stands on stage to declaim the story as she has rewritten it, Anderson places her body actively and conscientiously in a performing role. But, according to McClary, although Anderson's female body appears to us on stage, she uses vocoders to perform vocal androgyny so as to resist the frequent scrutiny often inflicted on the female body as spectacle.[62] McClary writes, "Laurie Anderson's music is multiply charged. . . . It is electronically saturated at the same time as it insists on the body."[63] This, she says, contrasts with the neutrality with which music is generally regarded—what she terms, its "neuter" enterprise.[64] In a similar-but-opposite tactic, I see Truax *denying* the visually corresponding bodies of his performers, so as to leave room for ambiguity. Like Anderson, Truax's work confronts music's presumed "neuter" presence by simultaneously insisting on the gendered voice while also raising questions as to how categories of gender are typically defined in musical contexts. Also, like Anderson, Truax breaks away from the compositional mold typically assigned to his gender: he is a male composer who embraces sexual imagery, but not by way of the common misogynist representations all too prevalent in the Western musical canon.[65]

This comparison of Truax's music to Anderson's shows the transitivity of the universalizing view. Both examples confront established tropes of gender perception *and* music, with some similar methods and some methods that are less so. Anderson transforms the visible female body with the androgynous voice, and Truax emphasizes the elasticity of the voice—and its transformative potential—in a habitual stretching and retraction of the common markers of gender in sound. In both works, the text, vocal quality, and role of the voice (whether supporting or supported) are inseparable from the music and the world in which that music sounds.

Additionally, *Song of Songs* draws also on the context of the composer's own oeuvre, as many of the work's so-called external associations have been made in Truax's other pieces. For example, *Beauty and the Beast* (1989) features the oboe d'amore and English horn in a story-telling role, similar to how the instrument in *Song of Songs* seems to elongate the lines of the story by serving as a mediating presence between the internal world of the speakers and the external context of live performance. This instrumental mediation is not unlike *Wings of Fire* (1996) and *Androgyne, mon amour* (1997), in which an erotic text refers to a "personified" musical instrument played in each piece respectively by a female cellist and a male double bass player—a comparable relation to that which appears also Rønsholdt's works.[66] The personified instrument thus enriches the ineffable resonances of "abstract" music with the emotional realm—the listener's inner psychology—as when in *Song of Songs* the pitches of the processed background and live foreground cross (1:31 [i, 1:28 score]), or when the instrument is synchronized with the electronic part (5:28 [ii, 1:16 score] and 9:41 [iii, 0:44 score]). In *Basilica* (1992), Truax extends the sound of chiming bells through time-stretching so that their resonances invoke the human voice. Likewise, in the second movement of *Song of Songs*, the Monk's voice is overlaid with monastery bells recognizable from the previous movement and stretched to blur the distinction between the two sources and also to immerse them, together with the processed voices of the speakers, in the rich environmental soundscape of the context in which the work is imagined to take place.

CONCLUSION

Addressing gender in music can be a complicated task. On the one hand, we do not wish to reduce the complex relationships—both musical and erotic—emerging in *Song of Songs* by identifying its sounds with either woman or man. But, on the other hand, we threaten losing some aspect of the experience by omitting a discussion of gender and its implications for musical listening. According to Truax, it is often assumed that music's social level "include[s] all of the material aspects of the context surrounding the musical performance that only *indirectly* affect it," thereby situating the social context in excess of a work's musical relationships.[67] Conversely, his compositional strategy aims to maintain a connection between what is "musically interesting" and "the audience's cultural awareness."[68] What I find compelling about Truax's music is the way he draws attention to difference, gender identity, and sexual orientation, while simultaneously diminishing the centrality of normative categories, whether the binary man/ woman, gay/straight categories or the typical musical oppositions between melody and harmony, and between timbral profiles and harmonic spectra.

Song of Songs is significant in its use of the gendered voice to address and comment on the sexual fluidity of the characters. While the gender of each speaker may at first be associated with one or the other gender, man or woman, *Song of Songs* quickly teaches its listeners that such categories are neither reliable nor stable—not within the work and not in the world beyond. But it is not enough to acknowledge the blurred boundaries between hidden virtual sources and the live instrumentalist, or the frayed edges of the sample and its no longer recognizable processed counterpart. It is likely, though I cannot show this beyond a doubt, that Truax received acclaim for his "homoerotic" approach mostly because of his own gender, because he is a man. As we have seen in several chapters previously, women have also taken up electroacoustic composition to trouble the gender binary (e.g., Shields, Anderson), to blur distinctions between human and animal (e.g., Lockwood), and, as we will see in the next chapter, playing with pronouns of address to elicit associations that broaden the sexual orientations of respective speakers. Truax also benefits from his presumed neutral identity in other ways. By being a white man he could easily mask or deny his homosexuality to "pass" in situations that would be threatening to non-white non-men—for example, in the electroacoustic studio.[69] This is to say that without considering race, gendered categories are merely destabilized in the abstract.

The next three chapters interrogate the racial implications of such gendered presumptions in order to disentangle these idealized constraints as they perpetuate racial stereotypes of assumed sexual provocativeness and availability.

10 TLC'S "I'M GOOD AT BEING BAD" (1999)

TLC released their third studio album *FanMail* in February 1999—the year of the Y2K scare. Given the times, digitality stands out as one of the album's prominent themes. Critics hailed these computer enhancements as evidence of the group's maturation beyond their overt safe sex campaigns of the previous decade, even if computerized projections of the group's three performers appeared early on in the video for their iconic song "Waterfalls" (*CrazySexyCool*, 1994). With *FanMail*, the performers became branded with digitality.

    The album's cover flaunts an envelope icon commonly associated with email and centers three-dimensional computer animated portraits of the three performers, whose nicknames form the initialism: "T" for T-Boz (b. Tionne Watkins, 1970), "L" for Left Eye (Lisa Lopes, 1971–2002), and "C" for Chilli (b. Rozonda Thomas, 1971). Their blue-tinted skin recalls alien imagery from *La planète sauvage* (1973) to *Avatar* (2009). Musically speaking, the album enhances fandom through digital processing and overtly computerized voices. For example, the first and title track "FanMail" features a speedy percussive hi-hat pulse calling back to the trap music of their hometown Atlanta and thematizes abrupt cutting easily accomplished with digital editing in a Digital Audio Workstation (DAW)—a household tool for today's musicians, but still relatively novel at the time of *FanMail*'s release. A 2012 *Pitchfork* article articulates the album's novelty succinctly: "When we talk about TLC's current influence on a whole crop of web-minded, Tumblr-savvy, android-obsessed artists, we don't seem to realize how much we're talking about *FanMail*. . . . *FanMail was one of the first pop records to aestheticize the internet.*"[1] The album's total package—its lyrics, title, iconography, performance imagery, and sonic profile—all form an updated corollary to the traditional markers of both celebrity and courtship: adoring fans.

The "FanMail" track tells an email love story: "Everyday I think I'm gonna meet ya / Can't wait 'til the day I see ya." The author and recipient have not met face-to-face. And by the song's title, we might assume the fan in "FanMail" writes to the performers, TLC; they are, after all, the stars worthy of fandom. The author is gendered (sung by T-Boz), announcing "I'm still sittin' Miss Alone." But if delivered *from* a fan, the lyrics suggest a woman professing attraction to the email's intended recipients, TLC's three women, one possible queer hearing. And yet, because the song is performed from the group's perspective and directed toward listeners, an alternate hearing situates the artists themselves as courting their collective audience, or maybe one fan in particular.

Despite its timely subject, the album was comparatively less popular than their previous albums, even if it received eight nominations at the 2000 Grammy Awards. Still, *FanMail* remains present due in part to periodic revivals, innovations promoted by the group themselves as well as various tribute versions of songs from the album which serve the musical equivalent of fan fiction. Where theatricalization of internet communications may have appeared hokey at the turn of the millennium when concerns for industrial collapse were still rife, now, twenty years later, the clichéd audible and visible computer processing resonates with contemporary internet fatigue, recalling a more simplistic technological interface before machines claimed agential independence from human subordination on the path toward the AI-driven Fourth Industrial Revolution.

TLC's decline in popularity throughout the early 2000s coincided with the trio's release of their fourth studio album *3D* and the sudden death of one of the group's members, rapper, producer, and song-writer Lisa "Left Eye" Lopes (1971–2002), in a traffic accident in April 2002, just seven months before the album's release.[2] As nostalgic for their former celebrity as fans, the group's remaining members, Chilli and T-Boz, announced in 2012 media appearances that they would tour with a holographic image of their deceased member, Left Eye.[3] The revival came on the heels of Drake's 2010 release of his version of "FanMail," which the *Degrassi* heartthrob called "I Get Lonely Too"—the "too" here resurrecting a conversation with TLC. Drake's throwback revitalized TLC's fan base, piggybacking on the emerging nineties nostalgia of the 2010s. The #ThrowbackThursday hashtag and the final shift from CD to MP3 to digital streaming formed an ideal platform for TLC to reclaim their celebrity by capitalizing on a newly codified (and commodified) collective cultural reminiscence. In his investigation into *The Persistence of Sentiment*, musicologist Mitchell Morris has argued that musical throwbacks—what he broadly construes as "cultural retrospectivism"—harken back two decades as they amplify and extend previous trends. So, by 2010, a '90s girl group like TLC was due a reboot.[4] Drake led the way and, later, sibling boy band Hanson obliged, covering a number of TLC's songs; Ed Sheeran's 2017 "Shape of You" riffs on TLC's "No Scrubs" off the *FanMail* album; and, in 2019, rock band Weezer released a full-on cover of "No Scrubs." Also in 2012, rumors circulated that the group's remaining members, Chilli and T-Boz, would do a reunion tour to promote their fourth studio album, *3D*, first released in 2002.

In this chapter, I delve further into Morris's "cultural retrospectivism" through the lens of sexualized girlhood. I look at TLC's sample of Summer's 1975 hit "Love to Love You Baby" (chapter 6) on *FanMail*'s sixth track, "I'm Good at Being Bad." Though Summer, via litigation, eventually had the quotation removed from TLC's song, the sample's place on the album's initial release still provides an important segue to the contemporary politics involved in "sampling sexuality." I examine this sample as a conflicting culturally nostalgic artefact, as a sign of simultaneous fandom and celebrity, infantilizing as it adultifies. Such conflict, according to hip-hop scholar Tricia Rose, is typical of the "dialogue, exchange, and multidirectional communication" exemplified by Black women rappers, an aspect I examine below in TLC's intertwinement of rap and pop.[5] TLC's songs are certainly open to interpretations of performed sexual availability, but rather than demonstrate continuity between Summer and TLC via this imagined sexualized proximity, I use musical analysis to draw attention to how TLC pay homage to Summer's *composition* of the song's hook and verses, and not to her famous ad-libbed moaning. Interpolating the classic disco melody with *Fan-Mail*'s updated digital profile, TLC's "I'm Good at Being Bad" calls back on Summer's own projection of music-technological futurity in the '70s. TLC magnifies ambiguity through digitality, for example, by blurring reality and the virtual, or what is human and machine, and in this way redefines what *is* and what is *possible* in regards to performing womanhood. TLC's song "I'm Good at Being Bad" calls back on Summer in several ways. First in its identifiable quotation of the hook from "Love to Love You Baby." Second, the song employs melodic motives derived from Moroder's orchestration of the song. And third, the title itself calls back on Summer's 1979 album *Bad Girls* (Casablanca Records). This chapter prioritizes the accounts of the respective performers while contextualizing personal stories about sexuality and related musical performances within Black feminist and activist intellectual scholarship to investigate dialogic commonalities as well as departures between original and sampled context. In particular, I look at how the group updated sexual tropes from the disco of the 1970s while building a new legacy of musical and sexual empowerment.

TLC'S GENDER TRINARY

Where gender existed on a binary, TLC projected an image of *three* formed personas, epitomized by their "crazy" "sexy" "cool" moniker. Left Eye's raps contrast both with Chilli's sweet lyricism as with T-Boz's lower belting, which has been described by critics as her "husky come-on vocals."[6] A quintessential TLC sonority features T-Boz singing in parallel octaves or in harmony below Chilli, for example, in the choruses of "Diggin' on You" and "Red Light Special" both from the album *CrazySexyCool* (1994). This set up sounds consistently throughout the group's career even in their recent hit "Way Back" (feat. Snoop Dogg) from the group's third revival in 2018. As Nataki H. Goodall observes, "T-Boz and Chilli rap occasionally, but their primary role is in singing the bridges and choruses that frame Left Eye's raps."[7] That is, rap centers and distinguishes the songs of an otherwise pure pop girl group. According to Alexandria Lust who has written a thesis and the only academic work dedicated entirely to TLC's music, whereas T-Boz "commanded the role of lead singer," Chilli's lyrical higher register typifies "not only sensuality, but sensitivity" in songs like "No Scrubs" off the *FanMail* album.[8]

Left Eye's rapped refrains reproduce what Tricia Rose calls the "Bad Sista" trope of 1990's women rappers. Bad Sistas typically rap about themes of "heterosexual courtship, the importance of the female voice, and mastery in women's rap and black female public displays of physical and sexual freedom." To some extent, writes Rose, "female rappers are at least indirectly responding to male rappers' sexist constructions of black women."[9] For this reason, Bad Sistas might double down on particular stereotypes, and TLC retains three markers of stereotypical gender simultaneously. Yet, the relative stability of Chilli and T-Boz's respective singing roles together with Left Eye's rapping becomes compounded and confused by the multiple personae each performer suggests lyrically, often keeping the same vocal quality while exchanging pronouns and even self-referencing—as we saw in the above "FanMail" courtship email.

"I'm Good at Being Bad" opens with Chilli in an exaggerated state of euphoria. Enter the raunchy rap and we have a jarring contradiction between Chilli's mythical pop Cinderella and the rapping Bad Sista. The "Love to Love" quote then introduces a third ideal, mixing the two extremes—something perhaps appropriately captured by Donna Summer's nickname from *Rolling Stone Magazine*: the "Sexy Cinderella."[10] In this sense, TLC's vocals tug at the limits of what, in the 1990s, would have been socially acceptable as performing womanhood. In the previous chapter, I examined how, at the time, vocal quality became an important factor in determining audible gender identity. However, such determinations relied on *a priori* biases about the respective range spoken by either "male" or "female" voices. As a result, such empirical studies were biased from the onset toward deciphering gender on a binary basis. My analysis of *Song of Songs* showed that, aside from pitch, gender perception is also beholden to inflection and rate of change over time, and many potential factors that have not yet been given adequate attention. In seeking to understand what compelled TLC to sample "Love to Love," here again my analysis retains an assumption that the performers presented to fans as women, but TLC's performances do not reduce to a singular "female" identity. As Rose also explains, Bad Sistas confront and contradict "the logic of female sexual objectification."[11] If we hear the raunchy rap as an expression of the Bad Sista, Left Eye's presence introduces a third woman distinct from the other two, the lyrical and the belter. An alternative interpretation identifies Left Eye as the opposing figure to Chilli, while T-Boz's quote of Donna Summer seems to be responding flirtatiously to the other two singers: she loves to love Chilli's saccharine sweetness *and* the raunchy rapping. In this sense, Donna Summer's quote enters to disrupt the logic of objectification with expressive sexual agency. The sample embodies the Bad Sista mentality just as much as the rapping. Hence, TLC embolden a gendered trinary that is irreducible to any one form, since the performers do not conform to any certain interpretation for very long.

In a hearing that understands Left Eye's rap as a confrontation with "the logic of female sexual objectification," T-Boz's quote of Donna Summer would seem to represent what Rose observes as a contradiction between objectification and sexual freedom. But this hearing does not necessarily affirm the claim that women rappers are responding to men. Rather, the quotation sounds within a dialogue among, and simultaneous ode to, women, given Summer's compositional role in "Love to Love You Baby" detailed in chapter 6, which is compounded by TLC's choice to rerecord the track rather than sample Summer outright. Rose determines that more than merely responding, "black women rappers are in dialogue with one another, with male rappers, with other popular musicians (through sampling and other revisionary practices), with black women fans, and with hip hop fans in general."[12] TLC's quotation of Donna Summer therefore demonstrates dialogic engagement across racially, socially, chronologically, culturally, and sonically coded lines, such that "I'm Good at Being Bad" evinces the meeting point of sex and technology as multifaceted *vectors of difference*. In what follows, I detail how the two songs, Summer's "Love to Love You Baby" and TLC's "I'm Good at Being Bad" retain musical and sexual connections and other ways in which the songs diverge.

QUOTING "LOVE TO LOVE YOU BABY"

Common to hip-hop music, TLC's "I'm Good at Being Bad" interpolates borrowed material from Donna Summer's 1975 disco hit "Love to Love You Baby." Typical of allosonic quotation (when a sample is performed anew in the new context rather than incorporating prerecorded material), T-Boz borrows Summer's melodic hook "Ahhh Love to Love You Baby," singing the sample over the groove from earlier in the song.[13] This borrowed section, despite being identified in numerous sources for the lyrics as the "break," recurs three times in the song. Aside from that melody, transcribed in figure 6.1, which T-Boz sings an octave lower than Summer in TLC's "I'm Good at Being Bad" (1:20–1:41 and 2:56–3:17),[14] no part of the song uses recognizable themes from Summer's hit outright, though audible similarities exist. These similarities suggest that TLC's song was composed with "Love to Love You Baby" in mind, rather than the sample being appended as an afterthought to the already composed material.

In addition to the recognizable hook from "Love to Love You Baby," the two songs share some less apparent musical elements—(1) a hidden motive, (2) key relationships, and (3) a pedal tone on G.

Perhaps the closest musical similarity shared by the songs, after the vocal hook and the key, E-flat major, is the descending chromatic tetrachord (figures 6.2a-c). TLC's song opens with an accompanying acoustic guitar in an alternating-bass triplet figure which sounds against the duple hi-hat that sets up the song's meter. Figure 10.1 reduces the part to harmonic simultaneities. In its reduced form, notice the guitar's similarities to figure 6.2b from the "Love to Love" bridge. Both figures share (1) the sustained G and B-flat (in

Figure 10.1
Harmonic reduction of acoustic guitar accompaniment to verses in TLC's "I'm Good at Being Bad," featuring descending chromatic tetrachord.

Figure 10.2
Alternating descending and ascending minor triad arpeggiation, sustained ambient synthesized background from "I'm Good at Being Bad" (2:56–3:17).

different registers) and (2) the descending chromatic tetrachord, though the
tetrachords sound on different pitches.

Like Summer's "Love to Love," TLC's song also features a repeating
instrumental string arpeggio, transcribed in figure 10.2,[15] which sounds
continuously from Left-Eye's explicit rapped refrain and into the sample
of "Love to Love" in the break. That the arpeggio pulsates undisturbed
throughout several sections of the track becomes a uniting facet of the
song, an effect that secures what Robert Fink terms "'hypnotic' repetition"
familiar from various electronic musics, including disco and hip hop.

Though figure 6.2b and figure 10.1 share a pedal tone G and the melodic
tetrachord, the resulting harmonic simultaneities each outline progress
toward divergent ends. As mentioned in chapter 6, Summer's verse (figure
6.2b) leaves open harmonic interpretation inviting either E-flat major (to
repeat the bridge) or Cm7 (a refrain of the hook). In comparison, TLC's
accompaniment in figure 10.1 ends definitively on the latter, thus securing its
connection with identical pitches to the sample of the "Love to Love" hook
(figure 6.1), a likely source of inspiration for TLC's opening instrumentals
(figure 10.1). This level of musical overlap and simultaneous variation suggests that the sample was chosen before TLC completed the song and that the
group and its producers worked to create continuity between the two tracks.

Summer, nevertheless, objected to TLC's sample of her song. Given the
complex representations of identity I showed earlier in "I'm Good at Being
Bad," perhaps Summer disagreed with TLC's attempt to revive her image as
an idealized, indeed laughable, "Sexy Cinderella." TLC often uses humor to
amplify the anticlimactic mimicry of women's sexual satisfaction to assert distance between their performed roles and the women's actual sexual desires and
actions—distance Summer felt she had lost after releasing "Love to Love."
Through this lens, we might hear "I'm Good at Being Bad" as a parody of
Summer's classic hit.

In *Queer Tracks*, Doris Leibetseder draws on a wealth of postmodernist literature on parody in relation to gender subversion. Whereas both irony and parody may concern an original text and its imitation at a critical distance, "Many parodies do not render the original text ridiculous, but rather use it as a standard according to which they examine the contemporary more closely."[16] While irony's gaze is aimed toward the past, parody in this reading is a reflection of the present. Similar to parody, pastiche also reacts to an original text, but it

> does not have to be critical or comical. In post-modern debates, there are many contributions and differences of opinion regarding the different descriptions of pastiche and parody. Like parody, pastiche is the imitation of a peculiar or unique style, the wearing of a stylistic mask, speaking in a dead language. However, it is a neutral practice of this kind of mimicry without the underlying thoughts of a parody, without the satirical impulse, without laughter, without the constant latent feeling that something normal exists in comparison to the imitation that gives a rather strange impression. Pastiche is an empty parody that has lost its sense of humour.[17]

Looking to TLC's past reputation, we might situate "I'm Good at Being Bad" in the camp of parody, along with another song, "Kick Your Game."

The lyrics of "Kick Your Game" similarly lend themselves to multiple, if not conflicting, interpretations. In the words of critic Darren Levin:

> This song is set in the dating scene, where the scrutiny is so intense that voices and even personalities have become mixed up. [Watkins] narrates the story, but when Lopes seamlessly picks up the verse, the situation is complicated. Lopes gives off her usual bravado, yet she also sweet-talks herself using another persona. She appears to be mouthing the responses for both herself and an admirer, but it's hard to tell, since the change between characters is done without a break. Throughout the track, the singers use the same tone for different speakers—whether they're playing at being flirts, nagging partners or wise-ass suitors ("I'm just a n*gga that followed you to the coat rack").[18]

Such instances of gender swapping are frequent on the *CrazySexyCool* album and are therefore not far off from my girl-courting-girl hearing of their later track "FanMail." The song lends itself easily to a parodistic hearing because of the performers' humorous gendered "masking," as when Left Eye supposedly imitates her ex-boyfriend's clichéd courtship ritual in the parenthetical quote Levin adds above. Through this lens, we could also hear "I'm Good at Being Bad" as simply mocking Summer's hypersexualized image. Such a hearing coincides with the typical notion of "sampling sexuality," a facet of electronic music that appears to strip agency from hypersexualized performers, especially Black women.

MOCKING SEXUALITY

Barbara Bradby's lauded article "Sampling Sexuality: Gender, Technology and the Body in Dance Music" identifies pop music as "the twentieth-century cultural genre most centrally concerned with questions of sexuality."[19] Bradby argues that, similar to my conclusions in chapters 6 and 7, whereas women have sometimes appeared as contributors to musical traditions, rarely have they been celebrated as productive leaders or instigators of new modes of musical expression. She considers how dance music, derived from "the black musical genres of soul and rap," discounted the role of Black performers inhibiting them from gaining creative authority, since, for many feminists, the high-tech recontextualizaton of these voices marked them "as *inauthentic* performances—a representation of a performance as the expression of emotion, rather than the performance itself."[20] Following the "death of the author," feminists worked to unveil the very ideality of the author as implicitly male. His demise contributed to a double indemnity against women creators, who were never acknowledged properly to begin with. For white feminists, there was a notable shift with wider acknowledgment of the fact of their lesser societal value. For Black women, at least in the US and other colonized geographies post-slavery, their role was not determined solely in relation to men but also to white women, and often (underpaid or unpaid) labor centrally contributes to their public identity—after all, the Mammy wasn't defined by her own offspring but by her relation to the white children for whom she provided a labor of care, a facet of "[her] 'dysfunctional' sexuality and motherhood," to quote Tricia Rose.[21]

Importantly, for our context here, Bradby identifies the time from 1990 to 1991 as a time when dance music solidified "the division of voices . . .

between male rapping and female soul-singing, between male speech and female song, male rhythm and female melody"—a trend apparent already in the 1970s in songs like Summer's "Love to Love You Baby," where, as I have argued in earlier chapters, men controlled the electronic backdrop propping up the vocalist's one-dimensional image. According to Bradby, in mainstream chart hits like Snap's "The Power" and Black Box's "Strike It Up," "the female voice is being used to connote the expression of human emotions (anger, desire, love, pity), which have been emptied out of the monotonous, mechanical style of male rapping one hears in this music."[22] And yet, such expressions draw listeners' attention to performances by women as idealized sexual constructs, the monotony of a moaning that has no end in sight. As we saw in chapter 3, Ferrari could not or would not convince singers to enact his intimately envisioned performances. He would either hire actors or record women in their natural state. Similarly, as popular music scholar Jon Stratton points out, those who perform explicit electrosexual songs like the extended "Love to Love" track or "*Je t'aime . . . moi non plus*" are usually known as actors (Brigitte Bardot, Jane Birkin) or envision themselves as playing a part (Donna Summer)—it was practically the only way they could get through the song.[23] It is precisely this plastic-wrapped version of sex that TLC uses as the basis of their citation. If we hear "I'm Good at Being Bad" as a parody of Summer's "Love to Love," it's as if to say, "This is *your* idea of sex, and here is *ours*." "Girl groups" have been especially demeaned for their subordination in the capitalist hierar-chies of the commercial music industry, such that TLC could be inter-preted as taking up the roles of the "disco diva," "fembot" or the "robo-diva R&B," precisely the roles that its sample of Summer dismantles when the three members of the all-woman vocal group participate in *all* of the roles.

Bradby notes that sampling practice in dance music habitually omits the name of the original performer or tends to substitute the visible per-former's voice with other, invisible singers—much like the backup singers in the original "Love to Love," who, according to Neale, were surprised at how prominently they featured in the final cut. Such substitutions, argues Bradby, do not necessarily contradict feminist endeavors to attribute women's labor; rather,

In exposing the deception played by juxtaposing body and voice of two women, it contradicts that deception with a new form of honesty. This maneuver actually challenges the primacy of the *visual* in our everyday imagining of the body (which has been central to the feminist analysis of the representation of women), and the implication that the voice is somehow "disembodied." It both deconstructs our assumption of the singing voice as emanating from an individual rooted in a body we can *see*, and re-roots that expectation into plural *bodies*, or the female body seen/heard in different ways. Perhaps feminism is stuck here with the postmodern contradiction, between the need to piece together the female subject that appears as fragmented by male discourse, and the need to acknowledge that subject's, or those subjects', fluidity and internal difference.[24]

It is precisely this tension between multiplicitous facets of womanhood that is carried forth in each new sample of "Love to Love."

As much territory as Bradby's groundbreaking feminist call covers in popular music, the article's title "Sampling Sexuality" refers more to a generic, implicitly heteronormative notion of sexuality, one that has relatively little to do with sound *per se*. She discusses sexuality only at the level of embodiment (in distinction from situated performances)[25] and relies on gendered assumptions of how listeners engage with dance music. The article outlines popular norms from the early 1990s, the habits of collaborations between "female" singers and "male" producers, and between "female" singers and "male" rappers, homing in on a critique of the patriarchal construction of labor in dance music from a feminist perspective. Though the article examines musical tropes that reinforce perceptions of gender and race, it does not attempt to trace a lineage of women performers that is distinct from the male-centered canon, nor does it trouble the racist dichotomy between black and white that this chapter does. Importantly, in this chapter I also seek to clarify musical discourses surrounding gender by amplifying how sex sounds.

As noted earlier in this book, pop music scholars Corbett and Kapsalis determined that "[w]omen's pleasure is not usually seen though it is often heard, their satisfaction most often equated with the 'quality and volume of the female vocalizations.'"[26] While such analyses, from Bradby to Corbett/Kapsalis have identified the pervasiveness of the moaning trope in popular sound, these commonly fall into a trap of idealizing women's audible sexual satisfaction—indeed, note the pervasive use of the descriptive "female" qualifying such descriptions. "Female" is a term used to connote the biological sex in many species, not just humans. Like hen for chickens or feline for cat, the female of the species also carries its own designation—in the case of humans, the female of the species is a *woman*. Often women's sexuality is examined very much in clinical terms, an aspect evinced by titles like Corbett and Kapsalis's subtitle "The Female Orgasm" or Bradby's opposition of "male rapping and female soul-singing, between male speech and female song, male rhythm and female melody." The qualifying adjective "female" dehumanizes the subjects of analysis, distancing and alienating some idealized subject from any actual human presence. Distance imposed in this way, as in the recording process itself, sets up a forced separation that grants listeners and, indeed, specialist analysts the right to perform far-off clinical observations from the perch of feigned objectivity. And maybe this is exactly what these performers want.

One question to which this chapter returns repeatedly is whether "sampling sexuality" is indeed possible, or if the politics inherent in sampling as a citational practice might interfere with the original song's message by enacting a falsified barrier that reinforces distance. The objective of the electroacoustic game rests on a question of whether disacousmatization is even possible, whether there can be such a thing as a "copy" of any single original.

Like Donna Summer, "Left Eye" is given song credits on the track that samples "Love to Love," as is Beyoncé Knowles on her track "Naughty Girl" (2003) that also samples the hit. Summer is also given credit on the later tracks. Each song captures a moment of the "Love to Love" energy, and yet, with each incarnation the message is given new context. Though one could argue that this is just another instance of the transposable identities of sexual laborers in the commercial music industry, in simply drawing a lineage among Black women's voices, scholars who trace this phenomenon as a continuous and uncontextualized lineage overlook any variation between artists, indiscriminately grouping together "representations of Black women as sexually insatiable and as commodities," to recall Brian R. Sevier's description of the disco diva, less overtly modeled after the hypersexualized Jezebel stereotype used to justify Black women's enslavement.[27] Such insistence on heredity thus further stigmatizes an entire group of people based on *a priori* descriptors of some feigned ideal—which may very well be an accurate portrayal of their reception by some listeners.[28]

Contrasting this common narrative, Mark Anthony Neal envisions Summer's "Love to Love You Baby" as a song marking when a Black performer "transformed from sexual object into sexual subject."[29] Whether hyperbolic or not, this sentiment, expressed in his 1999 book, *What the Music Said: Black Popular Music and Black Public Culture*, certainly embodies Summer's *new* image as a representative icon of a Black musical culture distinct from sex. In Summer's account:

> Upon its release, "Love to Love You Baby" became an immediate, international symbol of . . . *something*, although I've never been quite sure of what. It was certainly more than just another sexy song, and it went on to become the anthem of seventies pop culture. Or so they tell me.[30]

But she was not entirely pleased with the sexual immediacy the label attributed to her:

> What was unique about "Love to Love You Baby" was that it created a powerful, feminine image that was unlike anything previously released in pop music. It was an image I regretted. I mean, who could live up to that? Not even Marilyn! . . . This was not a direction of image I was comfortable with. I felt that the sensuality minimalized my self-worth and made people think that I had prostituted my talent to advance my career . . . On the stage in Germany, I'd felt protected by the distance between me and the audience. . . . [31]

Distance imposed by the staged nature of performance, as Summer suggests, was a marker of her *un*availability—an aspect those marketing the sound recording attempted to strip away. Rather than fall back on the trope of exploitation in "sampling sexuality," what concerns me more are the "aesthetics of sampling" in which a perceived sexual deviance is at play. An absence of sex acts performed is perhaps the most radical example.[32]

Susana Loza's updated investigation into "sampling (hetero)sexuality," published eight years after Bradby's, in 2001, explains that once the electrified voice of the diva—the "posthuman diva"—becomes manipulated enough to draw attention to sex via mimicry, she begins to parody "the natural with a musical masquerade that mocks the fixity of femininity."[33] Perhaps something akin to the way TLC's *FanMail* "aestheticizes the Internet." Loza problematizes the open-ended fantasy that the "exaggerated peak[s]" of "diva house and porno-techno" actually communicate "a plethora of pleasure-drenched climaxes."[34] She argues, any hearing that invites fantastic speculation, what she calls "the mystery of the erotic,"

> becomes hopelessly entangled with the riddle of race in the freaky figure of the fembot. For the prototypical nympho-diva not only renaturalises the technological with her organic orgasm, she also biologically essentialises the mechanical with her sonic racial coding. In other words, the classic fembot is a black diva who simultaneously samples hetero-sexuality and resurrects race.[35]

Given Summer's constructed image, grounded in her iconic performance of "Love to Love You Baby," it is easy to imagine her as a template for the now commonplace racialized and hypersexual "fembot," in Suzanne Loza's terminology, or what Tom Briehan declares derogatively as the face of "robo-diva R&B."[36] But to reiterate Rose's observations from earlier, women rappers respond to rap music's stereotypical (hyper)masculinity as well as women's purported availability by contributing to this lineage in newly productive ways. The form of citation they use employs themes of simultaneous critique and empowerment, expressly feminine while overtly feminist. Importantly, TLC's sample isolates the only original creative matter attributed to Summer as its rightful composer. Their "Love to Love" sample has competing stereotypes side by side—lyrical/rapper/husky and simultaneously sweet/sexy/ Sexy Cinderella—it does not allow for reduction to a unified representation of Black women's desires. In this way, as I hear it, "I'm Good at Being Bad" reinvigorates the destructive image of the idealized diva with an embodied corporeality, one that is not easily reduced to a singular representation of sexuality—even among Black women.

The kind of allosonic quotation T-Boz employs is often dubbed derogatively as "karaoke" by hip-hop producers and rock fans, alluding to amateur live singing with prerecorded playback. Karaoke singers are not usually expected to have the lyrics memorized or even to contribute productively to a given song. Instead, karaoke is considered among its critics as a form of diluted entertainment, a replication of preexisting practice even less valued than a song's cover. Such an interpretation might lend itself to Leibetseder's notion of pastiche, "the wearing of a stylistic mask" or even "an empty parody that has lost its sense of humour."[37] The karaoke singer is not expected to have had any compositional role in creating the song and can therefore be understood as a musical extension of fandom. Comparing fandom to a form of sampling, media scholar Mel Stanfill depicts fans as immaterial laborers whose work is continually discounted:

> Fan work is laborious, but fans are rarely seen as laboring. Partially, fan activity is considered unproductive because it's considered illegitimate. The value of the fan product is seen as coming from the media source, not the work of transformation. Fan production is classified as derivative, and therefore derived, lesser, taking from the source; the value is understood as in the original and borrowed by the fan.[38]

Within this discourse, where sampling is envisioned as fandom or mere karaoke, it becomes all too easy to dismiss TLC's "Love to Love" sample. Indeed, as mentioned, very few scholars have taken to TLC's music, though they were at one point one of the most popular acts in the world. Reflexively, one could say TLC were too popular, but from this analysis, it might be more accurate to say that people still do not know what to make of their music. It's still unclear whether we can or even should take TLC seriously. In his reading of Liebetseder, Stan Hawkins determines: "Because of its duplicity, parody is spurious," and TLC certainly retain this ambiguity of style (parody/pastiche), rhetoric (humorous/serious), genre (pop/R&B/hip hop), and identity—whether straight and/or queer.[39]

T-Boz's dressed-up voice idiomatically sounds not as an anomaly but as a quintessential feature of TLC's brand. T-Boz conjures the essential CrazySexyCool energy: in control and sexually suggestive. What is queerly subversive about the role is also its queer appeal. In this sense, the mask is hardly a joke but nevertheless imitative, what Leibetseder identifies as pastiche but bordering also on camp. Because of this irreducible duplicity, parody becomes a marker of queerness: "Every parody is obviously hybrid and has at least two meanings."[40] Paraphrasing Dennis Denisoff, she writes: "the aim of the sexual parodist is not to modify preceding representations to bring them nearer to the fundamental truth, but rather to elicit a pleasure that arises through the questioning of the idea of an original gender."[41] In this sense, both the parodic, laughably imitative and the earnest so-called "neutral" imitation can be heard simultaneously.

At the moment of the famous "Ahhh Love to Love You Baby" line, both songs support their singers with string riffs that include G, thus solidifying a connection between Summer and TLC on the basis of timbre and a distant skeletal harmony. And where both songs enter into modulations between major and minor tonalities, Summer's moves between E-flat major and C minor, while TLC's between E-flat major and G minor—reversing a minor-third descent to the relative minor to, instead, ascend a major third harmonically. Interestingly, the "Love to Love" melody works in both major/minor key pairs, that is, in all *three* keys.

As mentioned, "I'm Good at Being Bad" opens with Chilli in an exaggerated state of euphoria. According to music theorist Fred Maus, alternating major and minor modes can hint at alternative realities in which meanings conveyed via lyrics and instrumentals contrast or contract. Within a queerly coded sexual "ambivalence," Maus recognizes in Pet Shop Boys's song "So Hard" that "[t]he mingling of major and minor [simultaneously], reflects the mix of hope and resignation," one possible hearing of the later pairing of the minor string riff in TLC's song with both Left Eye's raunchy rap and Chilli/T-Boz's saccharine fairy-tale sexual availability narrated in the song's three verses. Maus characterizes such ambiguous pairings with "exciting eroticism that resists control" since, absent the certainty confirmed by sex *acts*, no one hearing is set in stone.[42] In the song Maus analyses, whereas the major inflected lyrics may purport an idealistic calmness, the accompanying orchestral notes betray "the truth of the situation, the facts that the words try to suppress."[43] In TLC's "I'm Good at Being Bad," where the lyrics purport sexual availability (in Chilli and T-Boz's verses) and might even "school" listeners on the criteria of a good suitor (Left Eye's rap), the exaggerated contradiction between these respective portions in both lyrics and delivery alludes more to three *caricatures* of sexualized women—first, the mythical pop Cinderella with the happy ending and, second, the "Bad Sista" popularized among Black women rappers. The "Love to Love" quote then introduces a third ideal, mixing the two extremes with Summer's "Sexy Cinderella."[44]

Qualitative differences between the songs' respective string sequences point to divergent goals and, nevertheless, convey separate representations and, hence, potential critiques of sexual expression in music.[45] It is possible that Summer's inviting availability perhaps proved inspiration for Chilli's apparent reverie in her verse; whereas, Left Eye's raunchy lyrics contrastingly promote quite a different sexual perspective, maybe even a little off-putting, one that is not so much inviting a suitor as it is confronting one. These same lyrics, however, are also sensational, provocative, empowering, and dialogically engaged in contemporaneous discussions of women's sexual independence, discussions at the heart of a decidedly woman-centered hip-hop movement and roots of sexual politics that we see later celebrated in music by Rihanna and Beyoncé, to name two examples.

Knowing this, could it be a coincidence that, rather than the non-distinct moaning on which all other analyses of "Love to Love" focus, TLC instead made a conscious decision to sample music *composed* by Donna Summer? Both Summer's and TLC's respective songs can be heard as "going somewhere," yet, to confirm Fink's "recombinant teleology," neither achieves a decisive moment of arrival. The teleological scale is therefore reconceived from the classically valorized phenomenological experience of a singular tension-release arc for the duration of the piece to form concomitant pitch relations as audible links across motivic figures, musical sections, genres (disco and hip hop/pop), chronologically defined stylistic nuances, and even, arguably, distinctive sexual expressions.[46]

What happens when performers—and even listeners—refrain from the implied closure of sexually consummating acts? Can musical performances still retain the two previous stages, or does the entire structure of teleology-directed musical "desire" collapse? In Fink's hearing it does not. But where Fink identifies desire by its implied goal, even if the goal is never achieved, desire as it relates to drives and teleology could very well be understood via *agency*, via women's own desires to perform a certain way, to dress a certain way, and to even refuse sex. My hearing explores the limits of aural sex understood as a performance of non-consummation and even non-arousal on account of its denial of sex acts.

RACIALIZING ASEXUALITY

Perhaps counterintuitive to the opposition of liberal and conservative sexual politics articulated above is that while encouraging sex-positive involvement from young audiences, the group themselves mostly abstained from acts that, under law, would be considered sexual. Chilli publicly acknowledged her celibacy already in 1994 and she has reiterated this vow as recently as 2017. Chilli's reticence toward sex is often framed in contrast to the group's apparent encouragement of sexualized behavior. As Lust writes, "Chilli's opinions, ironically, are more conservative," which the author qualifies as a temporary setback.[47] And yet, what if performing sexuality in this way—by denying confirmation of sex acts—is a form of *radical resistance* representing multiple contradicting sexual interpretations? Black women are stereotypically expected to perform sex acts for others, and sexualized performances have typically been interpreted as a precursor to such acts. Yet, absent consummation of these acts, the hypersexual stereotype is neither advanced nor realized, not in performance and, in the case of asexuals, not in performers' real lives.

When interviewed in 1994 by Joan Morgan, T-Boz acknowledged her recent vow of celibacy, a commitment she adhered to long after.[48] In 2017, she told Nosheen Iqbal: "I wouldn't be a good girlfriend right now; I don't want to have sex with nobody."[49] T-Boz's reticence toward sex is often framed in contrast to the group's apparent encouragement of sexualized behavior; though, in contrast to a romantic asexual outlook, T-Boz identifies sexual availability as the central quality determining one's potential as a "good" romantic partner. Chilli has also been criticized in the media for behaving in a way perceived as contrasting to the group's pro-sex message. In 1994, *Vibe* described Chilli as "now celibate."[50] In qualifying Chilli's avowal to abstain from sex with phrases like "*now* celibate" or "I wouldn't be a good girlfriend *right now*," interviewers leave open the possibility that the performers will one day break the vow. They leave women publicly available, reinforcing that their actions, beyond musical performance, are weighed down by greater public servitude—like they owe it to listeners to be sexually available. But who are the listeners who have such expectations? One reasonable assumption would be that the expectation comes from heterosexual men, but revisiting feminist activist literature, we see, too, that women's expectations also resurface hypersexual stereotypes of availability. Even some Black women rappers themselves cling to a premise of sexual choice and tout their availability.

In an analysis of popular music videos, Rana A. Emerson equates youth and sexual allure with the performers' perceived sexual availability. Surveying the conformity of Black women's images in music videos from MTV, BET, and VH1 in 1998, Emerson explains, "the portrait of Black womanhood that emerges from the video analysis is flat and one-dimensional. . . . Black women performers are not allowed to be artists in their own right but must serve as objects of male desire." Ultimately, Emerson determines, "Pregnant women and mothers, as well as women older than 30, are not desirable as objects of the music video camera's gaze, reinforcing the sense that only women who are viewed as sexually available are acceptable in music videos."[51] I have no doubt that the "one-dimensional womanhood" Emerson identifies certainly characterized the music videos of the mid- to late-1990s. I do, however, question whether limiting her analysis only to images available in music videos may have left out part of the story.

Likewise, Tricia Rose problematizes early hip-hop performances by women using an example of Salt-N-Pepa's "Tramp," which criticizes heterosexual club courtship rituals that encourage men to deliver cheap pickup lines and pursue multiple partners by lying about their intentions. "Salt 'N' Pepa's parable defines promiscuous *males* as tramps, and thereby inverts the common belief that male sexual promiscuity is a status symbol. This reversal undermines the degrading 'women as tramp' image by stigmatizing male promiscuity."[52] Yet, Rose believes Salt-N-Pepa nevertheless continue to advertise these same frowned upon behaviors because they likewise do not offer an alternative way to approach women.

"Tramp" does not interrogate "the game" itself. "Tramp" implicitly accepts the larger dynamics and power relationships between men and women. . . . What of women's desire? Not only is it presumed that men will continue their dishonest behavior, but women's desire for an idealized monogamous heterosexual relationship is implicitly confirmed as an unrealized (but not unrealizable?) goal. In their quest for an honest man, should not the sobering fact that "most men are tramps" be considered a point of departure for rejecting the current courtship ritual altogether? . . . . "Tramp" is courtship advice for women who choose to participate in the current configuration of heterosexual courtship, it does not offer an alternative paradigm for such courtship, and in some ways it works inside of the very courtship rules that it highlights and criticizes.[53]

Part of TLC's novelty rests in how they in fact do confront this courtship ritual.

In particular, Left Eye often takes on the part of the male suitor—at least reviewers of their albums assume so, for example in *Vibe*'s recent anniversary review "20 Years Since *CrazySexyCool*," Kathy Iandoli explains, "The album brought their traditional breed of gender role reversals, as tracks like 'Creep' highlighted infidelity on behalf of the woman and 'Kick Your Game' schooled men on the proper way to approach a lady." Iandoli identifies TLC's audience as men and therefore sees only one way of understanding the gender dynamics at play. On Billboard's critic picks of the top "100 Greatest Girl Group Songs of All Time," TLC's "Creep" lands the second spot on account of its portrayed gender reversal. Staffer S. J. H writes:

Produced by Dallas Austin, the track, driven by a muted trumpet and T-Boz's husky come-on vocals, subverted what had become a traditional trope in R&B: women somberly crowing about their man's infidelity. The '90s saw a turning point where women sang about not only being unfaithful to their men, but doing it as an act of revenge. . . . "Creep" . . . set the groundwork, a pioneer in the adulterous subgenre.[54]

Here, the issue with presuming a male-centered audience is whether we identify this audience as TLC's envisaged suitors or as the main demographic of the group's *actual listeners*. Regarding the latter, it is curious, given their Billboard success where they are touted as one of the highest grossing "girl bands" of all time, to consider whether TLC's main audience was comprised of mostly men.[55] And, regarding the former, many songs on the album do not easily map onto discrete dialogue along a woman/man gender binary.

In 1997, Polly E. McLean conducted interviews with African-American and Latino youths about the relevance of music and sex. Respondents confirmed music accompanied many daily activities, including sex, and the study showed music as a significant educational tool teaching listeners about sex topics (safe sex, sex positions, birth control, STDs, and flirting behavior). Most notably:

> Music seems to provide these adolescents, who are both consumers and critics, with a way of thinking about their gender roles and contradictions and their ideas of constructing sexual expression. For example, among some female respondents T.L.C.'s message encouraging females to take charge of their sexuality had the opposite effect. Consequently, they were incensed over the suggestion that to be sexually assertive means to implore males ["Beg"], because these females had already inverted the traditional male-female role within a sexual context by being aggressive. Most surprising was how Latino males took T.L.C.'s song directed to females as desirable and empowering of their own male sexuality.[56]

The study argues that "at-risk" youth are neither homogeneous nor reticent in articulating their distinctive points of view. It was easy for some participants to distance themselves from "dis" lyrics when the rapper was not of their ethnicity, or the song was not targeting a group with which they identified (whether by gender, ethnicity, or sexual conduct). This would suggest that the rapper's identity and perceived target audience might influence how listeners relate to the music, which raises further questions about who we might speculate was TLC's target audience and who actually identified with the music.

SAFE SEX

TLC's safe sex promotional campaigns already defined the group's image from their earliest popularity. Their public "safe sex" advocacy and pop youth image, like a lot of youth-centered hip hop, pushed against the commonly advanced narrative of the 1980s: that sex was an "adult theme" and an activity that should be controlled, protected, shielded from youth, and enjoyed in private. From one of their first successful singles "Waterfalls," written by Left Eye, TLC have employed digital projection, replication, and enhancement to advance a pro-sex message to counter the common political and legal drive to control and correct United States citizenry—a citizenship to which only certain individuals are entitled.[57] At the helm of this message are Left Eye's consciously rapped lyrics and the signature condom she wore over her one eye, alongside TLC's iconic baggy clothing symbolic of 1990's "urban" street youth—at once both sexually elusive and alluring. In Nataki Goodall's view,

> T.L.C.'s debut album, *Oooooooohhh . . . On the TLC Tip* represents a major breakthrough in the expression of black female sexuality. A product of the sexual (r)evolution in female rap, it serves as a powerful, liberating assertion of black women's economic (Depend on Myself), psychological (Shock Dat Monkey), emotional (Baby-Baby-Baby), and sexual (Ain't 2 Proud 2 Beg) independence and their ability to dictate and take responsibility for the terms, the process, and the outcome of their own experiences.[58]

Oooooooohhh . . . On the TLC Tip (1992) clearly channeled contemporary hip-hop vibes in the performers' appearance and, in the vein of rap's "didactic function,"[59] offered sex-positive messages geared toward "girls" ("Shock Dat Monkey") by alluding to sexual situations rather than spelling out explicit scenarios and using language clean of expletives. TLC's later albums *CrazySexyCool* (1994) and *FanMail* (1999), though still retaining their youthful humor, raise issues surrounding AIDS and women's self-respect in heterosexual courtship with more descriptive language and profanity and thus appear to target more mature audiences—perhaps fans who grew up listening to them—but we cannot know for sure whether that's the case. Indeed, those albums and later also *3D* came embossed with Parental Advisory labels. As mentioned, the "Love to Love You Baby" sample only sounds on the explicit version of the album.

Spearheaded by wives of politicians (e.g., Tipper Gore) and elevating the stereotypical maternal instinct to protect children via a higher authoritative power (i.e., the US government), Parental Advisory labels imposed on musical media in the United States since 1985 warn consumers of explicit content (profanity and age "inappropriate" references to sex and violence). The warning labels emerged under the same logic of abstinence advanced by Nancy Reagan's "Just Say No" anti-drug campaign and the contemporaneous anti-pornography proto-feminist campaigns discussed in chapter 7. Abstinence, in legal and social discourses throughout the 1990s, was posed as a definitive solution to many vices, including drugs, heterosexual premarital sex, and homosexuality.

Conservative constituents touted abstinence as a form of maturity and accountability in opposition to such youthful indiscretions as (implicitly heterosexual) premarital sex and homosexual acts alike, each dismissed as a "phase" to be *overcome* in adulthood. Queer theorist Lauren Berlant frames this discourse within Tipper Gore's advocacy for Parental Advisory labels:

> Thus, when Tipper Gore places the words "Explicit material-parental advisory" on the title page of her book, we are to understand that her project is to train incompetent American adults to be parents, as a matter of civic and nationalist pedagogy. . . . Although all Americans are youths in her view, in other words, incompetent to encounter live sex acts or any sex in public, she also desperately tries to redefine "adult" into a category of social decay more negative than any national category since the "delinquent" of the 1950s.[60]

In this way, sex itself is touted as an achievement of adulthood, an act in which North Americans must not engage but something to which North American youth nevertheless ought to aspire.

Sexual maturity in such discourses is evinced by the ability to abstain from certain acts until lawfully, socially, and religiously permitted. Indeed, as argued in 1994 by Major Melissa Wells-Petry, US military attorney (paraphrased by Lauren Berlant), deviance is evidenced by actions:[61] "There is no possibility that a homosexual has an identity that is not evidenced in acts . . . the courts have proven that to be homosexual is to behave as a homosexual."[62] Such was her justification for prohibiting homosexual men from joining the military. According to this rhetoric, by not engaging in "acts of sexual alterity" (in Berlant's words), individuals are no longer classified by the state as being deviant. Abstaining is a legal limit, a stopgap and cure imposed upon non-normative sexuality, which serves to further erase "asexuality" from sexual discourses. Within abstinence rhetoric, abstinence is not a choice but an act of political and religious coercion no more than mere necessity.

In contrast to this outdated legal framing, refusal to engage in sex acts could be reframed along the lines of contemporary asexual discourses, as radical resistance to the social stigma and medical pathology assigned to those who abstain from sex. It is precisely this "anti-social" stigma, writes David Halperin, that defines music as "camp," even "queer."[63] Rather than failure to meet the "mature models of socialization," Stan Hawkins argues that "the aspirations of camp are a reversal of the conventional valences of style and appearance, where social identities are rejected as authentic or naturalized. Camp expression in pop therefore gains its leverage along these lines, in humorous forms of representation, such as parody."[64] Hawkins points to the example of Nicki Minaj, who plays "on the anxieties of straight masculinity by engaging in the rhetoric of same-sex play and 'different' sexual encounters" in her "raunchy" tracks.[65] TLC surely set a precedent for Minaj in this respect. But, where Hawkins identifies "masculinity" implicitly with men (hence the "straight" prefix), there is more to this parodic ambiguity when situating these choices within a history of sexual agency as a selective privilege, as a *property*.

Emphasis on choice has been a constant running theme in music performed by African-American women. Drawing on the dual intention of the "moan" as both a religiously spiritual and sexually allusive "cry," Angela Davis notes that "One of Bessie Smith's own compositions philosophically juxtaposes the spirit of religion and the spirit of the blues, and contests the idea of the incontrovertible separateness of these two spheres."[66] Davis goes on to examine songs that refer to moaning, including "Preachin' the Blues" (featuring the text "Moan them blues . . .") and "Moan, (You) Mourners," songs that "[turn] the blues into a spiritual discourse about love . . . revealing an interesting affinity with West African philosophical affirmations of the connectedness of spiritual and sexual joy."[67] Davis investigates cultural changes to the church's denunciation of sexuality as it was historically formulated in women's blues, being careful to emphasize the performer's choice in the matter. In her determination:

> the freedom to choose sexual partners was one of the most powerful distinctions between the condition of slavery and the postemancipation status of African Americans. In this sense, the incorporation by the black church of traditionally Christian dualism, which defines spirit as "good" and body as "evil," denied black people the opportunity to acknowledge one of their most significant victories.[68]

Most notably, the willingness to acknowledge women's agency—not as deterministically gendered *a priori* but by the fact that "women's blues were more extensively recorded than men's during the twenties" and hence more widely available—suggests that "women's blues influenced men's attitudes toward women even as they were influenced by the prevailing ideas of women as sexual objects rather than as subjects of their own experience."[69] Carried into the present framing of this chapter, we might read Davis as saying that, while women's music uniquely shaped how sexuality was heard in music overall, it is ultimately up to listeners to decide whether they are willing to acknowledge the performer's "desire" for music to convey, in Davis's words, "the affirmation of autonomous sexuality by women." Such willingness would allow musical performances to achieve an agential goal of not only performing availability but also of *refusing* to perform certain sex *acts*. In a word, performing unavailability.

Asexuality is neither sexual absence nor the wholesale opposition to sex. The "a-" prefix paints asexuality as an identity that expresses an orientation to sex that, like any other sexual orientation, does not apply uniformly to each person who claims it. In the context of TLC's hypersexualized youthful image, I want to claim asexuality as a form of safe sexual practice, a way of expressing sex via music and dance performances all while maintaining a safe distance. From the performing women's perspective, this distance is not a gap to be filled; there is certainly no room for suitors to advance. In this way, perhaps simply showing or talking about sex is not in itself inviting or suggestive of a woman's availability. The veil shields nonconforming performers.

For some individuals—such as those whose identities fall outside binary gender parity or the dominant racial/ethnic class—physical age is not the barrier to legitimacy, but rather a more fluid interpretation of maturity emerges from collectively governed social norms and expectations that bar or grant these individuals access to the rights extended to "adults"—such rights as marriage or pornographic and violent media, which come also with the right to privacy and protection from a nation's government. As Berlant explains in "Live Sex Acts," the right to privacy is given lawfully to those who achieve maturity, those who are sanctioned by the government as fully formed citizens. Laws are enacted to protect citizens. Laws bar or grant access to certain materials, habits, or acts, based on an acquisition and preservation of objects perceived as frozen or dead, and abstractly idealized over time. Included in such "objects" are also sex acts that must become legally legitimized by governing authorities.

Berlant contrasts "live sex acts" with the preserved and objectified acts of "dead citizenship," which "involves a theory of national identity that equates identity with iconicity. It requires that I tell you a secret history of acts that are not experienced as acts, because they take place in the abstract idealized time and space of citizenship."[70] Because abstinence presumes an absence of action, it can be mistaken for inaction. In this imagined ideal, the absence of sex acts is falsely equated with there having been no act performed at all, rather than an act of rejecting or refusing to act in a way that replicates "mature models of socialization," to recall Halperin. To state the implicit here, antisocial acts equate with public displays of affection (PDAs) whether sexual acts are performed or not. In Berlant's words, critics of a so-called deviant sexuality demonstrate that it is not sexual identity as such that threatens North America, which is liberal as long as sex aspires to iconicity or deadness, but suggest, rather, that the threat to national culture derives from what we might call sex acts on the live margin, sex acts that threaten because they are not private. They imagine they are protecting national culture by encapsulating all sex into one of the conventional romantic forms of modern consumer heterosexuality.[71] Make no mistake, abstinence is not celibacy; it is not a coerced or undesired denial of one's *actual* desires. Abstinence can be a certain and uncompromised refusal to engage sexually in any given scenario. Hardly a gap, silence, or abjection, abstaining confidently envoices resistance on a case-by-case basis.

Abstinence by choice is probably most frequently aligned with religious or spiritual fervor. When someone who is not particularly devout expresses anti-sex sentiments, there is an assumption the decision was not actually a choice but a consequence of one's inability to find a compatible partner or because one was once active but, after having been wronged by a lover, fears reliving the rejection, pain, and suffering—like in Kevin Lynch's "Abstinence the Musical" (2017), which the composer describes as being "about a main character who takes a vow of abstinence after a string of disappointing hookups, and empty lovers."[72] Lynch has since made a brand from the title.

Women who don't feel like having sex are especially pathologized, either valorized and idolized for their strong principles like the Virgin Mary, or heavily mocked, for example, the common "I have a headache" trope, as if sex couldn't possibly be about sex and there must be another underlying issue. Indeed, Patricia Hill Collins traces asexual pathology to the racial dichotomy between white women's infantilization and Black girl's adultification—both expressions of idealized girlhood:

> Girls constitute a related benchmark used to construct hegemonic femininity. Girls are allegedly pure, innocent, and sexual virgins. They should be unspoiled. Interestingly, representations of young women/girls within contemporary popular culture contain the contradictions currently plaguing views of young White womanhood. On the one hand, women are expected to aspire to a body type that approximates that of adolescent girls. The inordinate pressure placed on thinness within U.S. society advances a social norm that values youth. At the same time, these same inordinately thin adolescent girls are dressed as highly sexualized women within high fashion. Black women as sexualized, full-figured women become juxtaposed to the thin, young, fragile and increasingly ornamental and sexualized young White girls.[73]

Being sexually flamboyant commonly invites speculations, even criticisms, about a woman's behaviors, about what she *does* and not what she *does not*. This rhetoric became reinforced in the public consciousness in the 1960s when, as Breanne Fahs wrote, "sexual freedom for many women became synonymous with the freedom to have more sexual activity, partners, sexual positions, sexual speech, and physical pleasure."[74] In short, liberation became the cure for a lack or gap that always required one to consume more, newer, and better sex—like technology.

Around the same time, we see similar musical rhetoric emerge urging technological innovations.[75] Where novelty is equated with progressive liberatory sexual diversity, we might similarly position redundancy positively alongside asexuality, as mutually reiterative actions of self-sameness. Perhaps this is why Myra T. Johnson links "Asexual and Autoerotic Women" in the same breath in her landmark 1977 essay emphasizing women's right to choose sexual satisfaction absent any considerations of pleasing others.[76] Hardly a sign of dysfunction, asexuality exemplifies confidence in one's own desire to choose.

Today, Johnson's distinction is made by the designations "nonlibidoist and libdoist asexuals," where libdoist asexuals may enjoy romance, sexual desire, even fondling, with either mutable or immutable limits when it comes to engaging in particular sex *acts*.[77] People who identify as asexual recognize a broad spectrum of orientations under that banner, and not necessarily a total rejection of or aversion to sex, though some individuals do identify with this description.[78] Significantly, in contemporary discourses on asexuality, contrasting from the historical notion associated with racial oppression, rather than impose asexuality as a pathological impairment to be stigmatized or "cured," it is advocated for individuals to self-identify and, similar to other marginalized sexual orientations, to "come out" to gain acceptance and avoid being negatively stigmatized. In this way, asexuality aligns with minoritized sexualities under the LGBTQIA+ banner, which have also been theorized in the same way that having a visible or invisible physical and/or mental illness has been examined in Disability Studies.[79]

Abstinence, as often conveyed through liberal white feminist discourses, appears in opposition to sexual freedom, which writers like Berlant envision as the goal of securing legal rights to "sex law." However, if we view privacy as a right only for those on the so-called correct side of the law—which permits only monogamous (implicitly heteronormative) behaviors, as argued by Berlant and Gayle Rubin (see chapter 7)—then those to whom sex law does not extend must seek more discrete ways of securing privacy. One of the ways that Black women have done this within the US context is by vocally vowing against sex altogether. According to Collins:

> In a climate where one's sexuality is on public display, holding fast to privacy and trying to shut the closet door becomes paramount. Hine refers to this strategy as a culture of dissemblance, one where Black women appeared to [be] outgoing and public, while using this facade to hide a secret world within. As Hine suggests, "only with secrecy, thus achieving a self-imposed invisibility, could ordinary black women accrue the psychic space and harness the resources needed to hold their own in the often one-sided and mismatched resistance struggle."[80]

Collins contextualizes the need for secrecy around choosing to abstain from sex, since this choice was viewed as abnormal within the typical hypersexualization to which Black women were commonly susceptible. As an example, she quotes from Darlene Clark Hine's essay on "The Sexual Exploitation of Black Women," which examines the famous Anita Hill hearings, accusing Supreme Court nominee Clarence Thomas of sexual harassment, and the subsequent "backlash against affirmative action," as the period was commonly described.[81] By silencing Black women's voices surrounding their own sexual desires, under the auspices of group solidarity equally imposed by both proto-feminists as from Black men, each faction's aspiration toward uniformity, in effect, enacted racial and gendered suppression and, consequently, further marginalized Black women.

Rose observes that Black women rappers rarely consign themselves to the collective feminist movement, even if many rappers agree with some feminist premises.

> Given the identities these women rappers have fashioned for themselves, one might expect them to feel comfortable understanding themselves as feminists. And, as I mentioned earlier in the chapter, critical and journalistic writing on black women rappers implicitly and explicitly constructs them as feminist voices in rap. However, during my conversations with Salt, MC Lyte, and Queen Latifah it became clear that these women were uncomfortable with being labelled feminist and perceived feminism as a signifier for a movement that related specifically to white women.

. . . In MC Lyte's case, she remarked that she was often labeled a feminist even though she did not think of herself as one. Yet, after she asked for my working definition of feminist, she wholeheartedly agreed with my description.[82]

This explains one reason why the discourses collectively termed "Black Feminism," after Collins and others, have possibly been slow to take off in musicology and the study of Western art music. Feminism's implicit racial barriers preach equality between white women while ignoring the efforts of women of color or simply feigning *ignorance* to avoid acknowledging, while tacitly enforcing, Black women's sexual *difference*. C. Riley Snorton explains that ignorance can have at least three meanings relevant to how Black sexuality is represented in media and interpreted in reception:

> One, the state of not knowing, two, the act of wilfully ignoring, and, three, in the Black vernacular sense of describing something or someone who engages in socially and/or politically problematic activity. Ignorance, in the black vernacular, represents the very opposite of being politically correct; it is an affectively charged descriptor for those who act shamelessly . . . To be ignorant, and wilfully so, requires artistry, and . . . it may also be one tactic in negotiating stigma in representations of black sexuality.[83]

Outside of popular music, Black music studies, and maybe jazz, in Europe and North America, musicologists are generally ignorant of Black Feminism in all three senses: they willfully ignore and therefore do not know Black Feminist discourses, which then results in the problematic activity of continuing to advance white heteronormative discourses of sex and sexuality. Whereas electrosexual music in the limited white racial framing of part I, "Electroacoustics of the Feminized Voice," may appear one-dimensional and replicate racist tropes about Black women's availability—even when viewed through a queer lens—this is only one perspective, one reading of music history, and neither TLC nor Donna Summer really "fit" into that narrative. At the same time, these women may feign ignorance as a form of veiling to protect themselves from overeager suitors.

According to Collins, Black women's experiences, as uniquely defined by social, geographic, familial and personal context in the United States, are all factors that necessarily impinge on how they are actually treated in society, culture, and especially in the media. Collins writes in *Black Feminist Thought*: "Within U.S. Black intellectual communities generally and Black studies scholarship in particular, Black women's sexuality is either ignored or included primarily in relation to African-American men's issues," and this certainly holds within analyses, both musical and sexual, of Donna Summer's role as a performer in "Love to Love You Baby."[84] Fifteen years later, in *Black Sexual Politics*, Collins examined the musical particularities of this one-dimensional presentation of Black women's sexuality to conclude that musical media transfer and replicate systems of power that contribute to a common perception of Black women's sexual stasis. In her comparison, she names sexual stereotypes to later descend on "women of African descent who followed"—though each performer "may not have been aware of the power of the sexual stereotypes that were created in her image."[85] One of the points on which Collins's analysis hinges is the question of agency, of what say the performers had in constructing their own image (the image later collapsed into a collective "Black Sexuality"), raising a question about whose narrative about Black women is being sold—whether the imposed stasis or the uniform veil of anonymity.[86]

Indeed, even the juxtaposition of white and black can be problematic for analyses that seek to amplify Black women's agency:

> Long before the English explored Africa, the terms "black" and "white" had emotional meaning within England. Before colonization, white and black connoted opposites of purity and filthiness, virginity and sin, virtue and baseness, beauty and ugliness, and God and the devil. Bringing this preexisting framework with them, English explorers were especially taken by Africans' color. Despite actual variations of skin color among African people, the English described them as being *black*. . . . From first contact, biology mattered—racial difference was embodied.[87]

E. Patrick Johnson summarizes this history succinctly in the opening to his book *Appropriating Blackness*, writing: "The fact of blackness is not always self-constituting. Indeed, blackness, like performance, often defies categorization."[88] Collins, Johnson, and many other scholars, from critical race studies to Black studies to whiteness studies and those investigating how these interests converge in cultural studies and the wider humanities, insist that race is a social construct determined by how and when a subject is investigated, but especially, by *who* is performing the investigation. There is no inherent biological imperative or "natural talent" that secures an individual's road to success in one musical practice over another; rather, musical success is a matter of interpretation owing to a host of ways that musical labor becomes valued in reception. In the case of the continuity between Summer and TLC, this success is a matter of *succession* via what some listeners imagine as a sexual proximity extended via musical sampling. Contrastingly, in my view TLC utilizes the electrosexual veil to their advantage to insert just enough digital ambiguity to disrupt the myth of sampled sexuality.

Musical composition is one way of asserting a performer's agency. Summer's song has been repeatedly interpreted through the lens of sexual availability (for example, by Robert Fink and Jon Stratton), because of her explicit moaning. By turning our gaze to the music and minimizing the centrality of sexualized moaning—after all moaning is one of many common features in Black music—we gain new interpretations of performance, especially when throwing gender into question in the ways TLC relies on digital ambiguity. Performing ambiguity through the confluence of sexuality and digitality raises questions about what sex is and which sex acts have been performed. Performers can suggest availability with the agential goal of refusing to perform certain sex *acts*. Acknowledging an irreconcilable conflict between Chilli's sunny major key and the rapped "raunchy" lyrics promotes simultaneously competing interpretations and, therefore, suitors of various identities. There are no pronouns to indicate who the three are addressing in their lyrics and leaving open the recipient, as they do throughout the *FanMail* album, means they could be addressing fans, one another, or self-referencing. That these interpretations exist together points to the singers' collective notion of womanly performance. TLC's gender trinary isn't reducible to the contributions of any one of the group's three performers.[89] Sampling therefore challenges our staid definitions of gender and how these inform sexuality by throwing into question the "logic of oppression" and "the logics of femininity," where "I'm Good at Being Bad" evinces the meeting point of sex, gender, and technology as multifaceted *vectors of difference*.

What an Afrofuture looks like per Janelle Monáe: "It looks like an orgasm and the big bang happening while skydiving as Grace Jones smiles."[1]

The last chapter investigated boundaries between a person's claimed sexual orientation as distinct from whether they perform particular sex acts. This chapter continues to interrogate how orientation can be expressed in sound, how coded references like musical tropes work to create communal belonging between listeners, including other artists, by way of shared musico-sexual orientations. Janelle Monáe's (b. 1985) song "It's Code" from *The Electric Lady* (2013) features one such passage. This recurring theme, which I identify as the "wobble warp," is doubly encoded, hardwired, while also sending a message to similarly wired listeners. Where Donna Summer's "Love to Love" struggles with the laboriously synthetic, and TLC's sample of that song in "I'm Good at Being Bad" subtly repositions Summer's song through digital cultural retrospectivism, Monáe's concept albums display explicit Afrofuturist flights of optimistic fantasy rooted in Black iconography and double meanings that flit across the "sonic color line."

The Electric Lady presents Suites 4 and 5 of Monáe's multi-album adventure, interfusing the artist's dreams with themes from Fritz Lang's *Metropolis* (1927).[2] A prequel to *The ArchAndroid* (2010), the album tells the origin story of Monáe's loosely fictional alter ego, android Cindi Mayweather, or Electric Lady No. 1. She explains: "I started to think of a world where there were more electric ladies, there was a new breed of women."[3] Where the electroacoustics of the feminized voice isolate sound merely as reflexive iconography for the patriarchal fantasy of power and control, what Laura Mulvey has deemed "the phantasmagoric space conjured up by the female body," Monáe's style-traveling pastiche of blues, R&B, hip hop, pop, and many more electronic currents reimagines the relationship between the human, the voice and the electronic. This chapter defines the "wobble warp" as a "quare" expression and coded mode of communication between this breed of women in Monáe's song "It's Code."

"IT'S CODE"

The lyrics of "It's Code" tell a common story of scorned love. Cindi sings of her love for human Sir Anthony Greendown in a world where relations between humans and androids are forbidden. The "code" of the song's title plays on the word's two meanings: (1) computer/genetic code; (2) communication of coded messages. In the song's verse Mayweather/Monáe sings: "Oh baby, it's code / I want you to hold me, and love me until I want no more."[4] A common interpretation of this passage imagines that "love" is programmed or encoded into the android. Mayweather is on the run, forced to leave her love, but obviously she still cares for him. The song also features "code" of another sort, a message to her lover.

Harmonically, Monáe's opening verse oscillates stepwise between Gm9 and Am7 on successive downbeats for close to forty-five seconds. Melodically (as transcribed in figure 11.1a), Monáe's verse begins with a scalar ascent (labelled Motive 1) B-flat C D E F, skipping to A, then "filling" in the "gap" by landing back on G slightly before the downbeat of the next measure (labelled Motive 2). Again, Monáe repeats the slightly varied opening melody.

At the end of the first verse, the harmony breaks away distinctively from its neighboring Gm9-Am7 oscillation for the first time. The passage begins harmonically as before on Gm (minus the seventh), but soon shifts to a 3–4–5 root progression (C D E) delaying the arrival of 1, Am7. Conditioned as we were for almost a minute that Am7 will succeed the Gm7 chord, we expect Am7 on the downbeat of the next bar with the word "played," but in this spot Am7 does not sound and in fact it does not arrive for an additional two bars. Instead, the lyrics and melody support harmonic

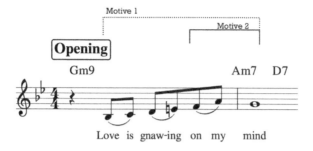

Figure 11.1a

Ascending hexachord B-flat—G (Motive 1) with leap followed by "gap fill" (Motive 2), opening Verse 1, Janelle Monáe's "It's Code" (0:12–0:17).

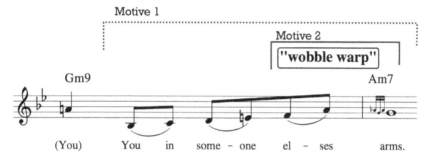

Figure 11.1b

Ascending hexachord (Motive 1) with leap followed by "gap fill" (Motive 2) and "wobble," continued Verse 1, Janelle Monáe's "It's Code" (0:59–1:04).

prolongation, as we hear the text "Cause when I turn back around" together with the 3–4–5 "turnaround" figure that mirrors this text.

"You" finally arrives emphatically on the downbeat in the backing vocals, represented with parentheses in figure 11.1b. The sustained A precedes the lower B-flat, as before, preparing for a reiteration of Motive 1, which proceeds as before through C D E F, skipping to A, and then finally back to G on "arms" (Motive 2). This time, however, G is muddled by a murky vibrating "wobble" hinting at A-flat. This is the only time in the song we hear this melodic hesitation, which importantly is preceded by a "gap fill." Together, (1) the scalar lead in, (2) "gap fill," and (3) punctuating pitch "wobble," form what I'm calling the "wobble warp" trope.

To distinguish the "wobble warp" from other reminiscent sections of the song, I refer to two other instances of Motive 1, one immediately after the "wobble warp" in the pre-chorus, and once again in the instrumental interlude thereafter.

In the pre-chorus (figure 11.2), Motive 1 ascends as before, this time on different pitches through a B-flat minor scale, C D-flat E-flat F. Then, leaping as before to C before returning to B-flat, now 3 in G minor, the motive prepares our ears for Motive 2, but no "wobble" sounds. The passage leads to a 6–4–1 harmonic root progression back to G minor. Although the pre-chorus version recurs throughout the song, the "wobble warp" version only sounds once in the vocals.

In the instrumental interlude leading out of the chorus (figure 11.3) we again hear preparation for the "wobble warp," beginning on the word "You," recalling figures 11.1a and 11.1b, but the lyric trails off into a light jazz flute rendition of the melody omitting the melodic "wobble," evading the expected resolution. I want to suggest that the gap fill in figure 11.1b, so

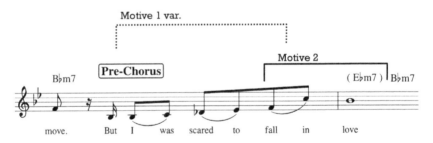

Figure 11.2
Ascending scale suggesting a variation of Motive 1, with leap followed by "gap fill" lead up to a "wobble" that never arrives, pre-chorus, Janelle Monáe's "It's Code" (1:11–1:15).

Figure 11.3
End of chorus through arrival on "You" preparing for Motive 1 but evaded by instrumental interlude, Janelle Monáe's "It's Code" (1:39–1:46).

audibly unique to the rest of song, is a sonic trope linked to the song's title: "It's Code." Not coincidentally, the song's chorus touts "Baby it's code" over and over on a descending tetrachord, an inversion of Motive 1. As we will see, encoded in the music is a key to hearing *quarely* across multiple contemporary R&B/hip-hop crossover songs. But code for what?

BEING QUEER, HEARING QUARE

The "wobble warp" trope sounds in songs that could be said to embody a queer Black aesthetic. It's not that the songs are somehow representative of queer Black individuals, but rather the "code" performs a wink at those who get it—at those who have been acquainted with its symbols through repeated exposure. For this reason, transcriptions of these momentous occasions are only so helpful, listeners who really get what the trope is about acquire this understanding by internalizing the trope via repeated hearing, sometimes, themselves performing it by singing along with recordings or independently. Queerness thus becomes affective in the way people in the know hear and pick up on queer significance to those with common ground. As Gayle Murchison explains, Janelle Monáe's music "provides examples of quare black music and the way black music (or any listener) can be 'hair-yuh' and 'quare-yuh' (i.e., 'here [or, to pun, hear] and queer,' in African American Vernacular English . . .)."[5] Whereas, some may hear the "wobble warp" simply as vocal vibrato, others may hear its electronic resonances with the Minimoog (discussed later), and still others may associate *quarely* with its cross-genre reach. All this is to say that affect is context dependent. Situational knowledge is important in figuring out what qualities are taken on or expressed by a given trope and *to whom*.[6] My aim is to highlight a quare bent on the "wobble warp" and to theorize the trope as more than mere figuration.

Queerness as a classification (musical or nonmusical) is in itself a term full of contradiction. In reclaiming "queerness" from its derogative historical connotations, queering can arise through resistance and opposition to oppression, as a refusal to conform—as we note from its comparisons to parody and antisocial behavior in the last chapter. A contradiction herein is that, collectively, this nonconformity also becomes something of a unifying descriptor, uniting through irreconcilable difference—Sedgwick's minoritizing outlook from chapter 9. The "wobble warp" embraces this collective nonconformity by hinting at regularity in the figuration's sound, in its dwelling at the porous bounds between genre, chronology, and technology—whether acoustic or electric.

Homosexuality is not criminalized in the United States, yet many queer folks still seek solace in the secrecy of coded messages and furtive glances, especially as anti-discrimination laws have remained in flux.[7] Dance music enthusiasts and scholars alike have examined how experiencing Electronic Dance Music (EDM) collectively fosters a strong communal bond among those listeners and dancers. EDM emerged simultaneously as both a musical and social phenomenon, with historical ties to underground or private dance spaces in disco clubs and urban house parties (chapter 6), as well as the more public venues of block parties and the Black church. Underground music and dance clubs have long served as uniting safe havens for the LGBTQIA+ communities in geographic regions ranging from Berlin to Nigeria, Chicago to Singapore. In many parts of the world there remain laws criminalizing same-sex dancing and even certain ways of looking, let alone physical contact that could be interpreted as sexual.[8] Private musical spaces afford safety to those individuals faced with violence and excluded elsewhere for expressing a nonconforming gender identity or for participating in sexual activities that would have been denied or suppressed as queer or deviant in spaces in which heteronormative behaviors were expected. Safe havens like discotheques, according to Brian Currid, could therefore replace the biological bonds of blood with an alternative, more transient notion of family.[9] One is not necessarily born into this family, but rather one subscribes voluntarily by way of inter-musical participation, a mode of belonging with possible resonances in the secondary, more public arenas above.

Although queer hip-hop artists often rap about sexuality,[10] their music is not necessarily sexually explicit in the way that "pornorap" or "dirty rap" are.[11] Those genres, for instance, are sexually connotative on account of graphic lyrics and music videos with explicit visuals more than any specific musical associations (see, for example, videos by CupcakKe or RoxXxan).

Self-described queer rappers or hip-hop artists like Monáe, Stas THEE Boss (discussed in the next chapter), Guayaba, or Frank Ocean, tend to draw their roots from so-called consciously rapped lyrics, using interpolative and intertextual associations across musical streams to advance sexual content via multiple constitutive dimensions, whether lyrics, music, technical processing, quotation, and any number of mechanisms operating simultaneously, such as the multiple modalities in which Monáe expresses her "code." Though explicit lyrics can certainly also express queer sentiments—for example, when trans rapper Mykki Blanco in "Wavvy" declaims topless, bedazzled in belly dancer jewels: "I scalp these haters . . . I cut throats." What is usually meant by "queer" in this context, in the words of C. Riley Snorton, is a "politics of disidentification and disavowal" to the heteronormative nuclear structure.[12] And, as I wrote in the introduction to part II: "'norms' of performance come to influence expectations of how audiences receive codes, whether sexual, musical, or racial, since norms dictate whether codes are perceived as facilitating or disrupting the performance framing." In the case of quare codes, norms of "disidentification and disavowal" are formed and passed on via sonic tropes like the "wobble warp." Though music theory writ large still perpetuates the marginalization of Black (musical) aesthetics from academic music-theoretical discourses,[13] the designation of a "quare" outlook instead forms a cohesive analytical orientation from which to understand and interpret the queer underpinnings that link musical expressions.

Francesca Royster's *Sounding Like a No-No? Queer Sounds and Eccentric Acts in the Post-Soul Era* uniquely examines what's "quare"-ly audible in different popular electronic musics, most notably in music by Prince, Stevie Wonder, Grace Jones, and Janelle Monáe.[14] Royster draws from Imani Perry's imagery on alienation and Tricia Rose's optimistic hip-hop imaginary to theorize "eccentric" and "quare" elements in Stevie Wonder's music, writing:

> While Wonder's *Journey through the Secret Life of Plants* doesn't sound at all like the dynamic, pared down aesthetic of much early rap, both emerge out of a context where "social alienation, prophetic imagination, and yearning intersect," in Tricia Rose's words. Despite the characterization of *Journey* by many of its critics as (only) airy or otherworldly, we might, in particular, connect Wonder's exploration of return in this project, in addition to its other

themes, to rural landscapes and knowledge, and to Africa, as a response to the deindustrialization and burned-out city landscapes that have also influenced hip-hop graffiti, breakdancing, rap, and DJing.[15]

Royster relishes in the joy of Wonder's success, his forward-sounding synthesizer-driven music links to hip hop through Afrofuturist flights of fancy. Yet, she cautions also that an artist's aspirations to break through are limited by gendered notions of nonconformity:

> In popular culture (and often played out in everyday life), black male genius has its limits concerning where it can go and what it can do. For one thing, the term genius in the world of black music is almost always used synonymously with male (with the exception of Billie Holiday, thanks to Farah Jasmine Griffin), and its most embraced and best-known models are reassuringly and zestfully masculine, deeply loyal and embedded in the world of men. Black male musical genius is often connected to sex and the seductions of power. . . . The stories we tell about black male genius are bigger than life, mythological.[16]

Thus, concludes Royster: "black geniuses are exceptions to the rules of racism," an exceptionalism that does not extend to Black women.[17]

Hip-hop scholar Gwendolyn Pough summarizes society's stereotypical representations of Black women as loud, gesticulating, and angry.[18] Royster's chapter on Grace Jones shows how she defies and resists such labels, her fans characterizing the performer as "'Alien Grace. Detached Grace. Frozen Grace.' She is often seen as an emblem of cold steel androgyny."[19] Through this lens, Jones come to exude antisocial characteristics. A mature woman's self-assured standoffishness contrasts so starkly with the angry Black woman as to make her *quare*.

From the heteronormative perspective, standoffishness commonly characterizes lesbians who are perceived as avoiding the sexual complacency and prowess society often expects of women. Tying this popular image of insubordinate women to an instrumental and choral composer, musicologist Elizabeth Wood revealed hidden messages in music by lesbian composers. Ethyl Smyth was said to have communicated cryptically through music as a way of warding off unwanted scrutiny and "to escape, deny, and transcend gender."[20]

In this sense, Black women's insubordination is doubly articulated via sex/gender/sexuality as well as race, both of which become identified as alien and detached characteristics—an affect they attempt to capture in sound by encoding nonverbal messages via performative gestures. The refrain: hair-yuh and quare-yuh.

The "wobble warp's" coincidence with music by queer-identifying performers of electronic music and in queerly identified spaces marks it historically as both electronic and queer while remaining an unnegotiable facet of Black experiences within these spaces, hence *quare*. The trope's gendered and racial origins are inseparable from the "wobble warp" as sounding effect and therefore essential to its hearing and its transhistorical and intergeneric confluence.

ENCODING QUARENESS IN "MANY MOONS"

Monáe's "It's Code" is track 12 of *The Electric Lady* and the first song of Suite V (after the Overture). Set in a future which forbids interspecies relationships between androids and humans, the series follows Mayweather through her love affair with the human Sir Anthony Greendown, the ensuing trial, and the verdict sentencing her to disassembly. Alongside futuristic themes of robots, time and space travel, and assembly-line capitalistic production, Monáe's albums also convey an allegory for Black oppression via imagery, drawing on blackface minstrelsy, slave auction, commoditized sex, and miscegenation (as conveyed via the law against interspecies coupling)—imagery that provokes critique and simultaneous reclamation.

Traveling back to the first installment, Gayle Murchison proclaims Monáe's *Metropolis* series "a successor to [Patti] Labelle and other Afrofuturist funkateers and funkmeisters." Echoing Mark Dery's foundational definition of Afrofuturism, *Metropolis* is a "speculative fiction that treats African-American themes and addresses African-American concerns in the context of twentieth century Technoculture."[21] However, she distinctly emphasizes women's contributions to an Afrofuturist genealogy whose scholarly scope does not historically feature many women, whether this failing is from the typical neglect or from an aesthetic vantage, in that Afrofuturist theorists have failed to recognize Afrofuturist traits as these are expressed by women.[22]

Not only does Murchison's analysis enrich definitions of Afrofuturism with a feminist perspective, but her contribution also highlights queer themes in this music thus augmenting the analytical intersectional breadth.[23] Murchison echoes Black feminist theorists like Patricia Hill Collins and bell hooks to offer "quare feminist critique" as a mode of musical theorization that "troubles the way in which the music industry offers black female bodies on a continuum from eroticized to hypersexualized."[24] Following Murchison, I acknowledge a deferential power distribution in how music is used to create sexual tension. I hear certain sounds like the "wobble warp" trope expanding how we witness queer musical belonging expressed sonically, where the trope becomes a way of communicating *quare*ly to those in the know at the same time as music, in all its ephemerality, easily evades the scrutiny of those seeking to sever the continuity of these connections. Specifically, Murchison's analysis identifies "quare emancipations" in gestures and symbols particular to the "Afrofuture" as articulated by Monáe's performances. In Dery's words from 1994, it is "perplexing" that

> so few African-Americans write science fiction, a genre whose close encounters with the Other—the stranger in a strange land—would seem uniquely suited to the concerns of African-American novelists. . . . African Americans, in a very real sense, are the descendants of alien abductees; they inhabit a sci-fi nightmare in which unseen but no less impassable force fields of intolerance frustrate their movements; official histories undo what has been done; and technology is too often brought to bear on black bodies (branding, forced sterilization, the Tuskegee experiment, and tasers come readily to mind).[25]

Since Dery's proclamation, we observe many more Afrofuturist contributions. Situating *quare*-ness in proxy to Afro-diasporic themes of being alien and alienation, Murchison credits Monáe with "appropriat[ing] visual symbols of white masculinity, wealth, and power to call out and push back against oppression," honing in on the scene of the "annual android auction" from the short film released for "Many Moons" (https://www.youtube.com/watch

?v=Yy-ugv9kxG0) from Monáe's debut album and Suite I of the collection, the *Metropolis* EP (2007).[26]

Whereas Monáe's fictional future forbids relations between androids and humans, this particular scene in "Many Moons" highlights dimensions of class oppression by showing how laws become enforced as a means of empowering some by way of the oppression of many others. The Black androids in the auction scene are not prized merely for the value of their labor or skills but for their visible attractiveness, even if interspecies relations are forbidden. One white human couple sits in the audience seeking to purchase a suitable mate (slave?). The couple's lascivious gaze and visible excitement features in several shots. Murchison describes the moment they decide to purchase (Monáe as) Suzie Scorcher (3:26–3:40):

> The wife whispers in her husband's ear. As Mayweather [singing] asks, "Are you bold enough to reach for love," we observe a *quare visual exchange* between the white couple and the black droid, Scorcher, a scene that is followed by Mayweather's rap, modeled on Madonna's "Vogue."[27]

What Murchison refers to here as a "quare visual exchange" is the mutual acknowledgement between the couple and the droid about possibly engaging in illicit activities of interspecies mingling, which are "queer" because the activities transgress established societal norms. The exchange becomes *quare* when taking into account a non-diegetically coded meaning arising from uneven power dynamics between the auctioned and the buyers. In this regard, Francesca Royster points to Monáe's android persona Cindi Mayweather, the auction master in "Many Moons":

> Monáe's creation, the android Cindi Mayweather is another eccentric tactic used to address issues of power specific to the music industry: the dehumanization of the commercial marketing of black performance, the ways that capitalism manages to appropriate the underground, and the always present push back of that underground to keep creating. . . . The auction is the site of multiple exchanges of power and desire, and we watch while the androids are traded between men and women competing for power and visibility. . . . "Many Moons" captures the rebellious energy of black quare musical performance, but it also speaks to the power of black performance to meet and produce the demands of pleasure seekers, sometimes to the point of their own destruction.[28]

"Quare," in the ways both Royster and Murchison refer to Monáe, is not merely a sexual "orientation" but an "eccentric" aspect of identity, in the sense that E. Patrick Johnson means when he writes in *Black. Queer. Southern. Women*: "While folks may have whispered about them, their dalliances with the same sex did not necessarily make them 'lesbian' as much as it did just another eccentric whose membership in and contributions to the community outweighed their sexual behavior."[29] This notion of eccentricity is uncompromisingly intersectional, as pointillistic in individual instances as it is historical in its adherence to communal integrity.

In the closing of my introduction to part II, I interrogated the limited and limiting white supremacist spectator perspective. I examined the one-sided perception by which performers become enslaved via a lens that elevates the white human couple's intentions and restricts Black experiences to a collective singular existence rooted in histories of alienation and/as racialized oppression. The "eccentric" breaks away from this mold, to "push back" with agency, exerting desire and a will to insist on private and personal boundaries, whether sexual or otherwise. In this sense, the "code" is both encoded, and hence predetermined, as well as formed via ever-changing linguistic, gestural, visual, and sonic exchanges. The transmorphography of the "wobble warp" is one such articulation.

Where previously I examined how distance reinforces imbalances between the empowered and oppressed, codes become tools for radical resistance in opposition to presumed racialized norms. In this sense, "code-switching" has been used by wanderers in the diaspora as a means of distantiation, for security and protection. Monáe says she created Cindi Mayweather as a voice through which she could express frustrations with and vulnerability about social injustices, like racial inequality and sexual oppression, to those for whom these messages had empowering effects. However, in a 2018 interview for *New York Times Magazine*, Monáe divulged why she abandoned her alter ego in recent albums like *Dirty Computer*:

> The public . . . doesn't really "know Janelle Monáe, and I felt like I didn't really have to be her because they were fine with Cindi." When Prince died in April 2016, she started to rethink how she would present herself. "I couldn't

fake being vulnerable. In terms of how I will be remembered, I have anxiety around that, like the whole concept about what I'll be remembered for."[30]

Thinking back to Donna Summer and TLC, who, as we have seen, leaned into the sexual ventriloquy of the electrified, Monáe recognized the problems of this pattern and identified a solution that could dissolve the stereotypical iconography of "robo-diva R&B," one that seems perhaps counterintuitive: being real. Monáe's recent performances aspire to collapse the distance between her real and performed selves. Who, anyway, can really separate the art from the artist?

In the next chapter, I delve deeper into the musical resonances of the "wobble warp's" codes as they resound through electrosexual currents of not-so-distant pasts and futures.

12 THE WOBBLE WARP

THEESatisfaction's "Enchantruss" (2012), Janelle Monáe and Esperanza Spalding's "Dorothy Dandridge Eyes" (2013), and Stas THEE Boss's "Before Anyone Else" (2017)

This chapter expands on the last to include more examples of the "wobble warp" trope. I will show how this audible signifier draws together electronic musical strands from different times and artists, who on account of this sound, can be grouped in a way that enhances a shared collective identity irrespective of marketed genre or musical style. The "wobble" effect familiar from Moog synthesizers—itself a likely imitation of vocal vibrato and guitar effects—has become so prevalent within contemporary R&B and hip-hop electronic crossovers, that, more than a trope in a single genre or style, the motive may be understood as a broader, overarching "topic," in the music-theoretical sense.[1] In this chapter, I examine the "wobble warp" as a queerly encoded intergeneric topic. Staying the path of this book's history of electrosexual currents, the "wobble warp" is a punctuating, vibrating motive. On its own, its quintessential oscillation between two adjacent pitches could be taken as incidental, as afterthought even. However, this particular vibration often sounds as ornate punctuation to significant musical and textual passages. In terms of tonality, it can reconfigure the key of the passage by introducing tonally adjacent pitches or hint at distant harmonies quickly approaching.

In the previous chapter, I suggested that the punctuating "wobble" in Janelle Monáe's song "It's Code" signaled quare significance because of how it subverts expectations normalized and conditioned from earlier in the song. Once the "wobble" arrives—even emphatically on a downbeat—it warps expectations, raising some hesitations both musically and in how it coincides textually with the lyrics. In that song, I identified the "wobble warp" with three successive criteria: (1) a scalar lead in, (2) a "gap fill," and (3) a punctuating pitch "wobble." In this, the final chapter, I expand the reach of possibilities to propose that this sliding chromatic wobble is coded to function as a sonic trope calling back on music by THEESatisfaction, Grace Jones, Betty (the group responsible for the theme song to HBO series *The L Word*), and others. To my ears, the coincidence of the "wobble trope" in so many coded situations suggests a type of "quare sonic exchange" similar to the "quare visual exchange" Murchison identifies in her analysis of Monáe's *Metropolis* EP, discussed in the last chapter. As I showed in that chapter, Monáe performs the wobble vocally. In this chapter I point to instrumental and electronic instances as well.

As detailed in the case study below, I borrow the term "wobble warp" from a review of THEESatisfaction's first album *awE naturalE* (2012), in which journalist Carrie Battan describes the track "Enchantruss" as "alien-like, wobble-warped."[2] The encompassing description jumped out at me for its familiar resonance with how I experienced much of the duo's electronic/hip-hop oeuvre. Battan's hearing of "alien" features likely references its Afrofuturist bent, drawing on synthesizer-driven music from Sun Ra, Stevie Wonder, and J Dilla. In the poesy of Kodwo Eshun,

> The music of Alice Coltrane and Sun Ra, of Underground Resistance and George Russell, of Tricky and Martina, comes from the Outer Side. It alienates itself from the human; it arrives from the future. Alien Music is a synthetic recombinator, an applied art technology for amplifying the rates of becoming alien. Optimize the ratios of eccentricity. Synthesize yourself.[3]

Black futuristic imperatives are central to the uniforms, paraphernalia, and overall *telos* of hip hop, straddling the line between alien beings and alienated subjects. For Imani Perri:

> Life on the margins of postindustrial urban America is inscribed in hip hop style, sound, lyrics, and thematics. Situated at the "crossroads of lack and desire," hip hop emerges from the deindustrialization meltdown where social alienation, prophetic imagination, and yearning intersect. Hip hop is a cultural form that attempts to negotiate the experiences of marginalization, brutally truncated opportunity, and oppression within the cultural imperatives of African-American and Caribbean history, identity, and community.[4]

THEESatisfaction's music is situated canonically within the particular brand of hip hop emerging from the Pacific Northwest, but because the vocalists often mix their own beats, their respective practices also touch on a broadly defined notion of electronic music production. I examine the wider impact and significance of this intrageneric crossover later in this chapter, but before delving into why the "wobble warp" matters, it is important that I first provide a few more examples to nail down a definition of the trope from which to later examine its topical potential as "queer sonic exchange" operating within and against the electrosexual oeuvre.

THE "WOBBLE WARP" TROPE

The "wobble warp" is related both in pitch and melodic contour to a figure known as the "gap fill" scheme, a melodic leap filled in by step in the opposite direction. This common technique can be traced back millennia in many kinds of music, recognized among music theorists and psychologists alike as a figure that builds melodic momentum as it subverts expectations. Although there are disagreements about whether listeners intuit or anticipate the gap-fill mechanism, its archetypal appearance in many styles of music certainly secures it as a trope that crosses multiple genres, music-compositional approaches, and time periods.[5] It is precisely this indiscriminate pervasiveness that marks the gap fill as a trope.

In his book, *Interpreting Musical Gestures, Topics, and Tropes: Mozart, Beethoven, Schubert*, Robert Hatten explains: "Troping in music may be defined as the bringing together of two otherwise incompatible style types in a single location to produce a unique expressive meaning from their collision or fusion." He compares a localized trope to a more global topic, writing: "Topics are style types that possess strong correlations or associations with expressive meaning; thus, they are natural candidates for tropological treatment."[6] Since THEESatisfaction's music features the most frequent and apparent use of the "wobble warp" across several songs and albums, I present the group at the core of a network of influence between motivic instances of the "wobble warp." With THEESatisfaction at the center, one imagines other musical instances of the "wobble warp" (including Monáe's "It's Code") drawn to THEESatisfaction, lingering at the boundaries of this group and its affiliated genres like protons around a nucleus. Though there are obvious intergeneric connections between these songs, it is still important to situate each song, to contextualize the trope historically and socially in each case. As we will see, whereas Monáe's and Michael Jackson's music hovers on the cusp of Motown, R&B and pop, J Dilla would mostly be understood as rap or electronic, Sun Ra as funk, and, as mentioned, THEESatisfaction lingers somewhere in the margins between rap, R&B, hip hop, pop, and electronic.

Table 12.1 lists several tracks with an identifiable "wobble warp." The artists are aesthetically linked via the trope in their intergeneric electronic inclusivity as well as their reception as queerly presenting musicians, and so I only include a few relevant tracks from each of these artists. The furthest column to the left provides the title, album, and release date of each track in chronological order with distinctive text and timing for the particular instance. In four adjacent columns, I indicate with a check mark whether the passage includes a wobble warp, scalar pattern (not only hexachordal), and/or gap fill. Since examples may convey the three motivic aspects but not concurrently or only a portion of these aspects in cases where the elements are more disparate, I indicate timings for a specific motivic aspect in the appropriate column. For example, though parallel thirds are pervasive in many kinds of tonal music, they occur in such high incidence with the "wobble" that I included them as a fourth characteristic in the "wobble warp" trope. Again, not all songs display all four motivic aspects but, in order to be included, the instance conveys enough of a resemblance to others like it—at least to my ears.

The chart is not meant to be exhaustive of all artists whose music may be related; rather, first, the chart represents the pervasiveness of these tropes in THEESatisfaction's music, with the secondary purpose of showing its historical and intergeneric extensions to other artists. Demonstrating chronological extensions may hint at a historical lineage of sorts, as in predecessors who may have inspired the artists musically (whether consciously or not). The chart only features tracks that present the motives both vocally *and* electronically, since either/or does not encompass the necessary uniqueness of how the motives permeate these tracks.

Table 12.1

Songs featuring a "wobble warp"

Track Details	Wobble Warp	Parallel Thirds	Scalar Pattern	Gap Fill
Grace Jones "Pride" (1978)	✓ (Instrumentals)			
Salt-N-Pepa "Shoop" (1993), throughout in sample of The Sweet Inspirations "I'm Blue" ("Shoop-eh-doop" motive)	✓	✓		✓
Salt-N-Pepa feat. En Vogue "Whatta Man" (1993) (opening "Yeah")	✓	✓	✓ (electric guitar riff sample of Linda Lyndell "What a Man")	✓
Alanis Morissette "Front Row" (1998)		✓	✓	
Betty, "The Way That We Live" (2005) (Theme from *The L Word*) (0:15–0:26 vocals)		✓ (0:15–0:26 vocals)	✓ (0:15–0:26 vocals)	✓ (0:26–31 chromatic vocals)
THEESatisfaction "Enchantruss" *awE naturalE* (2012)	✓ (main vocals, punctuating ends of each phrase with gap fill)	✓ (loop throughout)	✓	✓ (loop throughout)
THEESatisfaction "Extinct" *awE naturalE* (2012)	✓	✓	✓	
THEESatisfaction "Deeper" *awE naturalE* (2012) (1:03–1:20)	✓	✓	✓	
THEESatisfaction "God" *awE naturalE* (2012) (1:10–1:13)	✓	✓		
THEESatisfaction "Earthseed" *awE natural* (2012)	✓		✓	
THEESatisfaction "Sweat" *awE natural* (2012)	✓			
Janelle Monáe "It's Code" *Electric Lady* (2013) (1:00)	✓ (1:07–1:09)	✓	✓ (0:20–0:24)	✓ (0:24–0:29)

(Continued)

Table 12.1
(continued)

Track Details	Wobble Warp	Parallel Thirds	Scalar Pattern	Gap Fill
Janelle Monáe/Esperanza Spalding "Dorothy Dandridge Eyes" *Electric Lady* (2013)	✓		✓	✓
THEESatisfaction "Prophetic Perfection" *EarthEE* (2015)	✓			
THEESatisfaction "No GMO" *EarthEE* (2015) (0:44ff)	✓			
THEESatisfaction "Fetch/Catch" *EarthEE* (2015) (2:44ff)	✓			
THEESatisfaction "EarthEE" *EarthEE* (2015)	✓ (0:35)			✓ (0:04–0:22 background vocals)
THEESatisfaction "I Read You" *EarthEE* (2015) (2:00)		✓		
THEESatisfaction "WerQ" *EarthEE* (2015)		✓		
SassyBlack "Discovery of Self" (2019)	✓	✓	✓	

In writing this chapter, I hesitate to expose this code, to threaten a queerly cordoned-off hip-hop aesthetic and possibly invite others to exploit it, take it away, appropriate it. But as with any code, once discovered, it modulates, shifts positions, becomes altered such that it can never be fully decipherable. And maybe there's good reason to share a portion of our codes of conduct, to expose a thread for others who may wish to follow suit.

THEESATISFACTION'S "ENCHANTRUSS"
AWE NATURALE (2012)

As mentioned, the term "wobble warp" comes from "Enchantruss" (https://www.youtube.com/watch?v=5jXceebT-c4). The song opens with a four-beat clip of electronically processed polyphonic singing in a series of three "gap fills" created by an alternating figure: a descending third + ascending second + ascending third/fourth + descending second. This figure is looped throughout, in the sense that it becomes a repeated electronically processed figure and, metaphorically, the intervals throw our ears for a loop. A visual representation of these successive intervals might suggest a three-dimensional trefoil knot, looping in one direction, turning in on itself, looping back out and then in on itself again for the third time. The riff serves as one of the backbeat layers for the remainder of the track and is recognizably sung by THEESatisfaction's two members. When the track ends, we hear the unprocessed version, so it is easy to make the connection from the singers' voices to the underlying backbeat.

When the main vocals enter, each phrase is punctuated with a slight slide—a miniature "wobble" and characteristic facet of THEESatisfaction's music overall. A pointed moment in the song occurs at the line "can be a tricky thing" (0:45–47). The full wobble on "THI-ing" shifts the syllabic emphasis as well as seemingly dividing a monosyllabic word in two, which consequently also elongates the phrase rhythmically to disrupt or queer its metric accent.

JANELLE MONÁE AND ESPERANZA SPALDING'S "DOROTHY DANDRIDGE EYES" *THE ELECTRIC LADY* (2013)

Turning to a collaboration between Janelle Monáe and fusion jazz bassist and vocalist Esperanza Spalding we find the song "Dorothy Dandridge Eyes," track 18 on *The Electric Lady*. Like "It's Code," lyrical themes in the song find resonance with sexual encoding in the form of coded glances or multiple "quare visual exchanges," to apply Murchison's term for this gesture. This time, after a brief instrumental introduction, the song *opens* with the coded "wobble warp" which is later thematized again in the song's chorus.

Monáe's opening melody from "Dorothy Dandridge Eyes," transcribed in figure 12.1, evinces the familiar scalar ascent and slight chromatic "wobble" from earlier examples punctuating its first phrase, transcribed on the first system. The second phrase, transcribed on the second system, moves into two successive gap fills, one filling a gap from C to E, and then another filling in the gap from G to C, substituting the opening A with an A-flat

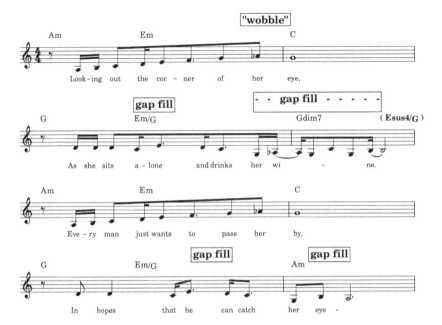

Figure 12.1
Janelle Monáe and Esperanza Spalding, "Dorothy Dandridge Eyes," opening vocal line (0:20–0:37).

familiar to us from, and therefore reminiscent of, the wobble in the previous line.

Harmonically in the bass, the song first outlines an A minor triad, A E C, in system 1. Then, in system 2, the bassline sits on a G pedal the entire time only providing a hint of the A at the end of that phrase which is subtly couched in the inner voicing of an Esus4 (G A B E). In system 3, the next phrase begins again on A providing temporary relief, and this time the phrase itself resolves prominently following the second gap fill, G to B, to finally land on A at the end of system 4. Lingering around A, as this phrase does, but never giving it until the end of the verse on the word "eye," the notes come to mirror the text—men try to catch Dorothy Dandridge's eye, or A. A is code for "eye" contact, resembling again the "quare visual exchange" common to the "wobble warp." Though Dorothy Dandridge herself was not known for her quare engagements, she and her sister were raised by two women who encouraged and trained them as performers. Is it possible then that the famous smolder in her eye was entrained by her mother(s)? Less speculative is the way Monáe and Spalding caress the hypothetical woman in the story, as the chorus touts: "She's got Dorothy Dandridge eyes, / And you love her." This coupled dance of the voices, Spalding's bass, and the electronic synthesizer filling in harmonies, intermingle suggestively—entrancingly.

BEYOND THEESATISFACTION

I've already mentioned the audible trefoil knot in THEESatisfaction's "Enchantruss," from the *awE naturalE* album. A number of other tracks on the duo's next album *EarthEE* (2015) also feature the "wobble warp." For example, the title track "EarthEE" has a recurring wobble throughout, punctuating every phrase in the song with semitone slippage and a more explicit "wobble warp" sounding in parallel thirds in the song's chorus (1:12–1:30). The album's opening track, "Prophetic Perfection," also plays with the gap fill instrumentally, sounding G F F-sharp (0:09–0:11).

THEESatisfaction disbanded in 2016 after the couple stopped being involved romantically, but we still hear remnants of the "wobble warp" in each performer's independent projects. For example, Stas THEE Boss's song "Before Anyone Else," from the album *S'Women* (2017) opens with a descending hexachord from G# to B (figure 12.2), which continues as a loop throughout the song, obscuring the tonal hierarchy to the extent that it is unclear whether the passage centers on E or B.[7] As we see in figure 12.3, Stas double's her own rap in thirds, and phrases are often punctuated with a "wobble." One such punctuated phrase sounds significantly on the words "sad queers," a sentiment that resonates with Monáe's allegorical yoking of "quare" interests to both queer sexual expression and, simultaneously, to a history of Black oppression, eccentricity, and excellence (or genius).

HIP-HOP PRODUCTION AS ELECTRONIC PRACTICE

The two most recognizable traits of hip-hop music are the rapped lyrics and a continuous beat to and against which the lyrics respond—though many have theorized about the order in which rap music is composed.

Figure 12.2
Descending hexachord loop throughout Stas THEE Boss's song "Before Anyone Else,"
S'Women (2017).

Figure 12.3
Parallel thirds punctuated by a wobble on the words "sad queers" in Stas THEE Boss's song
"Before Anyone Else" (1:06–1:08), *S'Women* (2017).

Relatively early on, Robert Walser called attention to discourses on the academic categorization of hip hop and, subsequently, its undertheorization as a valid and valued musical practice. Walser critiqued distinctions between the terms *lyrics* and *non-lyrics*—the latter a catchall term, coined by critics dismissive of hip hop's musical merits, to describe *everything else* apart from textual words. The *non-lyrics* can emerge either from original composition or musical borrowing, often created by precise technical manipulation (sampling and, on occasion, synthesis) to enhance precomposed music, text, or sounds, whether recognizably or not. For example, Donna Summer's "Love to Love" hook preceded the *non-lyrical* electronics, whereas its sample in "I'm Good at Being Bad" and instrumental variations presumably came before TLC's new lyrics in the verse and chorus.

The process of first producing beats and then adding lyrics is one touted by rapper Busta Rhymes in his relationship to producer J Dilla, and, according to music theorist Kyle Adams, is a common occurrence in rap music composition.[8] Responding to Adams, Justin Williams argues that many rap songs are written more collaboratively, exchanging tips and notes between producer and rapper—something akin to Donna Summer's professional relationship with Giorgio Moroder examined in chapter 6.[9] This latter theory would certainly appear to be the case in THEESatisfaction's music, produced either by one of the two performers, Stasia "THEE Boss" Irons (b. 1985) or Catherine "SassyBlack" Harris-White (b.1986), but also producers external to the group, for example, Erik Blood (also associated with Shabazz Palaces). THEESatisfaction's songs often feature a beat composed of the duo's voices, suggesting that the background beat and the vocals in the forefront are compositionally intertwined.

Referring to the term *rap music*, Adams points out that definitions are mixed and difficult to generalize:

> Such a definition will need to be more specific than the statement that rap primarily contains lyrics that are spoken or chanted rhythmically, since that could include a variety of musical styles from around the world. It will also need to be more general than my implication that rap always includes sampled music that is pre-recorded before the lyrics are written, since, as Williams has pointed out, that excludes a vast and important body of "rap music."[10]

In 2000, a decade prior to Adams's determination of hip hop's ambulatory relationship between lyrics and beat, music theorist Adam Krims proposed a taxonomy of genres based in ethnographic research and music analysis. According to Krims's taxonomy, THEESatisfaction's music most likely fits into the category of *jazz/bohemian rap*, which Krims characterizes as a diffuse and eclectic genre in terms of its musical inspiration, hailing from live or sampled jazz instrumentals to smoother, rounded percussion beats, and "consciously" rapped lyrics.[11] Krims describes the lyrics as generally less aggressive than *reality rap* (what in the press is often derogatively termed *gangsta rap*),[12] though the subject matter of both can touch equally on social consciousness. Writing in 2004, Gwendolyn Pough acknowledges that, beyond subject matter, "even the musical component of Hip-Hop has grown and encompasses more than just rap music. We can now talk about 'rock/rap,' 'Hip-Hop soul,' 'rapso,' and 'Hip House.'"[13] Indeed, THEESatisfaction's aesthetic has been referred to variably in reviews as "electronic psychedelic soul,"[14] "psychedelic space-rap/jazz,"[15] "hippie-hop,"[16] and "hologram funk" (the last two terms used mostly to describe SassyBlack's independent work), descriptions that might place the music in proximity to Krims's nomenclature, but also evidence of increasingly diffuse permutations of styles encompassing music with rapped lyrics.[17]

It is crucial at this point to note that groups Krims identifies within the "jazz/bohemian rap" style are made up predominantly of men—De La Soul, Jungle Brothers, A Tribe Called Quest, Black Sheep, The Roots, Organized Konfusion, the one exception being the Black Eyed Peas.[18] Alienation as an Afrofuturist concept also resonates with Krims's description of the "incomprehensibility" of the "jazz/bohemian rap" genre, with its complex instrumental backing and conscious social attentiveness—all features of music discussed in this chapter, from Janelle Monáe, THEESatisfaction, and Stas THEE Boss. Distinctly from Krims who minimizes the gendered make-up of jazz/bohemian's "college-boy" designation, which he says "probably [refers] less to any official demographic than to a perceived projection of artistic arrogance among some of this genre's devotees," I find it crucial to emphasize women's omission from his taxonomic category.[19] Although Krims attempts to counter the gendered "college-boy" stigma, the examples he provides under the jazz/bohemian heading are primarily from men, an omission I attribute equally to a known habit of minimizing women's roles in hip-hop history, while giving merit to the gendered implication that perhaps the music I explore here—despite its thematic, aesthetic, and even geographic parallels—nevertheless emerges from slightly varied motivations. The music in this chapter touches on various existing genres, but, having been created by queer-identifying women, situates itself intergenerically—in and between genres. In this sense, the music analyzed here does not fit neatly within Krims's "genre system," though it shares some similarities with the intellectual proselytizing cited in his description. We could linger on reasons why some might find aspects of this music "incomprehensible," as Krims says, given the previous discussion of the necessarily encoded obscurantism that serves as a security measure for *quare* communication.

Though an emphasis on the musical qualities of the genre may not distinguish Krims's groups from those I examine here—groups comprised entirely of women—gender is certainly important to flag, not merely because the vocal qualities of the respective rappers differ in pitch and inflection, but also because women rappers advance different values as is their music received distinctly from men's. According to Pough, "As the most visible element of Hip-Hop culture currently, rap sets the tone for a lot of what the dominant society recognizes as Hip-Hop culture. Having women's voices represented via Hip-Hop in the larger public sphere opens the door for a wealth of possibilities in terms of the validation of the Black female voice and Black women's agency."[20] Indeed, Cheryl Keys, Tricia Rose, and Gwendolyn Pough have respectively devoted significant analyses of music by women rappers as distinct from men, though these analyses focus more heavily on the subject matter of the lyrics and visual imagery in videos, hardly touching on the music's production qualities.[21] This gap may have arisen partially because the producers for many women rappers in the 1990s and early 2000s were men—exceptions include Missy Elliott, whose producer credits on hers and other artists' albums helped establish her distinctive reputation, and Sugar Hill Gang's producer Sylvia Robinson. Another reason, as Nancy Guevara, Keir Keightley, Christina Verán, Gwendolyn Pough and others have suggested, is that women hardly figure in the history of electronic music because sound engineering has been coded historically as a masculine pastime. Put simply, women's contributions have been ignored both in hip hop and in electronic music production histories.[22] Regardless of the reason, this chapter continues, in the vein of these authors' revisionist histories, to fill this gap by examining music performed and produced by women. Though hardly encompassing, the issues arising here around sound and its signification may well carry over into analyses of broader repertoire, whether music described as rap, queer, jazz/bohemian, or otherwise. Indeed, in his introduction to a special issue of *Music Theory Online* on the subject of Kendrick Lamar's *To Pimp A Butterfly*, Philip Ewell writes: "Because of rap's general lack of functional harmonies or other traditional music theoretical parameters, music theory must conceive of new ways for analyzing the music."[23] Thus, genre studies and music analysis, alike, become enriched by the variety of analytical approaches this music invites.

Perhaps some readers would rather skip this debate about musical style and genre entirely, with the view that pinning down genre is not the most important aspect in music analysis. However, I would argue to the contrary that unless some boundaries are defined to situate this music historically, socially, aesthetically, and geographically, the analysis would be lacking in sufficient context to prove convincing. Whereas some readers may agree that nailing down aesthetic and compositional process is important in analysis, these same readers may not be convinced that the artists' race, gender, or sexuality plays any significant role in musical reception. I will therefore address each of these in turn in the paragraphs below.

Before rap music grew to its current mainstream popularity in the United States, prior to the 1980s, when music by Black musicians incorporated electronic elements, musical categories and labels took a cue from racially segregated music marketing strategies.[24] As Guthrie P. Ramsey notes, "most of the genres were historically marketed and mass mediated in the culture industry as 'race records,'" and forms of Black electronic musical creation were often situated within rock 'n' roll music, jazz, and rhythm & blues, for music prior to the 1970s, where disco and hip hop continued in the same vain thereafter—a history I examined in part in chapters 6 and 10.[25] It is no coincidence that these genres are considered part of the musical "vernacular" in the academy, even if jazz was conceived as an experimental art form (often implicitly distinct from the presumed neutrality of "experimental music" more broadly).[26] Though gospel, as a form of religious expression, does not neatly align within this narrative of Black musical forms, Alisha Lola Jones writes of how "Historically black colleges and universities (HBCUs) have and still privilege Eurocentric music education over black vernacular music education." She provides an example of gospel composer Reverend Richard Smallwood, who continued a practice of performing gospel, soul, and R&B on HBCU college campuses "despite threats of expulsion."[27] Amy Coddington has traced the desegregation of popular music charts, once exclusive of rap and hip hop, through their 1980s incorporation into generic pop music charts. Today, rap and hip hop are hailed as the most dominantly listened to genres within the entire pop spectrum at large.[28] In short, although Black musicians have always had a hand in music's innovative electronic technologies, the domain of electronic music is not easily reconciled either within commonplace academic textbooks on electronic music or in how histories of Black music are typically narrated by music critics.

When one believes that there exists a unified Western art musical canon, vernacular music and electronic music alike both appear tangential. Electronic music forms a niche in the history of music—a niche within which women remain minoritarians, and Black women, with few exceptions, all but disappear. Obviously, if one instead looks to commercial music and pop culture as the canon, then rap music and electronic dance music are heavily represented in recent accounts, both academically and commercially. Where electronica and hip hop alike are genres with origins in underground, urban, live performance, the early 2000s witnessed a boom of mainstream popularity for both, for example, in the characteristic timbre, groove, and build common to the EDM "soar" dominating pop music charts, like Psy's "Gangnam Style" (2012). Robin James identifies this "maximalist" period with the thrill and pleasure of greater risk represented by the motto "You only live once"—YOLO. In contrast, the period after 2016, coincides with an apparent hesitancy, where musical aesthetics mimic a wider sociopolitical and economic anxiety over risk management.[29]

James observes that this shift in economic, political, musical, and psychological outlook, paved the way for a greater appreciation of musical remixability, presumably because it limited musical innovation to incremental variations of familiar favorites.[30] Perhaps this wider context provides one explanation for the reason composers in the early 2010s would have once again become interested in the electrosexual tropes, such as the climax mechanism and the feminized voice, as we hear in Hodkinson and Rønsholdt's *Fish & Fowl* (chapter 5). Whereas earlier in the twentieth century composers mainly collected concrete samples to recreate electrosexual atmospheres, Hodkinson and Rønsholdt pick up on this trend of remixability, revitalizing a familiar trope within the palette of their own individual compositional styles. James also identifies this turn to individualism as a symptom of how recent neoliberalism measures value, moving away from quantification towards more speculative, affective, and "qualitative forms of rationality"[31]—still measured and controlled, not from external force but leveled internally and individually. This may be one reason that, despite *Fish & Fowl*'s production links to pop culture, its sampled lineage remains yoked to experimental electroacoustic scenes.

With increased popularity of musical remixability and "chill," comes "ever more privatized listening."[32] Giving Billie Eilish's "Bad Guy" (2019) as a recent example, James demonstrates how privatized listening through headphones has given way to an emerging valuation of music that presses up against the surface, as if occurring within the listener's intimate personal head-space. James ties Eilish's music to "early 20th century microphone technology [which] birthed a new vocal style as singers no longer needed to forcefully project their voices to be heard."[33] This microphonic utility combined with its desired "intimacy," easily recalls Ferrari's anecdotal experiments from chapter 3. Yet, where Ferrari, Normandeau and others manipulate this utility for their own personal usage, "bedroom artist" Eilish does not fabricate or co-opt an unknown "other" place to elicit pleasure or relay as her own; rather, she invites listeners into her space, to share her actual bedroom. With music "apparently made with earbuds in mind," she encourages listeners to utilize her own sounds of intimacy in creating a room of their own.[34] Close-miking in "Bad Guy" and other songs by then seventeen-year-old Eilish aims to elicit ASMR (an autonomous sensory meridian response) in susceptible listeners, amplifying mouth sounds, clicking, breathiness, and other perceptual associations to physical closeness.[35] This global attention to private listening began earlier in the characteristically Northwest hip-hop sound of THEESatisfaction, and their predecessors Shabazz Palaces, Digable Planets, and others. A regional, intimate invitation coupled with home remixing, extending from hip hop onto the international stage.

Debuting in 2008, with their first releases on Bandcamp, THEESatisfaction gained wider popularity through a 2011 collaboration with jazz rap inspired Shabazz Palaces, a hip-hop duo formed in Seattle by Ishmael Butler, formerly of Digable Planets named in the "bohemian rap" section of Krims's typology. With James's sociocultural assessment of the rise of "chill," it is easy to hear how THEESatisfaction rose to fame. Seattle's hushed hip-hop beats lend to wind-down listening, and their detailed musical textures beg especially close-up listening. Where electroacoustic music at one point examined the reach of the acoustic horizon, for example Takemitsu's *Vocalism A·I* (1956) or Normandeau's *Jeu* (1989), both of which play with panning to the far reaches of audible space, more recent intimacies magnify sounds in their closest proximity. In this range, acousmatic pleasure aligns with the rise in ASMR-inducing videos.

Looking to gender-specific instances, James observes that, whereas the 1970s saw women audibly revolting against the old adage "little girls should be seen and not heard," by the mid-2010s, this rhetoric reached its apex.[36] *Fish & Fowl* was composed in 2011, at the end of this trend. Figure 5.1 shows the spectrogram completely maxed out for the majority of the piece. We might understand this fashionable "maximalism" as one expression of Hodkinson's feminist-inclined "sonic writing." *Fish & Fowl* settles into its cold hushed tones only to set off its climactic arrivals, making them sound even more charged. Still, the peaks coincide with the normalized way of experiencing the electrosexual—as a "protagonist," to use Rønsholdt's word. Even when we are told who is performing the music, the technique of sexualizing through splicing alters how we come to hear the music. The idealized feminized voice cannot break from its tropic significance, even when evading the singular climax mechanism or when incorporating more risqué sounds, like those of whips, stiletto heels, or heartbeats.

This maximal risk to both culture and hearing becomes a force of unrestrained expression—loud, deafening shouting. *Fish & Fowl*'s protagonist could be reacting to and shielding herself from (potential) harm. She slaps, she hits, and she fights back. Even her voice rings forth, not chopped or cut up as in so many other layers of this work and others explored in part I. For once, she is uninterrupted. With this in mind, Shields's *Apocalypse* pushes against all allegedly conceded boundaries—against languages, chronologies, genres, instrumentations, volumes, sonic barriers, ranges, genders, and, of course, also sexual expressions, sexualizations, and sexual experiences. During this period, the 1990s to early aughts, women's voices carried the maximalist trend: "the mainstream media explicitly praises women for their loudness, especially when that loudness is the sound of feminists busting out of the misogynist chains that once held them back."[37] Yet, when risk became a growing factor, when women risked tipping the scales, making real change, musically, politically, socially, maximalism became negatively characterized by and large for precisely the qualities that made this music stand out. As a wall of sound, it came to be characterized as noise, as dissonant. This dissonance emerges when elements are perceived as qualitatively too strong and therefore over-represented. Maximalism could only be remedied with moderation, as James says, with "moderation as an ethical ideal."[38] And maybe this is the reason these electrosexual works have not caught the same attention as these composers' other works. *Apocalypse* is thoroughly feminized in the ways it grates against established patriarchal norms in the US, and, therefore, remains quintessentially and inexcusably feminist within the US context even now. There are too many roles and too many styles. The penis is too big. Too many orgasms. Overall Shields is just too present. The piece simply attempts to accomplish too much, and, as a result, it has never been performed and never will be.

In response to soaring maximalism, musicians and producers came to moderate loudness, rescinding the climax to maintain a perceived aural balance or "chill." Listeners use music to sooth their nerves, music absent the dissonant tensions of hypersexualized and racialized tropes. Think, for example, of Taylor Swift's neutered *Evermore* or *Folklore*, both from 2020. Critics have ascribed an audible "emotional distance" to these albums, all in moderation.[39] James observes that because of a gendered and racialized double standard, compounded by the historical polarized extremes used to describe Black women—being too loud, either *hyper*-sexual or *a*-sexual— they have not been accorded the same access to moderation. "Society both values and devalues loud women and feminized loudness: women can be loud when their voices rationally sync up with the rhythms of white supremacist patriarchy and amplify its signal, but this loudness is dangerous when their voices don't rationally sync up with those rhythms and introduce dissonance and noise."[40] Electrosexual music does not merely respond to evident social, political, and historical misogynoir; electrosexual music pushes these currents right up to the surface. When revisiting music's sexual themes, we must take care not to limit the dynamic range of this music, to delimit its nuanced interpretation by bracketing out seemingly extraneous or ill-fitted materials.

James hears the subdued hush of Solange's music as a moderating vehicle and an acceptable medium through which to convey the harsh truths of injustice and oppression addressed in her lyrics. Similarly, Stas's "Before Anyone Else" delivers reminisces of the queer melancholy, dressed in a monotonous, moderating drone. The "wobble" signals the song's only melodic departure, which, even in its limited range, significantly punctuates its difference, its eccentricity. While artists from Janelle Monáe to THEESatisfaction embrace sexual expression, recent artists neither reinforce nor counteract the established electrosexual tropes—(1) the climax mechanism; (2) the feminized voice; or (3) their canonization. In fact, quare codes like the "wobble warp" do not even engage with earlier maximalist framings. Strict boundaries between hip hop and electroacoustic culture, upheld on either side of the divide, ensure that these two musical lineages, while similar in production technique, do not mingle in terms of content or even context really.

In many ways, this book reinforces the dominating perspective of European and North American musical histories at the expense of trends elsewhere that might not fit as neatly into this electrosexual narrative. For example, Japanese Francophile Toru Takemitsu's *Vocalism A·I*, or "Love," from 1956, employs a man's voice alongside a woman's—one of few electroacoustic examples to do so and the only electrosexual example that I have found to play with the perceived gendered distantiation I have described throughout this book.[41] In another context, the Middle East has recently drawn international attention for its hip hop and especially trap music, characterized by an aggressive lyrical delivery and boosted bass.[42] Though trap first emerged in Atlanta in the early 1990s, at a time when such bombastic beats would have been compared to equally raucous rap on the West Coast of the US, today, within the musical backdrop of the maximal–banal divide, this contrast—this dissonance—seems out of place, almost contradicting or even resisting "the West's" current musical narrative.[43] Within these many pan-global dividing musical and contextual hearings, the halting chill of American chart-topping pop emerges as yet another means by which to distance and bracket off those whose music already occupies the margins. Still, rather than dismiss these examples as ancillary, we can recognize such outliers as flickering glimpses across a shifting veil of coded and later newly encoded subtlety.

Conclusion

Since the 1990s, musicologists exploring sexuality have written of the teleological drive to climax in music, a phenomenon often aligned with masculinist sexual aggression identified by the pulse, intensity, and form of the music—for example, Susan McClary's foundational analysis of Beethoven reviewed in my introduction. This analysis caused much consternation among musicologists, not merely because it sullied the reputation of this presumed disembodied music by insisting on its marked bodily signification but, I would argue, more so because the analysis "engendered" the sounds *with masculinity* at a time when sex sounds were so often presumed to be quintessentially feminine. Shortly after McClary's analysis, Corbett and Kapsalis published their observations on the prevalence of "The Female Orgasm in Popular Sound." They declared that, like in pornography, women's sexual pleasure in popular music equated with the "quality and volume of the female vocalizations"—something we hear throughout this book. By 2022, I thought we would know better.

By now, most music scholars and creators should know to recognize McClary's infamous "beanstalk" looming over the façade of feminized sexual pleasure. We know that sound cannot be anything other than socially and contextually dependent, and that when women are not consulted about a phenomenon that so critically impacts them, we ought to be suspicious, concerned even. We know how centrally "The Tyranny of the Female-Orgasm Industrial Complex" figures into recent music, its histories and theories—especially those concocted in North America and Europe.[1]

I first stumbled onto the subject of this book in graduate school, in a class on analysis of contemporary music. The professor asked that we analyze a piece composed in the last ten years. I was clueless about how to discover new music—Twitter wasn't a thing and my Google searches only brought up pop songs. I ended up searching through Naxos Music Library, which is where I encountered a piece of music that I just couldn't figure out: Hodkinson and Rønsholdt's *Fish & Fowl*. This entire project emerged in answer to the fateful questions that piece raised for me: What are those sounds? Who is making them? How do I make sense of what I hear?

I am fully willing to admit that what motivated me then differs significantly from my present motivations for this book. Back then I experienced a jolt when I thought of this music and its mysteries, a feeling most aptly described in Suzanne Cusick's hallmark essay, "On a Lesbian Relationship with Music": "I am in search of a union with that music, and I am most alive when I find it. *That* union is more like the supposed thrill of sex."[2] But Cusick also factored a measure of kinship into this feeling, her own sensed distance between "the pleasure of sex with a woman," which is more "'like it' than anything I have experienced with a man."[3] In this electrosexual research, I, too, have come to better understand my own sensations, to acknowledge that some pieces work to arouse pleasure in me better than others. Regrettably, for my own mental soundness, music that deliberately worked to incite sexual associations tended to fall short of the mark. The gimmicks of the climax mechanism and the feminized voice have been all-too-present, too persistent to have drowned out the happy accidents of this work.

Back when I began this work, I didn't believe it when I read Hannah Bosma's report that when voices sound in electroacoustic music, or, as she called it, electrovocal music, they are most commonly voices sounded by women in works composed by men.[4] Bosma tallied the numbers, looking through piles of electroacoustic recordings or "digging in crates," to use the common musical nomenclature. Yet, I didn't want to believe. In fact, I was motivated by my beliefs, by my disbelief in these gendered imbalances. I set out to show that women, too, have contributed, that women, too, compose electroacoustic music, electrovocal music, and what I have now called electrosexual music. And they have, which is how I discovered Alice Shields's music and Annea Lockwood's music—remarkable for the use of their own voices. But I was wrong. Almost all the electroacoustic works I encountered that sought to allude in some way to sex feature women's voices. Looking only to electroacoustic music, as Bosma and I did, there can be no argument that this music is composed overwhelmingly by white men; European men, like the French Ferrari, and North American men, like the Canadian Normandeau. Bosma was right, and still, I don't want to believe it. I doubted her as I doubted myself.

This book is a record. That is, the book records a heritage of sexual imaginings in recent music from the path to discovery through this inevitably pessimistic resolve. I write in honor of McClary and Bosma who already showed us the trends years ago and whose predictions of the harmful consequences of such tendencies we continued to ignore. This book revisits calls to action in electroacoustic music from Andra McCartney, Elizabeth Hinkle-Turner, Tara Rodgers, and Robert Gluck, who repeatedly criticized the genre's insularity owing to national, geographic, gendered, ethnic and racial exclusions.[5] It seems to me necessary and also futile to recall their criticisms, since even today, when searching among the entries for "electroacoustic" in academic databases, such insularity persists.[6] And yet, this impervious racist misogyny—its permanence—is precisely why this book needs to be. Because alongside these important interjections, the book also records the staid and stagnant narratives of electroacoustic hegemony. This book stands as a record of the electrosexual catalog, whose definition as I finish these last pages is already in flux.

At the risk that readers will scrutinize and therefore set aside this book like its predecessors, here are some takeaways: (1) there is yet to be a plausible theory confirming the occurrence of sex acts in music; nevertheless, (2) even when no bodies are visible, music takes on set gendered roles based in the expectations set up from listeners' (including composers') prior experiences of hearing sex; (3) these roles are difficult to dismantle because of expectations listeners hold of music from particular eras, even as time lapses.

Although we do not have tools to pinpoint precise sexual qualities in music, the tropic normalization of certain sonic features causes correlations between works to emerge such that composers may be inspired to replicate certain tropes from elsewhere, and we analysts welcome these reappearances into our broader theories. Perhaps Strauss's "Symphonia Domestica" (with its famous lovemaking scene between the composer and his wife) inspired this book's featured "Symphony," Schaeffer and Henry's *Symphonie pour un homme seul*, with its notable fourth movement "Erotica" (chapter 1). We might interpret Schaeffer and Henry's *Symphonie* in the philosophical sense of "sounding together," where each movement contributes to the sense of balance necessary to constitute a sound man, *homme seul*. As symphony, the work captures sonically all of the acousmatic features veiled by the physical form, and articulates the moderate "soundness" of man in all his sensibilities. If we imagine the *Symphonie* as two harmonious halves, where the first six movements complement the last six, then, "Erotica's" inverse complement comes in the movement "Apostrophe." Indeed, we hear complementary sounds in both. Both movements feature women's laughter, where "Erotica" uses this voice as fundamental to or grounding the affective drive of the presumed sexual encounter. In "Apostrophe," the cackling woman also features noticeably, but more instrumentally, and where "Erotica" omits the audible man's voice leaving us to presume his role, "Apostrophe" features both characters in dialogue. "Apostrophe" opens with the woman's hushed singing. Soon she is greeted by the man. Each time the two speak, the conservation ends in a punctuating flourish: the woman's voice rises in a sudden, shrill outburst. Though the words are hardly intelligible (I hear "actually no" and "*physiment*"), we might interpret the woman's outburst as an emotional response to their conversation. This unexpected "outburst" so captivates our attention because in comparison to the conversational exchange in the middle register at a medium volume, it sounds contrastingly, with a relatively wide range, a very high register and very loud. Comparing this movement to "Erotica," which is relatively static, "Apostrophe" is quite spirited. Its piano also features the characteristic "um-pah" but is confronted each time by frenetic, dissonant chaos, a pairing comparable to its human counterparts. And compared to the dynamic envelope of "Apostrophe," the stasis of "Erotica" presents controlled movement—one implication of "man," as I suggested in chapters 1 and 12. Another understanding also stems from the position of the "Erotica" movement as one part of a whole. Sex is not an independently verifiable concept.

Electrosexual music does not necessarily replicate the trope of the insatiable feminized voice. Enmeshed in the "electroacoustic game" of whether or not to refer, many hearings use the trope as a baseline from which to deviate and resound otherwise. These pieces rely on feminized associations to confirm the essentializing trope of women's sexuality, while also playing with the voice's gendered associations to disturb firm binaries. Still other pieces do not include the voice or women at all, but merely allude to the trope, while subverting its implicit heterosexual gaze. Barry Truax's 2004 work for percussion and electronics *Skin & Metal* is written for a shirtless performer dressed in all leather, complete with chest harness, gloves, and wrist- and neckbands. Though *Skin & Metal* does not use the voice, pulsation and the climax mechanism still feature prominently in this ode to homoerotic S&M (and the recorded material provides more avenues for discussing consent in performance, as well). There are also many works by women to which the composers attribute sexually explicit content, including: Annea Lockwood's *Tiger Balm* (1970), Alice Shields's electronic opera *Apocalypse* (1991–1994), Pauline Oliveros's *Skin* (1991)[7] and her soundtrack to Annie Sprinkle's film (1992), as well as Juliana Hodkinson and Niels Rønsholdt's jointly composed *Fish & Fowl* (2011). These works differ from the stereotypical sampled sexuality evident in electrosexual works like "Erotica" from *Symphonie pour un homme seul*, *Presque rien avec filles*, *Jeu de langues*, and others, because their respective creators had intentions to attend to consent. Lockwood, Shields, Oliveros, and Hodkinson/Rønsholdt take care to identify and give attribution to the sounds and the individuals who utter these sounds. They still play the electroacoustic game, but in these instances, everyone involved knows they are playing. Neither is there a question whether they are of consenting age.

Earlier chapters in this book demonstrated the potential harm, even violence performed by sound cutting and splicing but, as these examples show, sampling can be done with care as well as attention and attribution to a sample's source and cause. TLC paid homage to Donna Summer as a sexually allusive predecessor while adding their own voices to the commentary on sexual availability, exploitation, and appropriation. Similar to Truax's music, gender—even when stereotyped as TLC presents it—need not become reduced to a man/woman binary. Unlike Summer/Moroder, Shields, or Truax, the cumulative compositional approach of TLC's "I'm Good at Being Bad" shows a process shared by the instrumentals and vocals, a creative non-hierarchy where neither receives more credit for how the song is put together. Such cumulative collaboration became standardized as studios moved into private homes, and vocalists took on the work of producing their own music—as we observe in THEESatisfaction's music.

Even recently several works have utilized electronic tools to enhance musico-sexual experiences where participants and musical creators willingly collaborate. Miya Masaoka (b. 1958) writes of the "Vagina as the Third Ear," as a unique and significant site of musical engagement explored in her work "Vaginated Chairs" (2017).[8] Inspired by Pauline Oliveros's theory of deep listening, the piece invites participants to explore pitched sensations through their nether regions as mediated through chairs vibrating at different frequencies. "Vaginated Chairs" capitalizes on music's embodied connections, moving beyond ears as *de facto* sites of musical sensation while engaging directly with sexual stimulation. Conversely, Jen Kutler's album *Disembodied* (2019) deliberately attends to the veil between source/cause and effect.[9] Each track on the album features the work of a different artist, "a series of audio pieces generated by the vibrations and movements captured by an electronic ring worn on the finger of a feminine spectrum body while bringing themselves to orgasm." Whereas Masaoka seeks to engage directly with the audience in their live participation, Kutler moves to sonify veiled distance by extracting data points from her masturbating subjects with the goal to "de-objectify" the body through "de-sexualization experiments." The vibrational data collected from Kutler's participants is abstracted, first, into data points in a spreadsheet, and then again when Kutler assigns seemingly unrelated timbres and textures to these data via the Pure Data programming language. Where composers of early electrosexual music—Schaeffer, Henry, Ferrari, even Normandeau—skirt consent and thereby threaten their audible subjects with sexual coercion, more recent composers attend to the consent of performers and listeners alike to offer an alternative hereditary lineage extending at least since the Sprinkle–Oliveros collaboration to the present and even toward futures of sounding sex.

Masaoka describes her "Vaginated Chairs" in technical terms as "sonification," a domain usually reserved for realizations of scientific data in sound. However, recognizing the historical tendency to situate sound as *mere* representation of some more tangible metric, the composer leans into contemporary critiques of science and technology that understand "intelligence" not as an independently verifiable metric, but rather, like art and sex, as "interpretation" and "expression" of social, cultural, historical, and political forces. Where electronic and computer hardware facilitate large-scale automation of certain sonifying practices, the technology of this "artificial intelligence" largely depends on parameters that were selected and weighted manually by humans, as does Kutler's disclosed manipulation of the tracks on *Disembodied*.

It is by now common knowledge that electronic music can inherit gendered associations through the voice via imitation and replication, a form of mimesis that arouses listener expectations and familiarity with common tropes. This book shows that these frequent allusions to feminized vocalizations and the climax mechanism no longer refer merely to *extra*musical sexual situations; rather, the sexual tone in this music is cultivated from a common contour, duration, pulse, and musical form that has ultimately become a *convention*, in essence, *canonized* as sexual expression in music.

While showing this confluence of composing sex in electronic music, I also showed that there exist many diverse engagements with sex within these conventions, including, of course, potentially diverse *responses* to the music from listeners, who are also sexed, gendered, racialized, socialized, and enculturated. Combining several analytical methods based on musical commonalities highlights similarities across categories or genres of electronic music, such that we can redraw a disciplinary axis of inclusion that insists on sexuality as a crucial and robust paradigm at the intersection of music, society, and technology.

But the work is not finished. There are many areas due further examination. For example, returning to eroticism and exoticism in orchestral music to analyze compositions from Strauss to Zemlinsky from any other perspective than their white male creators. Another area ripe for inquiry examines electroacoustic works that are not explicitly sexual via the electrosexual tropes mapped out throughout this book—that is, the pulse, gait, periodic and harmonic development, rising pitch level, and envelope of the teleological "climax mechanism" commonly heard in both sex and music. Computer algorithms still cannot compose music idiomatically, and software for analyzing music (e.g., using spectrograms or music information retrieval [MIR]) still requires a guiding human hand. Musical creation, analysis, and listening are therefore still mostly informed by human judgments and values. According to contemporary critiques of artificial intelligence, one cannot determine "intelligence" without admitting the criteria of which intelligence is a measure.[10] And, similarly, sex is an expression and one of many human traits. Measuring sex becomes a relative measure of traits like any other, of personality, physical fitness, mental acuity, and so on. The time is therefore ripe to turn a spotlight on the long-accepted norms of electronic musical composition that rest on shaky ethics regarding gender and sexuality, and their interlocking racializing and marginalizing tropes, to question assumptions grounding a domain whose compositional standards have been determined by a relatively small and homogeneous demographic. The expansive breadth of electronic possibilities awaits.

Appendix A: Alice Shields, *Apocalypse*, Track List

Scene (Liner Notes)	Title	CD Track	Page in Libretto
Part I			
1	Sacrifice	1	1
2	The Land of the Dead	1	3
3A (3)	Conception	1	5
3B (4)	On the Dark Plain	1	5
3C (5)	Push	1	7
Part II			
4A (6)	The Sea	2	9
4B (7)	Approach	2	11
4C (8)	First Greeting	3	11
4D (9)	Dialog 1	3	13
4E (10)	Sea Dance 1	3	14
4F (11)	Second Greeting	3	15
4G (12)	Dialog 2	3	16
4H (13)	Sea Dance 2	3	18
4I (14)	Third Greeting	3	19
4J (15)	Dialog 3	3	20
4K (16)	Sea Dance 3	3	21
4L (17)	Looking	3	22
4M (18)	Corpse	3	23
4N (19)	Here	4	24
4O (20)	The Flower	(omitted)	25
4P (21)	Truth	5	26

(Continued)

Scene (Liner Notes)	Title	CD Track	Page in Libretto
Intermission			
Part III			
4P-2 (Continuation)			28
5A	The Forest	(omitted)	28
5B	This is Good	(omitted)	28
5C	Prince of Rain	(omitted)	29
5D	Mantra 1	(omitted)	30
5E	Supplication 1	(omitted)	30
5F	Mantra 2	(omitted)	32
5G	Supplication 2	(omitted)	33
5H	Mantra 3	(omitted)	35
5I	Supplication 3	(omitted)	35
5J	Undulating 1	(omitted)	38
5K	Forward 1	(omitted)	38
5L	Undulating 2	(omitted)	38
5M	Forward 2	(omitted)	39
5N	Undulating 3	(omitted)	39
5O	Forward 3	(omitted)	39
5P (21)	Shiva	(omitted)	40
6A (39)	On the Dark Mountain	6	41
6B (40)	Dialog 1	6	41
6C (41)	First Naming	6	43
6D (42)	Dialog 2	6	44
6E (43)	Second Naming	6	45
6F (44)	Dialog 3	6	46
6G (45)	Third Naming	6	47
6H (46)	Final Question: Aeon	7	48
7A (47)	The Approach	(omitted)	49
7B (48)	Soliloquy	(omitted)	50
7C (49)	The Arrival	7 Retitled: "The Great Mother's Revenge"	51

Scene (Liner Notes)	Title	CD Track	Page in Libretto
7D (50)	Dismemberment Dance	8 Retitled: "Dismemberment and Eating"	52
8A (51)	Regrets	(omitted)	52
8B (52)	Someone, I Say, Will Remember Us	9	58
9A (53)	Invocation	(omitted)	59
9B (54)	The Epiphany of the Clouds	(omitted)	60
9C (55)	Someone Spoke of Your Death	10	61
Part IV			
10A (56)	Reactions	10	64
10B (57)	Beginning of the End	(omitted)	65
10C (58)	Apocalypse Song	11	65
11A (59)	Heat Drum	12	66
11B (60)	Organ Screaming	13	66
Epilogue			
12 (61)	The Dawn Wind	14	79

Appendix B: Works Incorporated in *Fish & Fowl* Listed with Instrumentation and Approximate Duration, and Organized Chronologically

Table A.1

Works by Niels Rønsholdt

Composition	Instrumentation
"Torso," scene from *Triumph*, a micro opera (2006)	woman's voice, clarinet, double bass, percussion, electronics, in collaboration with Signe Klejs [9 minutes]
HammerFall (2006)	piano, saxophone, percussion (including horse whip, hand thrown fire crackers, wine glasses for breaking, small balls made of paper, small stones/pebbles), w. optional lighting [8:30 minutes]
Die Wanderin (2007)	violin, piano, percussion, audio playback (footsteps and ambient chords), w. optional video [10 minutes]
Scores and sound files available at www.nielsroensholdt.dk/ and through publisher Edition•S	

Table A.2

Works by Juliana Hodkinson

Composition	Instrumentation
In Slow Movement (1994)	flute, clarinet, violin, cello, piano, guitar, percussion [14 minutes]
sagte er, dachte ich (1999)	flute, clarinet, viola, cello, piano, guitar, percussion [10 minutes]
what happens when (1999)	soprano, bass recorder, guitar [6 minutes]
Why Linger You Trembling In Your Shell? (1999)	violin and percussion with egg-shells, down feathers, and table-tennis balls [10 minutes]
Harriet's Song (2001)	singing woman violist and percussion (hanging objects such as chimes, keys, a transparent freezer-bag filled with milk, a small music box, metal chains) [10 minute]
Scores available through publisher, Edition Wilhelm Hansen. Juliana Hodkinson's originals scores, sketches, project folders, and digital documentation of all the works have been archived at the Royal Library in Copenhagen since 2016.	

Notes

INTRODUCTION

1. Aryelle Siclait and Madeline Howard, "Guess How Many Rihanna Songs Are on This Sex Playlist," *Women's Health*, August 4, 2020, https://www.womenshealthmag.com/sex-and-love/a19935020/sex-playlist/; Natalie Maher and Nicole DeMarco, "40 Songs That Probably Aren't on Your Sex Playlist (but Should Be)," *Harper's Bazaar*, May 16, 2020, https://www.harpersbazaar.com/culture/art-books-music/a23538483/best-sexy-songs/; Sophie Harris, "The 50 Best Sexy Songs Ever Made," *Time Out New York*, January 18, 2021, https://www.timeout.com/newyork/music/best-sexy-songs; The Editors, "72 Songs to Add to Your Sex Playlist Right Now," *Cosmopolitan*, August 6, 2020, https://www.cosmopolitan.com/celebrity/news/songs-to-have-sex-to.

2. Jacob Smith, *Vocal Tracks: Performance and Sound Media* (Berkeley: University of California Press, 2008).

3. Simon Emmerson, *Living Electronic Music* (New York: Routledge, 2007); Deniz Peters, Gerhard Eckel, and Andreas Dorschel, eds., *Bodily Expression in Electronic Music: Perspectives on Reclaiming Performativity* (New York: Routledge, 2012).

4. Margaret Schedel, "Electronic Music and the Studio," in *The Cambridge Companion to Electronic Music*, 2nd ed., ed. Nick Collins and Julio d'Escrivan (Cambridge: Cambridge University Press, 2017), 25–39, https://doi.org/10.1017/9781316459874.004; Keir Keightley, "'Turn It Down!' She Shrieked: Gender, Domestic Space, and High Fidelity, 1948–59," *Popular Music* 15, no. 2 (1996): 149–177.

5. Schedel, "Electronic Music and the Studio," 26.

6. Francisco Kröpfl, "Electronic Music: From Analog Control to Computers," *Computer Music Journal* 21, no. 1 (1997): 26, https://doi.org/10.2307/3681212.

7. Jennifer Iverson, *Electronic Inspirations: Technologies of the Cold War Musical Avant-Garde* (Oxford, New York: Oxford University Press, 2019), 24.

8. Luciana Galliano, *Yogaku: Japanese Music in the 20th Century* (Lanham, MD: Scarecrow Press, 2002), 166.

9.   Mark Dike DeLancey, Rebecca Neh Mbuh, and Mark W. DeLancey, *Historical Diction-ary of the Republic of Cameroon* (Lanham, MD: The Scarecrow Press, 2010), 65; Eileen Southern, *Biographical Dictionary of Afro-American and African Musicians* (Westport, CT: Greenwood Press, 1982), 30, http://archive.org/details/biographicaldict00sout.

10. James Andean, "Electroacoustic Mythmaking: National Grand Narratives in Electroacous-tic Music," in *Confronting the National in the Musical Past*, ed. Elaine Kelly, Markus Man-tere, and Derek Scott (New York: Routledge, 2018), 138–150.

11. Andean, "Electroacoustic Mythmaking," 145.

12. Frances Morgan, "Pioneer Spirits: New Media Representations of Women in Electronic Music History," *Organised Sound* 22, no. 2 (August 2017): 238–249, https://doi.org/10.1017/S1355771817000140.

13. Emmerson, *Living Electronic Music*, 75–77. In a 1952 letter to Pierre Boulez, John Cage recalls the thrill of meeting Pierre Schaeffer, though he does not regard *Symphonie* highly in its initial incarnation: "Merce choreographed part of the *Symphonie pour un Homme Seul* (a terrible piece) for a festival at Brandeis University." John Cage, *The Selected Letters of John Cage* (Middletown, CT: Wesleyan University Press, 2016), 166.

14. Edward T. Cone, "The Creative Artist in the University," *American Scholar* 16, no. 2 (1947): 192–200; Milton Babbitt, "The Composer as Specialist [First Published as 'Who Cares If You Listen?']," *High Fidelity* 8 (1958): 38–40, 126–127; Pierre Boulez, "Alea," *Perspectives of New Music* 3, no. 1 (Autumn–Winter 1964): 42–53; for a critique of this position, see Susan McClary, "Terminal Prestige: The Case of Avant-Garde Music Composition," *Cul-tural Critique*, no. 12 (Spring 1989): 57–81.

15. Ellie M. Hisama, "Getting to Count," in *Music Theory Spectrum* 43, no. 2 (2021): 359, emphasis added.

16. Danielle Sofer, "Categorising Electronic Music," *Contemporary Music Review* 39, no. 2 (2020): 231–251.

17. Tara Rodgers, "Tinkering with Cultural Memory: Gender and the Politics of Synthesizer Historiography," *Feminist Media Histories* 1, no. 4 (October 1, 2015): 8, https://doi.org/10.1525/fmh.2015.1.4.5.

18. Andrew Hugill, "The Origins of Electronic Music," in *The Cambridge Companion to Elec-tronic Music*, 2nd ed., ed. Nick Collins and Julio d'Escrivan (Cambridge: Cambridge Uni-versity Press, 2017), 7.

19. Theodor W. Adorno, "The Form of the Phonograph Record," trans. Thomas Y. Levin, *Octo-ber* 55 (Winter 1990): 56–61.

20. Michel Chion, *Guide to Sound Objects. Pierre Schaeffer and Musical Research* (originally pub-lished as *Guide des Objets Sonores: Pierre Schaeffer et La Recherche Musicale*), trans. John Dack and Christine North (Paris: Editions Buchet/Chastel, 1983), 39, http://monoskop.org/images/0/01/Chion_Michel_Guide_To_Sound_Objects_Pierre_Schaeffer_and_Musical_Research.pdf.

21. Denis Smalley, "Defining Transformations," *Interface* 22, no. 4 (November 1993): 285, https://doi.org/10.1080/09298219308570638.

22. Brian Kane, *Sound Unseen: Acousmatic Sound in Theory and Practice* (Oxford: Oxford University Press, 2014), 129, 193.

23. Jean-Jacques Rousseau, *The Confessions of J. J. Rousseau, . . . Part the Second. To Which Is Added, a New Collection of Letters from the Author. Translated from the French. In Three Volumes . . .*, trans. Anonymous, vol. 1 (London: Printed for G. G. J. and J. Robinson, and J. Bew, 1790), 85–90.

24. Jennifer Lynn Stoever, *The Sonic Color Line: Race and the Cultural Politics of Listening* (New York: NYU Press, 2016).

25. Smith, *Vocal Tracks*.

26. Kane, *Sound Unseen*.

27. Rousseau, *The Confessions of J. J. Rousseau, . . . Part the Second*, 1:85–89.

28. Jean-Jacques Rousseau, *Seconde partie des confessions de J. J. Rousseau, citoyen de Geneve. Edition enrichie d'un nouveau recueil de ses lettres . . .*, vol. 4 (A Neuchatel: l'imprimerie de L. Fauche-Borel, imprimeur du Roi, 1790).

29. Sarah Kofman and Mara Dukats, "Rousseau's Phallocratic Ends," *Hypatia* 3, no. 3 (1989): 123–36.

30. Kane, *Sound Unseen*, 109.

31. Kane, *Sound Unseen*, 110.

32. Kane, *Sound Unseen*, 109.

33. Waylon Lewis, "Beautiful Agony," September 19, 2013, http://www.elephantjournal.com /2013/04/beautiful-agony-8-songs-that-parallel-the-rhythmic-path-to-orgasm/.

34. The works Lewis mentions are Karl Jenkins, Allegretto from *Palladio*; Carl Orff, "O Fortuna" from *Carmina Burana*; Richard Wagner's "Flight of the Valkyries" (the orchestral prelude from *Die Walküre*, the second opera in the monumental *Der Ring des Nibelungen*); Aram Khachaturian, "The Sabre Dance" from the ballet *Gayane*; Hans Zimmer and Lisa Gerrard, "The Battle" from the film *The Gladiator*; Georges Bizet, "Habañera" from *Carmen*; Diana Ross, "Love Hangover"; and DJ Paul Van Dyk, "Sensation White 2004."

35. Susan McClary, *Feminine Endings: Music, Gender, and Sexuality* (Minneapolis: University of Minnesota Press, 1991), 112.

36. Susan McClary, *Feminine Endings*, 130.

37. Tia DeNora, "Music and Erotic Agency," *Body and Society* 3, no. 2 (1997): 55.

38. McClary subscribes to the feminist separatist agenda, precluding that the "simulation of sexual desire and fulfillment," which she finds pervasive in Western music, is evocative only of a violent and sexual aggression that is inherent to the *male* orgasm. McClary, *Feminine*

Endings, 126–127. On the feminist separatist agenda, see Jill Johnston, *Lesbian Nation: The Feminist Solution* (New York: Simon and Schuster, 1973).

39. Gayle Rubin, "Thinking Sex: Notes for a Radical Theory of the Politics of Sexuality," in *Deviations: A Gayle Rubin Reader* (Durham, NC: Duke University Press, 2011), 170.

40. Gayle Rubin, "Thinking Sex," 170.

41. Ruth Solie, ed., *Musicology and Difference: Music, Gender, and Sexuality in Music Scholarship* (Berkeley: University of California Press, 1993), 17.

42. Georges Bataille, "Kinsey, the Underworld and Work," in *Death and Sensuality: A Study of Eroticism and the Taboo. Translated by Mary Dalwood* (New York: Walker, 1962), 160. For an illuminating exploration of Bataille's resistance to Kinseyan methods, see James Mark Shields, "Eros and Transgression in an Age of Immanence: George Bataille's (Religious) Critique of Kinsey," *Journal of Religion & Culture* 13 (2000): 175–186.

43. Jerrold Levinson, "Erotic Art and Pornographic Pictures," *Philosophy and Literature* 29, no. 1 (2005): 229, https://doi.org/10.1353/phl.2005.0009.

44. Robert Normandeau, program notes to *Jeu de langues* (2009); Robert Normandeau, "Jeu de Langues," *electrocd*, 2012, https://electrocd.com/en/piste/imed_12116-1.3.

45. Morgan, "Pioneer Spirits."

46. Friedrich Nietzsche, *The Gay Science*, trans. Walter Kaufmann (New York: Random House, 1974), 60 quoted in Irigaray, "Veiled Lips," 104. The quotes of Nietzsche in Irigaray come from, and hence serve also as commentary on, Derrida's reading of Nietzsche. Jacques Derrida, *Spurs: Nietzsche's Styles*, trans. Barbara Harlow (Chicago: University of Chicago Press, 1979), quoted in Luce Irigaray, "Veiled Lips," 123n8.

47. Luce Irigaray, "Veiled Lips," 105.

48. Friedrich Nietzsche and Josefine Naukoff, *The Gay Science* (Cambridge: Cambridge University Press, 2001), 60.

49. Gilles Deleuze and Richard Howard, *Proust and Signs* (London: Continuum, 2008), 89.

50. Deleuze and Howard, *Proust and Signs*, 89–90.

51. Gilles Deleuze, *Difference and Repetition*, trans. Paul Patton (New York: Columbia University Press, 1994), 209.

52. Deleuze *Difference and Repetition*, 208, 209.

53. Marcel Proust, *A la Recherche du temps perdu* (Paris: Bibliothèque de la Pleide, 1988), II:622, quoted in Deleuze, *Difference and Repetition*, 89.

54. Hannah Bosma, "The Electronic Cry: Voice and Gender in Electroacoustic Music" (PhD diss., University of Amsterdam, 2013).

55. Derek B. Scott, *From the Erotic to the Demonic: On Critical Musicology* (New York: Oxford University Press, 2003), 19.

56. Alice Jardine, *Gynesis: Configurations of Women and Modernity* (Ithaca, NY: Cornell University Press, 1985), 119.

57. Jardine, *Gynesis*, 118.

58. This opinion is summarized in the "hearing-as" theory of expression in music, advanced primarily by Alan Tormey, *The Concept of Expression: A Study in Philosophical Psychology and Aesthetics* (Princeton: Princeton University Press, 1971).

59. Roger Scruton, *Beauty: A Very Short Introduction* (Oxford: Oxford University Press, 2011), 125.

60. Roger Scruton, "Representation in Music," *Philosophy* 51, no. 197 (1976): 273–287.

61. Soren Kierkegaard, "The Immediate Erotic Stages or the Musical Erotic," in *Either/Or: A Fragment of Life*, trans. Alastair Hannay (London: Penguin UK, 2004), 113–284.

62. Hans Maes, "Why Can't Pornography Be Art?," in *Art and Pornography: Philosophical Essays*, ed. Jerrold Levinson and Hans Maes (Oxford: Oxford University Press, 2012), 38–39.

63. Maes, "Why Can't Pornography Be Art?," 39.

64. Bernard Williams, ed., *Obscenity and Film Censorship: An Abridgement of the Williams Report* (Cambridge: Cambridge University Press, 1982), 8.2.

65. Susanne K. Langer, *Philosophy in a New Key: A Study in the Symbolism of Reason, Rite, and Art* (Cambridge, MA: Harvard University Press, 1949), 202.

66. Langer, *Philosophy in a New Key*, 202–203.

67. Stephen Davies, *Musical Meaning and Expression* (Ithaca, NY: Cornell University Press, 1994), 125.

68. Davies, *Musical Meaning and Expression*, 125–126.

69. Langer, *Philosophy in a New Key*, 209.

70. A number of recent publications address music's current intersections with sound. As Georgina Born describes in her edited collection: "The subject matter of the present collection congeals at the intersection of a series of related terms: music, sound, space, and how these phenomena have been employed to create, mark, or transform the nature of public and private experience." Georgina Born, *Music, Sound and Space: Transformations of Public and Private Experience* (Cambridge: Cambridge University Press, 2013), 2–3.

71. Michael C. Rea, "What Is Pornography?," *Noûs* 35, no. 1 (2001): 30.

72. See a video of a recent performance, https://www.youtube.com/watch?v=ZBgA84DoHEI&feature=youtu.be.

73. Linda Williams, *Hard Core: Power, Pleasure, and the "Frenzy of the Visible"* (Berkeley, CA: University of California Press, 1989), 95; John Corbett and Terri Kapsalis, "Aural Sex: The Female Orgasm in Popular Sound," *TDR (1988-)* 40, no. 3 (1996): 103, https://doi.org/10.2307/1146553.

74. "Pornography, n.," in *Oxford English Dictionary* (Oxford: Oxford University Press, June 2020), http://www.oed.com/viewdictionaryentry/Entry/148012.

75. Maes, "Why Can't Pornography Be Art?," 32.

76. Danielle Sofer, "Specters of Sex: Tracing the Tools and Techniques of Contemporary Music Analysis," *Zeitschrift Der Gesellschaft Für Musiktheorie [Journal of the German-Speaking Society of Music Theory]* 17, no. 1 (2020): n55, https://doi.org/10.31751/1029.

77. Catherine MacKinnon, "Feminism, Marxism, Method, and the State: An Agenda for Theory," *Signs* 7, no. 3 (1982): 515–544; see also Luce Irigaray, *This Sex Which Is Not One*, trans. Catherine Porter (Ithaca, NY: Cornell University Press, 1985), 176.

78. Andrea Dworkin, Catherine MacKinnon, and Women Against Pornography, *The Reasons Why: Essays on the New Civil Rights Laws Recognizing Pornography as Sex Discrimination* (New York: Women Against Pornography, 1985).

79. Simon Emmerson, "EMAS and Sonic Arts Network (1979–2004): Gender, Governance, Policies, Practice," *Contemporary Music Review* 35, no. 1 (2016): 21–31.

80. Hannah Alderfer, Beth Jaker, and Marybeth Nelson, *Diary of a Conference on Sexuality* (New York: Faculty Press, 1982).

81. Gayle Rubin, "Blood under the Bridge: Reflections on 'Thinking Sex,'" in *Deviations: A Gayle Rubin Reader* (Durham, NC: Duke University Press, 2011), 194–223.

82. Linda Williams, "A Provoking Agent: The Pornography and Performance Art of Annie Sprinkle," *Social Text*, no. 37 (1993): 117, https://doi.org/10.2307/466263.

83. Cady Lang, "President Trump Has Attacked Critical Race Theory. Here's What to Know About the Intellectual Movement," *Time*, September 29, 2020, https://time.com/5891138/critical-race-theory-explained/.

84. Lisa Nakamura, "Indigenous Circuits: Navajo Women and the Racialization of Early Electronic Manufacture," *American Quarterly* 66, no. 4 (2014): 920, https://doi.org/10.1353/aq.2014.0070.

85. Nielsen Music and Erin Crawford, "2017 Year-End Music Report: U.S.," January 3, 2018, 2, https://www.nielsen.com/us/en/insights/report/2018/2017-music-us-year-end-report/. In terms of consumer sales, hip hop already topped the charts in 1999, Justin A. Williams, *Rhymin' and Stealin': Musical Borrowing in Hip-Hop* (Ann Arbor: University of Michigan Press, 2013), 11.

86. Tricia Rose, *The Hip Hop Wars: What We Talk about When We Talk about Hip Hop—and Why It Matters* (New York: Perseus Book Group, 2008), 114–115.

87. Rose, *The Hip Hop Wars*, 115.

88. Mickey Hess, "Was Foucault a Plagiarist? Hip-Hop Sampling and Academic Citation," *Computers and Composition* 23 (2006): 280–295.

89. Gordon Colin and Michel Foucault, "Prison Talk: An Interview," in *Power-Knowledge: Selected Interviews & Other Writings—1972–1977* (New York: Pantheon Books, 1980), 52.

90. Sofer, "Categorising Electronic Music."

91. Sofer, "Categorising Electronic Music," 5; Robert J. Gluck, "Electroacoustic, Creative, and Jazz: Musicians Negotiating Boundaries," in *ICMC*, 2009. In personal communication, Robert Gluck, the author of this singular essay, explained that this was the last time he sought to present such cross-genre correlations to an electroacoustic audience, April 11, 2021.

92. https://electrocd.com/en/artistes/tous.

93. It is admittedly difficult to determine racial or ethnic origins by appearance alone. None of the featured composers indicate identifying characteristics in their biographies.

94. https://www.composerdiversity.com/.

95. Megan DeJarnett, "The ICD Internal Review Part 1: There's No Policy Like No Policy," *Megan DeJarnett* (blog), March 20, 2021, https://megandejarnett.com/2021/03/20/icd-internal-review-part-1/.

96. Kane, *Sound Unseen*, 110.

97. Smith, *Vocal Tracks*, 20.

98. Katherine Hayles, *How We Became Posthuman: Virtual Bodies in Cybernetics, Literature, and Informatics* (Chicago: University of Chicago Press, 1999); see also Joseph Auner, "'Sing It for Me': Posthuman Ventriloquism in Recent Popular Music," *Journal of the Royal Musical Association* 128, no. 1 (2003): 98–122.

99. Hayles, *How We Became Posthuman*, 2, 3.

100. Alexander G. Weheliye, "'Feenin': Posthuman Voices in Contemporary Black Popular Music," *Social Text* 20, no. 2 (June 1, 2002): 23.

101. Weheliye, "Feenin," 23.

102. Sofer, "Categorising Electronic Music."

103. Susan C. Cook, "'R-E-S-P-E-C-T (Find Out What It Means to Me)': Feminist Musicology & the Abject Popular," *Women and Music: A Journal of Gender and Culture* 5 (2001): 140.

104. Alexander G. Weheliye, *Phonographies: Grooves in Sonic Afro-Modernity* (Durham: Duke University Press, 2005), 7.

105. Guthrie P. Ramsey, *Race Music: Black Cultures from Bebop to Hip-Hop* (University of California Press, 2004), 96–97.

106. Sara Ahmed, *Living a Feminist Life* (Durham, NC: Duke University Press, 2016), 1.

107. Hannah Bosma, "Bodies of Evidence, Singing Cyborgs and Other Gender Issues in Electrovocal Music," *Organised Sound* 8, no. 1 (April 2003): 5–17, https://doi.org/10.1017/S135577180300102X.

108. Ahmed, *Living a Feminist Life*, 2.

109. George E. Lewis, "Improvised Music after 1950: Afrological and Eurological Perspectives," *Black Music Research Journal* 16, no. 1 (1996): 91, https://doi.org/10.2307/779379.

CHAPTER 1

1. Pierre Schaeffer, *In Search of Concrete Music*, trans. Christine North and John Dack (Berkeley: University of California Press, 2012), 48.

2. R. Murray Schafer, "Schizophonia," in *The New Soundscape: A Handbook for the Modern Music Teacher* (Ontario: Don Mills, 1969), 43.

3. Pierre Schaeffer, "La these naïve du monde: L'époche," in *Traité Des Objets Musicaux: Essai Interdisciplines* (Paris: Editions du Seuil, 1966), 265.

4. Brian Kane, "L'Objet Sonore Maintenant: Pierre Schaeffer, Sound Objects and the Phenomenological Reduction," *Organised Sound* 12, no. 01 (April 2007): 15–24, https://doi.org/10.1017/S135577180700163X.

5. Geoffrey Cox, "'There Must Be a Poetry of Sound That None of Us Knows . . .': Early British Documentary Film and the Prefiguring of Musique Concrète," *Organised Sound* 22, no. 2 (August 2017): 172–186, https://doi.org/10.1017/S1355771817000085.

6. Nicolas Vérin and Pierre Henry, "Entretien Avec Pierre Henry," *Ars Sonora* 9 (1999), http://www.ars-sonora.org/html/numeros/numero09/09c.htm.

7. Laura Anderson, "Musique Concrète, French New Wave Cinema, and Jean Cocteau's Le Testament d'Orphée (1960)," *Twentieth-Century Music* 12, no. 2 (September 2015): 198, https://doi.org/10.1017/S1478572215000031.

8. Alfred Charles Kinsey, Wardell B. Pomeroy, Clyde E. Martin, *Sexual Behavior in the Human Female* (Bloomington: Indiana University Press, 1953); Alfred Charles Kinsey, Wardell B. Pomeroy, and Clyde E. Martin, *Sexual Behavior in the Human Male* (Bloomington: Indiana University Press, 1948).

9. Smith, *Vocal Tracks*, 55–56, 54.

10. Smith, *Vocal Tracks*, 54.

11. Sylvie Chaperon, "L'histoire Contemporaine Des Sexualités En France," *Vintième Siècle: Revue d'Histoire* 75 (September 2002): 48n2; Georges Bataille, *L'Erotisme* (Paris: Les Editions de Minuit, 1957).

12. "Ecouter les partitions de Pierre Schaeffer n'a rien à voir avec la civilité musicale puérile et honnête. Il s'agit peut-être de découvrir un continent sonore aussi vierge que l'était l'Ile de Robinson Crusoe." Invitation for the École Normale concert, 1950, GRM Archives, quoted in Alexander John Stalarow, "Listening to a Liberated Paris: Pierre Schaeffer Experiments with Radio" (PhD diss., Davis, CA, University of California, Davis, 2017), 146.

13. Pierre Schaeffer, *La musique concrète* (Paris: Presses Universitaires de France, 1973), 22.

14. It is important to mention that these categories are not "human" and "non-human," as is written elsewhere, for example, in Peter Manning, *Electronic and Computer Music* (Oxford: Oxford University Press, 2013), 24. For Schaeffer the crux of the matter is sound *in relation* to man, and not the realm of possibilities implicit from a "non-human" context.

15. Schaeffer, *In Search of Concrete Music*, 55.

16. Marc Battier, "What the GRM Brought to Music: From Musique Concrète to Acousmatic Music," *Organised Sound* 12, no. 3 (2007): 195.

17. Daniel Teruggi, "Technology and musique concrète: the technical developments of the Groupe de Recherches Musicales and their implication in musical composition," *Organised Sound* 12, no. 3 (2007): 213.

18. Emmerson, *Living Electronic Music*, 67–68.

19. Teruggi, "Technology and musique concrete," 217.

20. The "initial series" refers to the first version of *Symphonie pour un homme seul* (1950). Schaeffer, *In Search of Concrete Music*, 58.

21. Gascia Ouzounian, *Stereophonica: Sound and Space in Science, Technology, and the Arts* (Cambridge, MA: MIT Press, 2021), 107.

22. A macron added to the "A" to indicate metrical emphasis.

23. Times refer to Track 4 of Pierre Schaeffer, *L'Œuvre musicale Volume 2*, INA-GRM ina c 1006–09 cd, 1990.

24. Listen to the 1950–1951 version of *Symphonie pour un homme seul* on Pierre Schaeffer, *L'Œuvre musicale*, vol. 2, INA-GRM ina c, 1990.

25. Italo Calvino, *Under the Jaguar Sun*, trans. William Weaver (New York: Harcourt Brace, 1988), 33–64, quoted in Adriana Cavarero, *For More Than One Voice: Toward a Philosophy of Vocal Expression* (Stanford: Stanford University Press, 2005), 2.

26. "Symphonie Pour Un Homme Seul (Béjart)," accessed July 15, 2019, https://www.youtube .com/watch?v=V8dCdQ3iTrc.

27. "Brigitte Bardot—Moi Je Joue," accessed July 15, 2019, https://www.youtube.com/watch ?v=hJ2mGqL1HEk.

28. Bardot's outro solo begins at 1:18 here, https://www.youtube.com/watch?v=hJ2mGq L1HEk.

29. R. Murray Schafer, "Schizophonia," in *The New Soundscape: A Handbook for the Modern Music Teacher* (Ontario: Don Mills, 1969), 43.

30. Schaeffer, *Traité Des Objets Musicaux*, 91.

31. Schaeffer, *In Search of Concrete Music*, 25.

32. Schaeffer, *In Search of Concrete Music*, 52.

33. Patrick O'Reilly, *Dancing Tahiti* (Paris: Nouvelles Editions Latines, 1977), 27.

34. O'Reilly, *Dancing Tahiti*, 27.

35. Linda Phyllis Austern, "The Exotic, the Erotic, and the Feminine," in *The Exotic in Western Music*, ed. Jonathan Bellman (Boston: Northeastern University Press, 1998), 27.

36. David F. García, *Listening for Africa: Freedom, Modernity, and the Logic of Black Music's African Origins* (Durham: Duke University Press, 2017), 75.

37. Daniel Defoe, *The Adventures of Robinson Crusoe* (London, 1862), 145. Defoe, *The Adventures of Robinson Crusoe*, 188.

38. Jardine, *Gynesis*, 219.

39. Jardine, *Gynesis*, 219.

40. Jardine, *Gynesis*, 219.

41. Statement by Tournier, quoted in Eugénie Lemoine-Luccioni, *Partage Des Femmes* (Paris: Editions du Seuil, 1976), 104.

42. Jardine, *Gynesis*, 221–222.

43. Gayatri Chakravorty Spivak, "Theory in the Margin: Coetzee's Foe Reading Defoe's Crusoe/Roxana," in *Consequences of Theory*, ed. Jonathan Arac and Barbara Johnson (Baltimore, MD: Johns Hopkins University Press, 1991), 155; Samir Amin, "Unequal Development: An Essay on the Social Formations of Peripheral Capitalism," trans. Brian Pierce, *ASA Review of Books* 4 (1978): 365, https://doi.org/10.2307/532256.

44. Spivak, "Theory in the Margin," 155–156.

45. Spivak, "Theory in the Margin," 157.

46. Spivak, "Theory in the Margin," 158; Jacques Derrida, *Margins of Philosophy*, trans. Alan Bass, reprint edition (Chicago: University of Chicago Press, 1984).

47. Spivak, "Theory in the Margin: Coetzee's Foe Reading Defoe's Crusoe/Roxana," 159, emphasis in the original.

48. Spivak, "Theory in the Margin," 159.

49. Deleuze and Howard, *Proust and Signs*, 50–51; Deleuze, *Difference and Repetition*, 106.

50. Gilles Deleuze and Félix Guattari, *A Thousand Plateaus: Capitalism and Schizophrenia* (Minneapolis, MN: University of Minnesota Press, 2007), 233; 244; 272.

51. Gilles Deleuze and Félix Guattari, *Anti-Oedipus: Capitalism and Schizophrenia*, trans. Robert Hurley, Mark Seem, and Helen R. Lare (New York: Viking, 1977); quoted in Jardine, *Gynesis*, 211.

52. Deleuze and Guattari, *A Thousand Plateaus*, 150. I return to Artaud and the colonial project in chapter 7.

53. Deleuze and Guattari, *A Thousand Plateaus*, 150, 151; quoted in Jardine, *Gynesis*, 212.

54. Jardine, *Gynesis*, 215.

55. Deleuze and Guattari, *A Thousand Plateaus*, 277; quoted in Jardine, *Gynesis*, 216.

56. Jardine, *Gynesisty*, 216.

57. Simone de Beauvoir, *The Second Sex* (New York: Vintage Books, 1973), 301.

58. Deleuze, *Difference and Repetition*, xix.

59. Jacques Lacan, *The Four Fundamental Concepts of Psychoanalysis: The Seminar of Jacques Lacan*, ed. Jacques-Alain Miller, trans. Alan Sheridan, Book 11 (New York: W. W. Norton, 1981), 165.

60. Elizabeth Grosz, *Volatile Bodies: Toward a Corporeal Feminism* (Bloomington: Indiana University Press, 1994), 165.

61. Emmerson, *Living Electronic Music*, 18.

62. Kane, *Sound Unseen*, 150.

63. Shanté Paradigm Smalls, "Queer Hip Hop: A Brief Historiography," in *The Oxford Handbook of Music and Queerness*, ed. Fred Everett Maus and Sheila Whiteley (Oxford: Oxford University Press, 2018), 5, https://doi.org/10.1093/oxfordhb/9780199793525.013.103.

CHAPTER 2

1. Rebecca Lentjes, "Surreal Conjunctions," *VAN Magazine*, July 26, 2017, https://van-us.atavist.com/surreal-conjunctions; Annea Lockwood, "Sound Mapping the Danube River from the Black Forest to the Black Sea: Progress Report, 2001–03," *Soundscape: Journal of Acoustic Ecology* 5, no. 1 (2004): 32–34.

2. Joel Chadabe, *Electric Sound: The Past and Promise of Electronic Music* (Upper Saddle River, NJ: Prentice Hall, 1997), 152.

3. Elizabeth Hinkle-Turner, *Women Composers and Music Technology in the United States* (Aldershot: Ashgate, 2006), 32.

4. Frank J. Oterion, "Annea Lockwood Beside the Hudson River," *NewMusicBox*, January 1, 2004, https://nmbx.newmusicusa.org/annea-lockwood-beside-the-hudson-river/.

5. Lentjes, "Surreal Conjunctions."

6. Hinkle-Turner, *Women Composers and Music Technology in the United States*, 32.

7. Angela Kane, "Richard Alston: Twenty-One Years of Choreography," *Dance Research: The Journal of the Society for Dance Research* 7, no. 2 (1989): 21, https://doi.org/10.2307/1290770.

8. Annea Lockwood, *Early Works 1967–82*, EM1046CD, 2007, compact disc.

9. John Young, "Source Recognition of Environmental Sounds in the Composition of Sonic Art with Field-Recordings: A New Zealand Viewpoint" (PhD diss., University of Canterbury, 1989), 479.

10. Lockwood likely refers here to Schaeffer's notion of *musique concrète*.

11. Christopher Hasty, "Segmentation and Process in Post-Tonal Music," *Music Theory Spectrum* 3, no. 1 (1981): 54–73.

12. Stephen McAdams and Albert Bregman, "Hearing Musical Streams," *Computer Music Journal* 3, no. 4 (1979): 26.

13. Dora A. Hanninen, "Associative Sets, Categories, and Music Analysis," *Journal of Music Theory* 48, no. 2 (2004): 150.

14. Young, "Source Recognition of Environmental Sounds in the Composition of Sonic Art with Field-Recordings: A New Zealand Viewpoint," 348.

15. Young, "Source Recognition of Environmental Sounds," 482.

16. Lockwood, *Early Works 1967–82*, 9.

17. Danielle Sofer, "Strukturelles Hören? Neue Perspektiven Auf Den 'idealen' Hörer," in *Geschichte und Gegenwart Des Musikalischen Hörens: Diskurse—Geschichente(n)—Poetiken*, ed. Klaus Aringer et al. (Freiburg: Rombach Verlag, 2017), 107–132.

18. Don Ihde, *Listening and Voice: Phenomenologies of Sound* (Albany, NY: SUNY Press, 2007), 32–33.

19. Ihde, *Listening and Voice*, 33.

20. Tara Rodgers, *Pink Noises: Women on Electronic Music and Sound* (Durham, NC: Duke University Press, 2010), 125.

21. This was at least one motivation behind the mid-twentieth-century faithfulness to high fidelity, which gave special attention to the recording's manner of presentation.

22. Lockwood, "Sound Mapping the Danube River from the Black Forest to the Black Sea: Progress Report, 2001–03"; Annea Lockwood, "What Is a River? Down the Danube by Ear," *Soundscape: Journal of Acoustic Ecology* 7, no. 1 (2007): 43–44. This is not to dismiss the erotic potential of hearing the rushing waters of the Danube; recall the intensely erotic glance exchanged between Count Friedrich and the girl in Josef Freiherr von Eichendorff's *Ahnung und Gegenwart* (1815).

23. Nothing in Young's transcription of his interview with the composer or subsequent notes on the piece suggest the composer's presence or masturbation, perhaps this was noted "off the record." Young, "Source Recognition of Environmental Sounds in the Composition of Sonic Art with Field-Recordings," 509.

24. Steven Connor, "Panophonia" (Pompidou Centre, February 22, 2012), http://www .stevenconnor.com/panophonia/panophonia.pdf.

25. Lockwood, *Early Works 1967–82*, 10.

26. Marion A. Guck, "Edward T. Cone's 'The Composer's Voice': Beethoven as Dramatist," *College Music Symposium* 29 (1989): 8.

27. Stephanie Jordan, "British Modern Dance: Early Radicalism," *Dance Research: The Journal of the Society for Dance Research* 7, no. 2 (1989): 9, https://doi.org/10.2307/1290769.

28. A. Porter, *The Financial Times*, August 15, 1972, quoted in Kane, "Richard Alston," 21–22.

29. A. Porter, quoted in Kane, "Richard Alston," 22.

30. Lockwood, *Early Works 1967–82*, 10.

31. Guck, "Edward T. Cone's 'The Composer's Voice': Beethoven as Dramatist," 9.

32. The example of (sex) act and actor maps onto Guck's analysis and her reluctance to make a definitive distinction between persona and act; Guck, "Edward T. Cone's 'The Composer's Voice': Beethoven as Dramatist," 10n6.

33. J. Griffith Rollefson, "Hip Hop Interpolation: Rethinking Autochthony and Appropriation in Irish Rap" (Department of Music Seminar Series, Maynooth University, November 23, 2018), https://www.maynoothuniversity.ie/music/events/griff-rollefson-hip-hop-interpolation-rethinking-autochthony-and-appropriation-irish-rap.

CHAPTER 3

1. Brigitte Robindoré, "Interview with an Intimate Iconoclast," *Computer Music Journal* 22, no. 3 (1998): 8.

2. Battier, "What the GRM Brought to Music."

3. Schaeffer, *Traité Des Objets Musicaux*.

4. Robindoré, "Interview with an Intimate Iconoclast," 11.

5. Eric Drott, "The Politics of Presque Rien," in *Sound Commitments: Avant-Garde Music and the Sixties*, ed. Robert Adlington (Oxford: Oxford University Press, 2009), 155.

6. In distinction from himself, Ferrari says that "Cage was not terribly concerned with sound, but rather with concepts." Robindoré, "Interview with an Intimate Iconoclast," 11.

7. Ferrari's four works under the heading *Presque rien* are: *Presque rien No.1—le lever du jour au bord de la mer*, *Presque rien No. 2—ainsi continue la nuit dans ma tête multiple*, *Presque rien avec filles*, and *Presque rien No. 4—La remontée du Village*, collected and reissued recently on, Luc Ferrari, *Presque rien*, GRM 2012—REGRM 005 et éditions MEGO—INA GRM, double vinyl discs. A more recent work is also titled, *Apres Presque rien* (2004).

8. Robindoré, "Interview with an Intimate Iconoclast," 13; Barry Truax, "Electroacoustic Music and the Soundscape: The Inner and the Outer World," in *Companion to Contemporary Musical Thought*, ed. John Paynter (New York: Routledge, 1992), 374–398; R. Murray Schafer, *The Soundscape: Our Sonic Environment and the Tuning of the World* (Rochester, VT: Destiny Books, 1977). For more on soundscape composition in practice, see the analysis of Barry Truax's *Song of Songs* in chapter 9.

9. Gilles Deleuze and Félix Guattari, *What Is Philosophy?*, trans. Hugh Tomlinson and Graham Burchell (New York: Columbia University Press, 1991), 164.

10. Deleuze and Guattari, *What Is Philosophy?*, 166.

11. Kane, *Sound Unseen*, 131.

12. Clara Hunter Latham, "Rethinking the Intimacy of Voice and Ear," *Women and Music: A Journal of Gender and Culture* 19 (2015): 125–132; Lacan, *The Four Fundamental Concepts of Psychoanalysis*; Mladen Dolar, *A Voice and Nothing More* (Cambridge, MA: MIT Press,

2006); Roland Barthes, "The Grain of the Voice," in *Image—Music—Text*, trans. Stephen Heath (London: Fontana Press, 1977), 179–189.

13. Italo Calvino, *Under the Jaguar Sun*, trans. William Weaver (New York: Harcourt Brace, 1988), 33–64; Cavarero, *For More Than One Voice*, 2.

14. Cavarero, *For More Than One Voice*, 3.

15. In distinction from Derrida, who delimits discussion to the framework imposed by his subject of critique, and Husserl, referring to presence only in the context of a speaker hearing their *own* voice and in turn reducing the voice to an internal (personal) dialogue, Cavarero prioritizes the mode of communication, sound, over the mode of reception, hearing, to express the phenomenon of hearing the voice as a form of collective participation. Cavarero, *For More Than One Voice*, 218–220.

16. A later publication goes to greater lengths to explain the difficulty of assessing *phoné* as a simultaneously embodying signifier, metaphorical representation, and disembodied sound. Kane envisions *phoné* in a tripartite juncture as a result of the many intersections and displacements of *logos* ("the content of [the voice's] utterances"), *echos* ("sound of the voice"), and *topos* (site of emission, whether source or perceived origin), all oriented and mediated by *techné*. Brian Kane, "The Model Voice," *Journal of the American Musicological Society* 68/3, "Colloquy: Why Voice Now" (2015): 671–677. Kane laments that musicological investigations, in order to avoid a reduction of voice to *logos*, remain stuck in the space between *echos*, and *topos*.

17. Kane, *Sound Unseen*, 156.

18. Scruton, "Representation in Music," 273–274.

19. "The vestige comes into focus when one considers the sensible remainder that persists after the idea has departed. . . . It describes the mark left behind after some event has occurred." Jean-Luc Nancy, *The Muses*, trans. Peggy Kamuf (Stanford: Stanford University Press, 1996), 96; quoted in Kane, *Sound Unseen*, 130, emphasis added.

20. Annie Goh, "Sounding Situated Knowledges: Echo in Archaeoacoustics," *Parallax* 23, no. 3 (July 3, 2017): 283–304, https://doi.org/10.1080/13534645.2017.1339968.

21. Anne Sauvagnargues, "Deleuze and Guattari as VJay: Digital Art Machines" (The Dark Precursor: International Conference on Deleuze and Artistic Research, Ghent, Belgium, November 10, 2015).

22. Connor, "Panophonia," 1; see also Steven Connor, *Dumbstruck: A Cultural History of Ventriloquism* (Oxford: Oxford University Press, 2000), 353.

23. Kane, *Sound Unseen*, 131, emphasis in the original.

24. Kane, *Sound Unseen*, 130.

25. Kane, *Sound Unseen*, 131.

26. Kane, *Sound Unseen*, 133.

27. Tim Hodgkinson, "An Interview with Pierre Schaeffer—Pioneer of Musique Concrète," *Recommended Records Quarterly*, 1987, 9; Drott, "The Politics of Presque Rien," 148.

28. Drott, "The Politics of Presque Rien," 153; Jacqueline Caux, *Almost Nothing with Luc Ferrari*, trans. Jérôme Hansen (Berlin/Los Angeles: Errant Bodies Press, 2012), 11.

29. Williams, *Hard Core*, 124.

30. The telegraphone, invented around 1900 by Vlademar Poulsen, was the first device capable of recording sound magnetically. "The Poulsen 'Wireless,'" *The Graphic: An Illustrated Newspaper*, Jan. 12, 1907, 62, https://archive.org/stream/graphicillustrat1907unse#page/n53/mode/2up

31. Drott, "The Politics of Presque Rien," 154.

32. Bosma, "Bodies of Evidence, Singing Cyborgs and Other Gender Issues in Electrovocal Music." In distinction from his electroacoustic works, Ferrari's text-based compositions remain relatively open to gender, for example, *Pornologos 2* (1971), a text score for private individuals or private groups.

33. Translation is taken from Ferrari's website, and includes all grammatical and typographical errors as published therein. http://www.lucferrari.org, first accessed April 29, 2015. The entry has since been updated with an English translation and a coy undated addendum: "New text: 'Suddenly the composer realizes that the title comprises the word "girls" and he decides to justify that by the presence of a truly present girl.'" http://lucferrari.com/en/analyses-reflexion/presque-rien-avec-filles/, accessed 4 December, 2018. Since Ferrari passed away in 2005, it is unclear where this addendum originates. The text was presumably added by Ferrari's widow, Brunhild Meyer-Tormin, who assumed control over his estate (hypothesized in personal communication with harpsichordist Mahan Esfahani).

34. Robindoré, "Interview with an Intimate Iconoclast," 15.

35. Caux, *Almost Nothing with Luc Ferrari*, 11.

36. Caux, *Almost Nothing with Luc Ferrari*, 52.

37. Emmerson, *Living Electronic Music*, 16.

38. Emmerson, *Living Electronic Music*, 23–25.

39. Leonard Meyer, *Style and Music: Theory, History, and Ideology* (Chicago: University of Chicago Press, 1989), 10; see also 345–349.

40. Kaja Silverman, *The Acoustic Mirror: The Female Voice in Psychoanalysis and Cinema* (Bloomington: Indiana University Press, 1988), 6.

41. John R. Searle, *Making the Social World: The Structure of Human Civilization* (Oxford: Oxford University Press, 2010), 45.

42. Searle, *Making the Social World*, 94.

43. Searle, *Making the Social World*, 95.

44. Searle, *Making the Social World*, 103. Searle insists that "acceptance or recognition" must always come together, so as to avoid equating acceptance with approval.

45. Searle, *Making the Social World*, 8.

46. Deleuze and Guattari, *What Is Philosophy?*, 180.

47. Malaika Fraley, "Berkeley: Renowned Philosopher John Searle Accused of Sexual Assault and Harassment at UC Berkeley," *East Bay Times*, March 23, 2017, https://www.eastbaytimes.com/2017/03/23/berkeley-renowned-philosopher-john-searle-accused-of-sexual-assault-and-harassment-by-former-cal-aide/.

48. Ahmed, *Living a Feminist Life*, 96.

49. Silverman, *The Acoustic Mirror*, 96.

50. On the reputability of this myth, see Kane, *Sound Unseen*, 45–72, in particular 46–45.

51. Kane, *Sound Unseen*, 150.

52. Kane, *Sound Unseen*, 212.

53. Kane, *Sound Unseen*, 213.

54. Kane, *Sound Unseen*, 149.

55. Kane, *Sound Unseen*, 150.

56. Kane, *Sound Unseen*, 199.

57. Ihde, *Listening and Voice*, 34.

58. Ihde, *Listening and Voice*, 35.

59. Ihde, *Listening and Voice*, 35.

60. Ihde, *Listening and Voice*, 37.

61. Searle, *Making the Social World*, 66.

62. Searle, *Making the Social World*, 69n4. Searle explored this notion more completely within an intentional context in *Intentionality: An Essay in the Philosophy of Mind* (Cambridge: Cambridge University Press, 1983).

63. Kane, *Sound Unseen*, 131.

64. Daniel Caux, "Les Danses Organiques du Luc Ferrari," *L'Art Vivant 43*, July 1973, 30. My translation.

65. John Hassard, "Representing Reality: *Cinéma Vérité*," in *Organization-Representation: Work and Organizations in Popular Culture*, ed. John Hassard and Ruth Holliday (London: Sage Publications, 1998), 41–66. In electronic music, producers do this as well. Alexander G. Weheliye calls this "sonic 'cinéma vérité' that depict the 'reality' of current technologically mediated life worlds.'" Weheliye, "Feenin," 33.

CHAPTER 4

1. Alexa Woloshyn, "Wallace Berry's Structural Processes and Electroacoustic Music," *EContact!* 13, no. 3 (2010), https://econtact.ca/13_3/woloshyn_onomatopoeias.html.

2. *Lieux inouïs* (Experientes DIGITALes, 1998), IMED-CD 9802. See also, https://www.youtube.com/watch?v=JVrTJvbiYmA.

3. Translation is the composer's as it appears in the album's liner notes, Ibid.

4. David Ogborn, "Interview with Robert Normandeau," *EContact!* 11, no. 2 (March 26, 2009), http://cec.sonus.ca/econtact/11_2/normandeauro_ogborn.html.

5. Alexa Woloshyn, "Interview with Robert Normandeau," *EContact!* 13, no. 3 (2010), http://econtact.ca/13_3/woloshyn_normandeau_2011.html.

6. Though Prokofiev names every character who appears on stage, the titles appear in name only in the score and libretto, such that the identities of various characters remain anonymous, for example, Croupier #1 and #2, Gamblers #1, #2, #3 (#4 is omitted), #5, #6, Reckless Gambler, Unlucky Gambler, Old Gambler, Sickly Gambler, Fat English, Tall English, "6 Players (2 Tenors, 2 Baritones, 2 Basses)," and "various silent roles." Danielle Sofer, "Confined Spaces/Erupted Boundaries: Crowd Behavior in Prokofiev's The Gambler" (MA Thesis, Stony Brook University, 2012), appendix.

7. Herbert J. Jenny, "Perotin's 'Viderunt Omnes,'" *Bulletin of the American Musicological Society* 6 (August 1, 1942): 20, https://doi.org/10.2307/829204.

8. The work was commissioned by the *Groupe de musique expérimentale de Bourges* (GMEB), France, and premièred on June 8, 1989 at the *Synthèse* Festival of Experimental Music. Robert Normandeau, "Jeu," *electroCD: The Electroacoustic Music Store*, accessed April 21, 2021, https://electrocd.com/en/oeuvre/14380/Robert_Normandeau/Jeu.

9. Henry Louis Gates, *The Signifying Monkey: A Theory of African-American Literary Criticism. Reprint Edition* (New York: Oxford University Press, 1989); Tricia Rose, *Black Noise: Rap Music and Black Culture in Contemporary America* (Hanover, NH: Wesleyan University Press, 1994), 62–98; Williams, *Rhymin' and Stealin'*; Joseph Glenn Schloss, *Making Beats: The Art of Sample-Based Hip-Hop* (Middletown, CT: Wesleyan University Press, 2014), 92–94.

10. https://electrotheque.com/oeuvre/28199.

11. Email correspondence with Paula Azguime, cofounder of the Miso Ensemble and its institution the *Música Viva* Festival, October 24, 2013.

12. A video recording of the roundtable, held in Portuguese, is available from *Música Viva*. Materials from the *Música Viva* Festival are archived by the Portuguese Music Research & Information Centre, last accessed March 26, 2015, http://www.mic.pt/index.html.

13. I am indebted to Piero Guimaraes for his Portuguese transcription and English translation of this broadcast.

14. Email correspondence with Robert Normandeau. Unless otherwise noted all correspondences are from October 9, 2013.

15. Robert Normandeau, email message to author, October 16, 2013.

16. Woloshyn, "Wallace Berry's Structural Processes and Electroacoustic Music," 5.

17. Woloshyn, "Wallace Berry's Structural Processes and Electroacoustic Music," 5.

18. Robert Normandeau in discussion with the author, June 10, 2014.

19. For some comic relief on a rainy day, take a look at the Wikipedia article on "cross-linguistic onomatopoeias," accessed June 9, 2015, http://en.wikipedia.org/wiki/Cross-linguistic_onomatopoeias.

20. Andrea Westlund, "Love and the Sharing of Ends," in *Twenty-First Century Papers: On-Line Working Papers from the Center for 21st Century Studies* (University of Wisconsin-Milwaukee, WI: Center for 21st Century Studies, 2005), 4, http://www4.uwm.edu/c21/pdfs/workingpapers/westlund.pdf.

21. Robert Normandeau, liner notes to *Pluie Noire, electrocd.com: The Electroacoustic Music Store*, last accessed on November 17, 2015, http://www.electrocd.com/en/oeuvres/select/?id=24606.

22. Robert Normandeau in discussion with the author, June 10, 2014.

23. Roland Barthes, *Le Plaisir Du Texte* (Paris: Éditions du Seuil, 1973), 15; translated in Elizabeth Locey, *The Pleasure of the Text: Violette Leduc and Reader Seduction* (Lanham, MD: Rowman, 2002), 21.

24. Locey, *The Pleasure of the Text*, 21.

CHAPTER 5

1. Juliana Hodkinson and Niels Rønsholdt, *Fish & Fowl*, Da Capo, 2011, compact disc.

2. Martin Nyström, *Dagens Nyheter*, 2011, https://www.dn.se/kultur-noje/skivrecensioner/scenatet-fish-fowl/. The review is written in Danish, English translations are the author's. *Fish & Fowl* includes several works from the composers' respective back-catalogues. These works are listed in appendix B.

3. See Norman Adams, "Visualization of Musical Signals," in *Analytical Methods of Electroacoustic Music*, ed. Mary Simoni (New York: Routledge, 2005), 13–28.

4. Niels Rønsholdt, email to the author, April 10, 2011.

5. In several of Salvatore Sciarrino's works for flute, including *All'aure in una lontananza* (1990) and *L'orizzonte luminoso di Aton* (1990), tempo is dictated by the rate at which the performer breathes.

6. Zach Herchen, Interview with Niels Rønsholdt, September 21, 2010.

7. Danielle Sofer, preface to Study Score "Erwin Schulhoff, Symphony No. 5" (Munich: Musikproduktion Jürgen Höflich, 2014); Judy Lochhead, "Hearing 'Lulu,'" in *Audible Traces: Gender, Identity, and Music*, ed. Elaine Barkin, Lydia Hamessley, and Benjamin Boretz (Zürich: Carciofoli Verlagshaus, 1999), 233–252; Sofer, "Specters of Sex."

8. Juliana Hodkinson, "Sonic Writing: A Vibrational Practice," *Seismograf/DMT* 31 (March 2014), http://seismograf.org/node/5502.

9. Hélène Cixous, "The Laugh of the Medusa," trans. Keith Cohen and Paula Cohen, *Signs* 1, no. 4 (Summer 1976): 875–893.

10. These themes resurface in part II.

11. Nyman, *Experimental Music*, 90.

12. Nyman, *Experimental Music*, 90.

13. Carolyn Abbate, "Music—Drastic or Gnostic?," *Critical Inquiry* 30, no. 3 (2004): 505–536.

14. Danielle Sofer, "Eroticism and Time in Computer Music: Juliana Hodkinson and Niels Rønsholdt's *Fish & Fowl*," in *Music Technology Meets Philosophy: From Digital Echoes to Virtual Ethos*, vol. 1 (ICMC-SMC 2014: 40th International Computer Music Conference joint with the 11th Sound and Music Computing Conference, Athens, Greece, 2014), 148–153.

15. I identify the end of the phrase here as I would in a traditional tonal context, by a change in character.

16. Deleuze, *Difference and Repetition*, 233.

17. Deleuze and Guattari, *A Thousand Plateaus*, 483.

18. Deleuze and Guattari, *A Thousand Plateaus*, 233.

19. Leon Botstein, *Alban Berg and the Memory of Modernism*, ed. Hailey Christopher (Princeton: Princeton University Press, 2010).

20. Botstein, *Alban Berg and the Memory of Modernism*, 342n103.

21. See "Desiring Woman, Becoming Other" in chapter 1 and "The Voice" in chapter 3.

22. See chapter 3.

23. Aristotle, *On Rhetoric: A Theory of Civic Discourse*, trans. George Alexander Kennedy, 2nd ed (New York: Oxford University Press, 2007), 197.

24. Aristotle, *On Rhetoric*, III.2, 1404b1, 197–198. Translation of ταπεινεν according to Christof Rapp, *The Stanford Encyclopedia of Philosophy* (Spring 2010 Edition), ed. Edward N. Zalta, s.v. "Aristotle's Rhetoric," accessed February 16, 2016, http://plato.stanford.edu/archives/spr2010/entries/aristotle-rhetoric/. See §3.

25. Francis Poulenc, "In Praise of Banality," in *Francis Poulenc: Articles and Interviews: Notes from the Heart*, ed. Nicolas Southon (Farnham: Ashgate, 2014), 28.

26. Poulenc, "In Praise of Banality," 28.

27. Giorgio Agamben, "Difference and Repetition: On Guy Debord's Films," in *Guy Debord and the Situationist International: Texts and Documents*, ed. Tom McDonough, trans. Brian Holmes (Cambridge, MA: MIT Press, 1995), 315–316.

28. Agamben, "Difference and Repetition," 318.

29. Agamben, "Difference and Repetition," 319.

30. Dolar, *A Voice and Nothing More*, 11.

31. Dolar, *A Voice and Nothing More*, 11.

32. An excerpt of the opening of "Torso" can be viewed on Niels Rønsholdt's YouTube chan-
nel, https://www.youtube.com/watch?v=M8LPYimkoho (1:05–2:35), last accessed April
22, 2021.

33. Michael Nyman, *Experimental Music*, 89.

34. https://www.dacapo-records.dk/en/recordings/hodkinson-fish-fowl.

35. Nyman, *Experimental Music*, 13.

36. A video of excerpts from the opera is available on YouTube, Niels Rønsholdt, "Triumph"
(filmed in Copenhagen, 2007), YouTube video, 5:43, posted November 3, 2008, https://www
.youtube.com/watch?v=M8LPYimkoho. The second scene in the video is from "Triumph."

37. Jens Voigt-Lund, "The Demon of Intimacy: About the Music," December 15, 2005, www
.nielsroensholdt.dk/uploads/2/3/2/1/23214662/demonofintimacy.rtf.

38. Voigt-Lund, "The Demon of Intimacy."

39. The sexual thrill of violence is not the same as enacting such violence. Those who watch
50 Shades do not necessarily want to enact such things themselves.

40. Niels Rønsholdt, email to the author, April 10, 2011.

41. Kathryn Kalinak, *Settling the Score: Music and the Classical Hollywood Film* (Madison: University of Wisconsin Press, 1992), 168. More on this "one-dimensionality" in chapter 10.
On the relationship between horror and pornography, see the discussion in chapter 8, as
well as Williams, *Hard Core*.

PART II

1. Deirbhile Brennan, "CupcakKe Tearfully Announces That She Is Quitting Music for Good,"
GCN, September 24, 2019, https://gcn.ie/cupcakke-tearfully-announces-quitting-music/.

2. Hermoine Hoby, "Ronnie Spector Interview: 'The More Phil Tried to Destroy Me, the
Stronger I Got,'" *Telegraph*, March 6, 2014, https://www.telegraph.co.uk/culture/music
/rockandpopfeatures/10676805/Ronnie-Spector-interview-The-more-Phil-tried-to
-destroy-me-the-stronger-I-got.html.

3. That I have not found evidence in electroacoustic circles of actual instances of physical
violence and abuse does not preclude these from occurring. Academics are only now turning attention to harassment and abuses within our own institutional ranks. Gerald Graff,
Professing Literature: An Institutional History (Chicago: University of Chicago Press, 1989).
Music is especially late in adopting this self-critical lens; Hisama, "Getting to Count."

4. Zakiyyah Iman Jackson, *Becoming Human: Matter and Meaning in an Antiblack World*
(New York: NYU Press, 2020), 122.

5.   Patricia Hill Collins, *Black Feminist Thought: Knowledge, Consciousness and the Politics of Empowerment*, 2nd ed. (New York: Routledge, 2000), 9.

6.   The term "Post-Bellum—Pre Harlem" came into use from an eponymous essay from 1931 by the African-American writer and phonographer Charles Chesnutt. Barbara McCaskill and Caroline Gebhard, *Post-Bellum, Pre-Harlem: African American Literature and Culture, 1877–1919* (New York: NYU Press, 2006); in relation to sound, see also Stoever, *The Sonic Color Line*, 163. According to Stoever, Chesnutt was known for his early "critique [of] white listeners' forms." Stoever, *The Sonic Color Line*, 160.

7.   Stoever, *The Sonic Color Line*, 136.

8.   Frederick Douglass, "The Color Line," *North American Review* 132 (June 1, 1881): 567–577.

9.   Robert Dixon and John Godrich, *Recording the Blues* (New York: Stein and Day Books, 1970), 17, http://archive.org/details/RecordingTheBlues.

10. Angela Y. Davis, *Blues Legacies and Black Feminism: Gertrude "Ma" Rainey, Bessie Smith, and Billie Holiday* (New York: Vintage Books, 1995), xii.

11. Stoever, *The Sonic Color Line*, 43.

12. Daniel Cavicchi, *Listening and Longing: Music Lovers in the Age of Barnum* (Middletown, CT: Wesleyan University Press, 2011), 7.

13. Stoever, *The Sonic Color Line*, 163.

14. Michael Rogin, *Blackface, White Noise: Jewish Immigrants in the Hollywood Melting Pot* (University of California Press, 1996), 5.

15. Pamela Brown Lavitt, "First of the Red Hot Mamas: 'Coon Shouting' and the Jewish Ziegfeld Girl," *American Jewish History* 87, no. 4 (1999): 253.

16. Lavitt, "First of the Red Hot Mamas," 255–256.

17. Lavitt, "First of the Red Hot Mamas," 253; Collins, *Black Feminist Thought*, 5.

18. Lavitt, "First of the Red Hot Mamas," 253.

19. Lavitt, "First of the Red Hot Mamas," 262.

20. Lavitt, "First of the Red Hot Mamas," 263.

21. Cavicchi, *Listening and Longing*, 170–171.

22. Cavicchi, *Listening and Longing*, 171.

23. Matthew D. Morrison, "Sound in the Construction of Race: From Blackface to Blacksound in Nineteenth-Century America" (PhD diss., Columbia University, 2014), 124.

24. Morrison, "Sound in the Construction of Race," 129–130.

25. Weheliye, "Feenin," 25–26.

26. Davis, *Blues Legacies and Black Feminism*, xii.

27. Dixon and Godrich, *Recording the Blues*, 17.

28. Hazel V. Carby, "It Jus Be's Dat Way Sometime: The Sexual Politics of Women's Blues," in *The Jazz Cadence of American Culture*, ed. Robert G. O'Meally (New York: Columbia University Press, 1998), 12.

29. I recently interviewed Shields for a series of podcasts recounting and celebrating her work as part of the project, "Unsung Stories: Women at Columbia's CMC." For more, visit https://www.unsungstoriescmc.com/podcast.

30. Barry Truax, "Homoeroticism and Electroacoustic Music: Absence and Personal Voice," *Organised Sound* 8, no. 1 (2003): 119.

31. Truax, "Homoeroticism and Electroacoustic Music: Absence and Personal Voice," 119.

32. Sheila Whiteley and Jennifer Rycenga, *Queering the Popular Pitch* (New York: Routledge, 2013), 256; Sarah Kerton, "Too Much, Tatu Young: Queering Politics in the World of Tatu," in *Queering the Popular Pitch*, ed. Sheila Whiteley and Jennifer Rycenga (New York: Routledge, 2013), 156.

33. Doris Leibetseder, *Queer Tracks: Subversive Strategies in Rock and Pop Music* (Farnham: Ashgate, 2013); Stan Hawkins, *Queerness in Pop Music: Aesthetics, Gender Norms, and Temporality* (New York: Routledge, 2015).

34. Rose, *The Hip Hop Wars*, 115.

35. Sofer, "Categorising Electronic Music."

36. Rose, *The Hip Hop Wars*, 115.

37. Guthrie P. Ramsey, "Who Hears Here? Black Music, Critical Bias, and the Musicological Skin Trade," *Musical Quarterly* 85, no. 1 (2001): 1–52.

CHAPTER 6

1. Donna Summer, "The Fresh Air Interview: 'Queen of Disco' Donna Summer," interview by Terry Gross, November 4, 2003, https://www.npr.org/2003/11/04/1491690/queen-of-disco-donna-summer.

2. Mark McLaughlin, *Disco: Spinning the Story*, 2005. This video is available online in two parts at https://www.dailymotion.com/video/x51r4ej.

3. Throughout this chapter, I refer to the 17-minute extended version of "Love to Love You Baby," DiscoSaturdayNightTV, *Donna Summer—Love to Love You Baby (Original Extended Version) Oasis Records 1975*, 2013, https://www.youtube.com/watch?v=VRI1yzn2ugA.

4. Lucy Neale in email communication with author, August 1, 2013.

5. Richard Buskin, "Donna Summer 'I Feel Love': Classic Tracks," *Sound on Sound*, October 2009, http://www.soundonsound.com/sos/oct09/articles/classictracks_1009.htm.

6. Timing based on the 17-minute extended version of "Love to Love You Baby," DiscoSaturdayNightTV, *Donna Summer—Love to Love You Baby (Original Extended Version) Oasis Records 1975*.

7. Timing based on the 17-minute extended version of "Love to Love You Baby," Disco-SaturdayNightTV, *Donna Summer—Love to Love You Baby (Original Extended Version) Oasis Records 1975.*

8. Robert Fink, "Do It ('til You're Satisfied): Repetitive Musics and Recombinant Desires," in *Repeating Ourselves: American Minimal Music as Cultural Practice* (Berkeley: University of California Press, 2005), 42.

9. Donna Summer and Marc Eliot, *Ordinary Girl: The Journey* (New York: Villard, 2003), 61–62.

10. Summer and Eliot, *Ordinary Girl*, 128.

11. Buskin, "Donna Summer 'I Feel Love.'"

12. Summer and Eliot, *Ordinary Girl*, 108, 110.

13. Summer and Eliot, *Ordinary Girl*, 112.

14. Patricia Hill Collins, *Black Sexual Politics: African Americans, Gender, and the New Racism* (New York: Routledge, 2004), 134. See chapter 10.

15. Summer, "The Fresh Air Interview"; Mikal Gilman, "Donna Summer: Is There Life after Disco?," *Rolling Stone*, March 23, 1978, 11.

16. Iris Carmon, "Steven Tyler's Teenage Girlfriend Tells Her Side of the Abortion Story," *Jezebel*, May 24, 2011, https://jezebel.com/steven-tylers-teenage-girlfriend-tells-her-side-of-the-5805190.

17. Charles L. Sanders, "Donna Summer," *Ebony*, 1977, 33–42.

18. Sanders, "Donna Summer," 36.

19. Some critical literature on the subject includes: Deborah Gray White, *Ar'n't I a Woman? Female Slaves in the Plantation South* (New York: W. W. Norton, 1985); Sander L. Gilman, "Black Bodies, White Bodies: Toward an Iconography of Female Sexuality in Late Nineteenth Century Art, Medicine, and Literature," in *Race, Writing, and Difference*, ed. Henry Louis Gates, Jr. (Chicago: University of Chicago Press, 1986), 223–261; Collins, *Black Feminist Thought.*

20. Brian Ward, *Just My Soul Responding: Rhythm and Blues, Black Consciousness, and Race Relations* (Berkeley: University of California Press, 1998), 426.

21. Joel Whitburn, *Hot Dance Disco 1974–2003* (Milwaukee, WI: Record Research Incorporated, 2004), 249.

22. Luis-Manuel Garcia shows the role sexuality played in solidifying the electronic dance music scene in the 1970s in New York, Chicago, and Detroit. See Luis-Manuel García, "An Alternate History of Sexuality in Club Culture," *Resident Advisor* (blog), January 28, 2014, https://www.residentadvisor.net/features/1927.

23. García, "An Alternate History of Sexuality in Club Culture."

24. Mitchell Morris, *The Persistence of Sentiment: Display and Feeling in Popular Music of the 1970s* (Berkeley: University of California Press, 2013), 36.

25. Richard Dyer, "In Defense of Disco," *Gay Left* 8 (Summer 1979): 23.

26. Judith Peraino, *Listening to the Sirens: Musical Technologies of Queer Identity from Homer to Hedwig* (Berkeley: University of California Press, 2006), 153.

27. Brian Currid, "'We Are Family': House Music and Queer Performativity," in *Cruising the Performative: Interventions into the Representation of Ethnicity*, ed. Sue-Ellen Case, Philip Brett, and Susan Leigh Foster (Bloomington: Indiana University Press, 1995), 165–196.

28. Robin James, "'Robo-Diva R&B': Aesthetics, Politics, and Black Female Robots in Contemporary Popular Music," *Journal of Popular Music Studies* 20, no. 4 (December 2008): 402–423, https://doi.org/10.1111/j.1533-1598.2008.00171.x; Eileen Hayes, Ingrid Monson, and Sherrie Tucker have respectively written of the problems of being "out" as a Black woman in musical communities. Hayes summarizes, "The neglect of black women in the scholarship of women's music is that gay and lesbian studies as well as popular music studies have tended to privilege the sexual identity disclosure of the artist." Eileen M. Hayes, "Black Women and 'Women's Music,'" in *Black Women and Music: More Than the Blues*, ed. Linda F. Williams and Eileen M. Hayes (Urbana: University of Illinois Press, 2007), 163. Ingrid Monson explains that Black women are not "out" in the same way as white women (or, implicitly, women of other ethnicities); Ingrid Monson, "Music and the Anthropology of Gender and Cultural Identity," *Women and Music: A Journal of Gender and Culture* 1, no. 1 (1997): 28; Sherrie Tucker, "When Subjects Don't Come Out," in *Queer Episodes in Music and Modern Identity*, ed. Lloyd Whitesell and Sophie Fuller (Champaign: University Press of Illinois, 2002), 293–310.

29. I am reminded of Brian Ward's quip: "Just about anyone seemed capable of piecing together a passable disco-by-numbers record." Ward, *Just My Soul Responding*, 426.

30. Tucker, "When Subjects Don't Come Out."

31. Ward, *Just My Soul Responding*, 426.

32. Backing vocals, no surnames provided on album.

33. Michael Veal, *Dub: Soundscapes and Shattered Songs in Jamaican Reggae* (Middletown, CT: Wesleyan University Press, 2013), 166, 171, 233.

34. James Arena, *First Ladies of Disco: 29 Stars Discuss the Era and Their Singing Careers* (Jefferson, NC: McFarland, 2013), 22.

35. Larry Crane, "Reinhold Mack: ELO, Queen, Black Sabbath & T. Rex | Tape Op Magazine | Longform Candid Interviews with Music Producers and Audio Engineers Covering Mixing, Mastering, Recording and Music Production," *Tape Op*, February 2011, https://tapeop.com/interviews/81/reinhold-mack/.

36. Watkins Electric Music (WEM) PA system, used by Sahara and Led Zeppelin.

37. Buskin, "Donna Summer 'I Feel Love.'"

38. L. Singerling, "Donna Summer, Queen of Disco," July 5, 2012, http://donnasummequeenof disco.blogspot.com/2012_07_05_archive.html.

39. Jon Stratton, "Coming to the Fore: The Audibility of Women's Sexual Pleasure in Popular Music and the Sexual Revolution," *Popular Music* 33, no. 1 (January 2014): 124.

40. Stratton, "Coming to the Fore," 124.

41. Summer and Eliot, *Ordinary Girl*, 102, 133.

42. For an annotated transcription of the themes in Summer's "Love to Love You Baby," see Fink, "Do It ('til You're Satisfied)," 57–58.

43. Mark Katz, "Sampling before Sampling: The Link between Dj and Producer," *Samples: Online-Publikationen Des Arbeitskreis Studium Populärer Musik* 9 (2010): 1–11.

44. Stratton, "Coming to the Fore," 126.

45. Fink, "Do It ('til You're Satisfied)," 42.

46. Fink, "Do It ('Til You're Satisfied)," 35. In this paragraph Fink revisits the argument of reductive analyses that map musical dualities easily onto gender and sexual binaries, in a similar vein to my analysis of Frith and McRobbie's "cock rock" in chapter 7.

47. Fink, "Do It ('til You're Satisfied)," 31.

48. Fink, "Do It ('til You're Satisfied)," 46.

49. Fink, "Do It ('til You're Satisfied)," 60.

50. Fink, "Do It ('til You're Satisfied)," 60.

51. Fink, "Do It ('til You're Satisfied)," 59.

52. Fink, "Do It ('til You're Satisfied)," 47.

53. Richard Dyer, *Only Entertainment* (London: Routledge, 1992).

54. Summer, "The Fresh Air Interview."

55. Summer and Eliot, *Ordinary Girl*, 133.

56. Alan Weitz, *Rolling Stone*, March 2, 1980.

57. Lucy Neale in email communication with author, August 1, 2013.

CHAPTER 7

1. I extend my gratitude to Alice Shields for providing me with the *Apocalypse* libretto and other documentation, and for taking the time to respond to my various questions via email and in person. Without her assistance this analysis would not have been possible. Alice Shields in private email communication with the author, December 27, 2013. Confirmed by Daria Semegen in private email communication with the author, August 30, 2016.

2. Alice Shields, "Liner Notes to *Apocalypse: An Electronic Opera*," *Composers Recordings* 647 (1993). Composers Recordings NXCR647, 1993, compact disc. The recording has since been reissued: Alice Shields, *Apocalypse: An Electronic Opera*, New World Records NWCR647, 2007, compact disc and other formats, https://www.newworldrecords.org/products/alice-shields-apocalypse-an-electronic-opera.

3. "Apocalypse Song," IV:10C, Alice Shields, *Apocalypse* (unpublished libretto, 1991–1994), 65. References to *Apocalypse* include part number in roman numerals and, following the colon, scene number according to the opera's libretto (published via the American Composers Alliance, https://composers.com/).

4. Alice Shields does not prioritize the Christian significance of these themes. My choice to begin from the text's Christian framing was motivated by my experiences presenting this research in Anglo-American settings, where questions from the audience overwhelmingly focused on Christian symbolism, presumably on account of the speakers' own points of reference. Portions of this chapter were presented at the "Deleuze + Art: Multiplicities | Thresholds | Potentialities" conference, cohosted by the School of Drama, Film, and Music at Trinity College Dublin, Trinity Long Room Hub Arts and Humanities Research Institute (April 8–10, 2016, Dublin, Ireland); and at the 14th annual plenary conference of the Society for Musicology in Ireland, hosted by Dublin City University (June 10–12, 2016, Dublin, Ireland).

5. Shields, email to author, December 27, 2013.

6. "Regrets," III:8A, Alice Shields, *Apocalypse* (libretto), 52.

7. Alice Shields was instructor and Associate Director of the Columbia–Princeton Electronic Music Center from 1964 to 1982, and Associate Director for Development of the Computer Music Center from 1994 to 1996.

8. I am indebted to Anna Giulia Volpato, a.k.a. Johann Merrich, for providing me with the original transcripts for this interview, which appears in Italian translation in Johann Merrich, *Le Pioniere della Musica Elettronica* (Milan: Auditorium Edizioni, 2012), 91.

9. Pril Smiley began working with Vladimir Ussachevsky at CPEMC in 1963. Daria Semegen studied with Ussachevsky from 1971 onward. Hinkle-Turner, *Women Composers and Music Technology in the United States*, 18–19, 92. For new documentation on the historic work of women composers at the CPEMC and the Computer Music Center, as it is known now, see unsungstoriescmc.com/. The Unsung Stories project, led by Zosha di Castri and Ellie Hisama, features podcast interviews, video recordings, and concerts.

10. Hinkle-Turner, *Women Composers and Music Technology in the United States*, 21.

11. Alice Shields in conversation with the author, July 25, 2014.

12. For a more extensive history of Alice Shields's role among her colleagues at the CPEMC, see Hinkle-Turner, *Women Composers and Music Technology in the United States*, 16–21.

13. Shields conversation, July 25, 2014.

14. Alice Shields, note to *Kyrielle, Soundcloud*, https://soundcloud.com/user-aliceshieldscomposer/kyrielle-for-violin-and-computer-music. The work appears on track 6 of Airi Yoshioka, *Stolen Gold*, compact disc (Albany Music Distribution Troy 1305, 2011).

15. Michel Foucault, "Sex, Power, and the Politics of Identity," in *Ethics, Subjectivity, and Truth*, ed. Paul Rabinow (New York: New Press, 1994), 163–164.

16. Occasionally, SHIVA's voice takes on a masculinized tone. In the New World Records recording SHIVA was sung by baritone Michael Willson in the Aeon and Organ Screaming scenes, and is sung elsewhere in the opera by Alice Shields.

17. *Apocalyse* has never been performed. Although in the early 2000s the Akademie der Künste in Berlin wanted to premiere the opera, this staging has yet to be realized. For a work posing comparable performance challenges that has been performed see *Mass for the Dead* (1993).

18. Andra McCartney, "'Creating Worlds for My Music to Exist': How Women Composers of Electroacoustic Music Make Place for Their Voices" (master's thesis, York University, 1997).

19. Emmerson, "EMAS and Sonic Arts Network (1979–2004): Gender, Governance, Policies, Practice."

20. Bosma, "Bodies of Evidence, Singing Cyborgs and Other Gender Issues in Electrovocal Music."

21. Keightley, "'Turn It Down!' She Shrieked: Gender, Domestic Space, and High Fidelity, 1948–59"; Emmerson, "EMAS and Sonic Arts Network (1979–2004): Gender, Governance, Policies, Practice."

22. Irigaray, *This Sex Which Is Not One*, 77.

23. Corbett and Kapsalis, "Aural Sex," 103.

24. "Heat Drum," IV:11A, Alice Shields, *Apocalypse* (unpublished libretto, 1991–1994), 66.

25. "Heat Drum," IV:11A, Alice Shields, 66.

26. Avanthi Meduri, "Bharata Natyam–What Are You?," *Asian Theatre Journal* 5, no. 1 (1988): 1.

27. Judith Lynne Hanna, *Dance, Sex, and Gender: Signs of Identity, Dominance, Defiance, and Desire* (Chicago: University of Chicago Press, 1988), 102.

28. The theory is attributed to Bharata, but Adya Rangacharya, the preeminent English translator of the *Natyashastra*, speculates that many authors have since "interfered" with Bharata's original manuscript. Adya Rangacharya, *Introduction to Bharata's Nātya-Śāstra* (Bombay: Popular Prakashan, 1966), 69.

29. Rangacharya, *Introduction to Bharata's Nātya-Śāstra*, 70–71.

30. Rangacharya, *Introduction to Bharata's Nātya-Śāstra*, 75.

31. Ranjana Thapalyal, "Sringara Rasa: The Feminine Call of the Spiritual/Erotic Impulse in Indian Art," in *The Sacred and the Feminine: Imagination and Sexual Difference*, ed. Griselda Pollock and Victoria Urvey-Sauron (London: I. B. Tauris, 2007), 137.

32. Meduri, "Bharata Natyam—What Are You?," 3.

33. Meduri, "Bharata Natyam—What Are You?," 4.

34. According to Meduri, Bharata's theory pertained more to the technical details of the arts, dance, music, and theater and their mutual signification. It was only in the eleventh century

that Abhinavagupta reinterpreted the *Natyashastra* to include more lofty transcendental psychologisms. Meduri, "Bharata Natyam—What Are You?," 4.

35. Pamyla A. Stiehl, "Bharatanatyam: A Dialogical Interrogation of Feminist Voices in Search of the Divine Dance," *Journal of Religion and Theatre* 3, no. 2 (2004): 282.

36. Meduri, "Bharata Natyam—What Are You?," 2.

37. Davesh Soneji, "Unfinished Gestures: Devadasis, Memory, and Modernity in South India" (Chicago: University of Chicago Press, 2012), 3. Soneji replicates the Madras Devadasi Prevention Bill in an appendix (Soneji, "Unfinished Gestures," 235–236).

38. Meduri, "Bharata Natyam—What Are You?," 4. One could compare this secularization and growing popularity to the Tai Chi and Yoga crazes that caught on in the West.

39. Stiehl, "Bharatanatyam," 282–283; Hanna, *Dance, Sex, and Gender*, 103.

40. Meduri, "Bharata Natyam—What Are You?," 9.

41. Stiehl, "Bharatanatyam," 285.

42. Soneji, "Unfinished Gestures," 3.

43. Partha Chatterjee, *The Nation and Its Fragments: Colonial and Postcolonial Histories* (Princeton: Princeton University Press, 1993), 136.

44. I am grateful to Rasika Ajotikar for emphasizing to me the importance of this point.

45. Shields, "Liner Notes to *Apocalypse: An Electronic Opera*," 3.

46. Shields conversation, July 25, 2014.

47. Deleuze, *Difference and Repetition*, 10.

48. Alan H. Sommerstein, *Aeschylean Tragedy* (London: Bloomsbury, 2010), 25.

49. Shields likely paraphrases Catherine Clément here, who writes, "In opera to love is to wish to die." For Clément, operas unfold as if in slow preparation for the death of the central female character; Puccini's fifteen-year-old Cio-Cio San sacrifices her body and her life for her American hero; Lucia is driven from madness to death, and Wagner's Isolde dies simply of love sickness. Catherine Clément, *Opera or The Undoing of Women*, trans. Betsy Wing (Minneapolis: University of Minnesota Press, 1988), 53–54.

50. On the "images and myths invoked in malestream discourses associated with electroacoustic music, in popular magazines, course texts and software," see McCartney, "'Creating Worlds for My Music to Exist,'" 43–76.

51. Shields, "Liner Notes to Apocalypse: An Electronic Opera," 7.

52. Shields, *Apocalypse* (libretto), 62. As the composer chose and composed the Greek text herself, it occasionally departs from strictly correct Greek grammar.

53. Alice Shields, "Patient and Psychotherapist: The Music," in *The Psychoaesthetic Experience: An Approach to Depth-Oriented Treatment*, ed. Arthur Robbins (New York: Human Sciences Press, 1989), 60.

54. Shields, "Patient and Psychotherapist," 58.

55. Shields, "Patient and Psychotherapist," 57.

56. Simon Frith and Angela McRobbie, "Rock and Sexuality," in *On Record: Rock, Pop and the Written*, ed. Simon Frith and Andrew Goodwin (New York: Routledge, 1978), 319.

57. Excerpt from review by *The Splatter Effect*, "What the Press Is Saying," Alice Shields's website, http://www.aliceshields.com/reviews.html.

58. For details on the Klangumwandler employed by Alice Shields, see Vladimir Ussachevsky, "Musical Timbre Mutation by Means of the-Klangumwandler, a Frequency Transposition Device," in *Audio Engineering Society Convention 10: Audio Engineering Society*, 1958. The instrument was first encountered by Ussachevsky and Otto Luening, whom Shields later assisted at the Columbia-Princeton Electronic Music Center, in 1955 on a trip to Baden-Baden while conducting "research on electronic music" funded by the Rockefeller Foundation. Otto Luening, "An Unfinished History of Electronic Music," *Music Educators Journal* 55, no. 3 (1968): 136–137.

59. Other examples include Joe Harriott and John Mayer's Indo-Jazz Fusions, the Mahavishnu Orchestra, and Turiyasangitananda (a.k.a. Alice Coltrane).

60. Anita Kumar, "What's the Matter? Shakti's (Re)Collection of Race, Nationhood, and Gender," *TDR* 50, no. 1 (2006): 72–95.

61. Janet O'Shea, "At Home in the World? The Bharatanatyam Dancer As Transnational Interpreter," *TDR* 47, no. 1 (2003): 182.

62. O'Shea, "At Home in the World," 182.

63. Artaud writes, "The actors with their costumes constitute veritable living, moving hieroglyphs. And these three-dimensional hieroglyphs are in turn brocaded with a certain number of gestures—mysterious signs which correspond to some unknown, fabulous, and obscure reality which we here in the Occident have completely repressed." Antonin Artaud, *The Theater and Its Double*, trans. Mary Caroline Richards (New York: Grove Weidenfeld, 1958), 60.

64. O'Shea, "At Home in the World," 177, emphasis added.

65. The Mexico City Policy was later reinforced by *Rust v. Sullivan* (1991) and *Planned Parenthood of Southeastern Pennsylvania v. Casey* (1992).

66. Barbara B. Crane and Jennifer Dusenberry, "Power and Politics in International Funding for Reproductive Health: The US Global Gag Rule," *Reproductive Health Matters* 12, no. 24 (2004): 128–137.

67. I borrow the phrase "sex acts understood as homosexual or queer" from Lauron Kehrer (private communication, May 1, 2018) to refer to what contemporaneously would have been designated "same-sex relations," a limited construction that implicitly omits trans and intersex identities, though these identities were certainly stigmatized in this politically conservative rhetoric. I thank K. E. Goldschmitt for helping me articulate this point (any errors are my own). Other "behaviors" condemned in this discourse, as outlined by Gayle

Rubin, include, "prostitution, transsexuality, sadomasochism, and cross-generational activities." Gayle Rubin, "Thinking Sex: Notes for a Radical Theory of the Politics of Sexuality," in *Deviations: A Gayle Rubin Reader* (Durham, NC: Duke University Press, 2011), 166.

68. Camille Paglia and Art, *Sex and American Culture: Essays* (New York: Vintage, 1992), 30.

69. Rubin, "Thinking Sex," 166, 167.

70. Theodor Adorno, "Sexual Taboos and Law Today," in *Critical Models: Interventions and Catchwords*, trans. and ed. Henry W. Pickford (New York: Columbia University Press, 1963), 77.

71. Chatterjee, *The Nation and Its Fragments*, 136.

72. Shields, "Liner Notes to *Apocalypse: An Electronic Opera*," 8.

73. Regrettably, a reconstruction of the old Irish texts would be nearly impossible, since the texts were likely partially composed by the composer's sister, a specialist of Celtic languages and literatures, who assisted in realizing the translated English text back into old Irish.

74. James January McCann, "Laoidh Chab an Dosáin" (master's thesis, Aberystwyth University, 2011), 65–66.

75. McCann, "Laoidh Chab an Dosáin," 66, line 23.

76. Some readings claim that she herself has turned into a man, matching her reversed role to his, but it's not clear to me from the text that her own transformation in fact takes place.

77. McCann, "Laoidh Chab an Dosáin," 1.

78. Seosamh Watson, "Laoi Chab an Dosáin: Background to a Late Ossianic Ballad," *Eighteenth-Century Ireland / Iris an Dá Chultúr* 5 (1990): 39.

79. Watson, "Laoi Chab an Dosáin," 38.

80. The "Dismemberment Dance" is alternately titled "Dismemberment and Eating" on track 8 of the album. Other nonvocal scenes featuring choreographed stage directions set to music include I:3, II:4B, 4L, 4M; III:4P-2, 5A, 5J, 5K, 7D, 9B, 10B, 11A; these occur primarily in part III of the opera and are mostly omitted from the CD.

81. Shields, "Liner Notes to *Apocalypse: An Electronic Opera*," 3. Shields notes also that she gained previous experience with manipulating belching sounds electronically while assisting Sam Shepard in the radio play *Icarus* (1966), though she did not recall her involvement in that project until after *Apocalypse* was completed. Shields, "Liner Notes to Apocalypse: An Electronic Opera," 5.

82. Shields, *Apocalypse* (libretto), 52.

83. Shields, *Apocalypse* (libretto), 57.

84. Shields, *Apocalypse* (libretto), 60.

85. Shields, *Apocalypse* (libretto), 59.

86. Shields, *Apocalypse* (libretto), 66.

87. Shields, "Liner Notes to *Apocalypse: An Electronic Opera*," 7.

88. In translation, Aristotle's attribution is to "the phallic songs," it should, however, be noted that Aristotle's original reads "*phaulika*," meaning trivial business, and not "*phallika*," of the phallic songs. Nevertheless, most English translations employ the latter, and this is likely the translation with which Shields was familiar. Aristotle, *Poetics*, trans. S. H. Butcher, IV:12 (London: Macmillan and Co., 1902), 1449a10–13.

89. Alice Shields, email message to author, December 27, 2013.

90. Frederic Jameson, "Of Islands and Trenches: Neutralization and the Production of Utopian Discourse," in *The Ideologies of Theory* (London: Verso, 2008), 386.

91. Frederic Jameson, "World Reduction in Le Guin: The Emergence of Utopian Narrative," *Science Fiction Studies* 2, no. 3 (1975): 221.

92. Milton Babbitt, "Twelve-Tone Rhythmic Structure and the Electronic Medium," *Perspectives of New Music* 1, no. 1 (1962): 51.

93. Iverson, *Electronic Inspirations*, 1.

94. Herbert Eimert, "What is Electronic Music?," *Die Reihe* 1 (1965): 1.

95. Author in conversation with Alice Shields, 25 July, 2014.

96. A gallery of photographs from the American Chamber Opera's staging of *Mass for the Dead* (1993) is available on the composer's website, http://www.aliceshields.com/massgallery.html.

97. Paul Attinello, "Performance and/or Shame," *Repercussion* 4, no. 2 (1996): 119.

98. Judith Butler, "Performative Acts and Gender Constitution: An Essay in Phenomenology and Feminist Theory," *Theatre Journal* 40, no. 4 (1988): 520.

99. Attinello, "Performance and/or Shame," 99.

100. Liz Wood and Mitchell Morris, "Flirting with 'Theory,' Flirting with Music: A Discussion in Advance of ForePlay," in *Gay and Lesbian Study Group (of the American Musicological Society) Newsletter*, 1994, 3–4.

CHAPTER 8

1. Pauline Oliveros, "eye fuck!" music to Maria Beatty and Annie Sprinkle, *Sluts & Goddesses Video Workshop: Or How to Be a Sex Goddess in 101 Easy Steps*, VHS (Beatty/Sprinkle Productions, 1992).

2. Perhaps of no coincidence, though tangential for the present concerns, *Pornosonic* (1990) is the title of an album by Don Argott, featuring the iconic Ron Jeremy. The album boasts "unreleased 70s porno music," though the tracks were actually newly composed. Don Argott, *Pornosonic*, compact disc (Mini Mace Pro, 1999), https://web.archive.org/web/19991013021409/http://pornosonic.com/.

3. Darshana Sreedhar Mini, "'Un-Sound' Sounds: Pornosonics of Female Desire in Indian Mediascapes," *Music, Sound, and the Moving Image* 13, no. 1 (September 11, 2019): 3.

4. http://sexecology.org/.

5. Martha Mockus, *Sounding Out: Pauline Oliver's and Lesbian Musicality* (New York: Routledge, 2008), 2, 62–63.

6. Hinkle-Turner, *Women Composers and Music Technology in the United States*, 31; 32.

7. Mockus, *Sounding Out*, 62.

8. Mockus, *Sounding Out*, 57.

9. Linda Montano, Annie Sprinkle, and Veronica Vera, "Summer Saint Camp 1987: With Annie Sprinkle and Veronica Vera," *TDR* 33, no. 1 (Spring 1989): 94, https://doi.org/10.2307/1145947.

10. Montano, Sprinkle, and Vera, "Summer Saint Camp 1987," 100.

11. Montano, Sprinkle, and Vera, "Summer Saint Camp 1987," 102.

12. Williams, "A Provoking Agent," 118.

13. Williams, "A Provoking Agent," 118.

14. Williams, "A Provoking Agent," 121.

15. Williams, "A Provoking Agent," 122.

16. Williams, "A Provoking Agent," 123.

17. Williams, "A Provoking Agent," 120.

18. Williams, "A Provoking Agent," 123.

19. Constance Penley, "Crackers and Whackers: The White Trashing of Porn," in *White Trash: Race and Class in America*, ed. Annalee Newitz and Matt Wray (New York: Routledge, 1996).

20. Holt N. Parker, "Love's Body Anatomized: The Ancient Erotic Handbooks and the Rhetoric of Sexuality," in *Pornography and Representation in Greece and Rome*, ed. Amy Richlin (New York: Oxford University Press, 1992), 91.

21. Annie Sprinkle in conversation with the author, July 28, 2020.

22. Johanna Drucker, "Review of Maria Beatty & Annie Sprinkle: *Sluts & Goddesses Video Workshop*," ed. Kathy High and Marshall Reese, *Felix* 1, no. 3 (1993), http://www.e-felix.org/issue3/Sprinkle.html.

23. Penley, "Crackers and Whackers."

24. Steven Ziplow, *The Film Maker's Guide to Pornography* (Portage, MI: Drake Publishers, 1977), 31–32; quoted in Williams, *Hard Core*, 127.

25. Williams, *Hard Core*, 128.

26. In personal communication, Sprinkle hinted that she would not typically have equated the Goddess persona with the "soft side" of sex, which goes to show how improvised performances can lead performers to say things they may later question, even regret. Sprinkle email, April 11, 2021.

27. Roger Hickman, "Wavering Sonorities and the Nascent Film Noir Musical Style," *Journal of Film Music* 2, no. 2–4 (2009): 166, https://doi.org/10.1558/jfm.v2i2-4.165.

28. David Rothenberg, "Three Ways Toward Deep Listening in the Natural World," *Sound American*, Artist Essays on Deep Listening, 7, accessed November 24, 2020, http://archive.soundamerican.org/sa_archive/sa7/sa7-artist-essays-on-deep-listening.html.

29. Williams, "A Provoking Agent," 127.

30. Williams, "A Provoking Agent," 127.

31. Williams, *Hard Core*, 122.

32. Williams, *Hard Core*, 147.

33. Williams, *Hard Core*, 124.

34. Williams, *Hard Core*, 146.

35. Williams, *Hard Core*, x.

36. Nadine Hubbs, "'Jolene,' Genre, and the Everyday Homoerotics of Country Music: Dolly Parton's Loving Address of the Other Woman," *Women and Music: A Journal of Gender and Culture* 19, no. 1 (2015): 72, https://doi.org/10.1353/wam.2015.0017.

37. Patricia Juliana Smith, *The Queer Sixties* (Hoboken: Taylor and Francis, 2013), 105–126.

38. Jeffrey A. Brown, "Class and Feminine Excess: The Strange Case of Anna Nicole Smith," *Feminist Review* 81 (2005): 74–94.

39. Corbett and Kapsalis, "Aural Sex," 103.

40. Williams, "A Provoking Agent," 128.

41. Williams, "A Provoking Agent," 126.

42. Williams, "A Provoking Agent," 126.

43. Penley, "Crackers and Whackers," 312.

CHAPTER 9

1. Barry Truax, "Composing with Real-Time Granular Synthesis," *Perspectives of New Music* 28, no. 2 (1990): 123, 132.

2. Truax, "Composing with Real-Time Granular Synthesis," 123.

3. Barry Truax, "Sound, Listening and Place: The Aesthetic Dilemma," *Organised Sound* 17, no. 3 (2012): 193–201.

4. Truax, "Sound, Listening and Place," 193.

5. Barry Truax, "The Inner and Outer Complexity of Music," *Perspectives of New Music* 32, no. 1 (1994): 186; Barry Truax, "The Aesthetics of Computer Music: A Questionable Concept Reconsidered," *Organised Sound* 5, no. 3 (2000): 125.

6. Barry Truax, "Editorial: Context-Based Composition," *Organised Sound* 23, no. 1 (April 2018): 1, https://doi.org/10.1017/S1355771817000218.

7. Truax, "Editorial: Context-Based Composition," 2.

8. The oboe d'amore and English horn on the recording are played by Lawrence Cherney, who also commissioned the work for Soundstreams, Barry Truax, *Song of Songs* (1992), on *Song of Songs: Computer and Electroacoustic Music by Barry Truax* (Burnaby, Canada: Cambridge Street Recordings, 1994), CSR-CD 9401. Computer graphic images were designed by Theo Goldberg. The CD's cover features one of these images. The score is available at the composer's website http://www.sfu.ca/~truax/songs.html; and the composer has made the complete documentation of the piece, including source sounds, processing, production score, live score, and spectrograms, available on the Documentation DVD-ROM: Barry Truax, *Documentation DVD #3* (Cambridge Street Recordings, 2010). The composer has since made these materials available on his website, http://www.sfu.ca/~truax/videos.html. I thank Barry Truax for providing me with all these materials.

9. Throughout this chapter I use the terms "male" and "female" when referring to human beings only when deeming it necessary to retain the contemporaneous terminology commonly invoked regarding sex and gender in the 1990s. Where possible, I update these terms, using man and woman, respectively, to refer to "male" and "female" human beings. I interrogate these terms further in the next chapter; in the meantime, for this chapter I will avoid gendering the subjects in situations where neither a man nor woman can be definitively named.

10. Truax, "Homoeroticism and Electroacoustic Music," 119.

11. The complete text is available on the composer's website, Barry Truax, *Song of Songs*, www .sfu.ca/~truax/songtxt.html (accessed March 14, 2016). In an interview I conducted with Barry Truax on June 12, 2014, in Berlin, the composer informed me that the text was adapted from the King James translation of the Bible.

12. Truax, "Homoeroticism and Electroacoustic Music," 119.

13. Susan McClary, "This Is Not the Story My People Tell: Musical Time and Space According to Laurie Anderson," in *Feminine Endings: Music, Gender, and Sexuality* (Minneapolis: University of Minnesota Press, 1991), 132–147.

14. Leslie C. Dunn and Nancy A. Jones, *Embodied Voices: Representing Female Vocality in Western Culture*. (Cambridge, England: Cambridge University Press, 1994); Barthes, "The Grain of the Voice."

15. Dunn and Jones, *Embodied Voices*, 1–2.

16. Cavarero, *For More Than One Voice*, 106–107; Peraino, *Listening to the Sirens*, 1–67.

17. Barbara Bradby, "Sampling Sexuality: Gender, Technology and the Body in Dance Music," *Popular Music* 12, no. 2 (1993): 158. See also Laura Mulvey, "Visual Pleasure and Narrative Cinema," *Screen* 16, no. 3 (October 1, 1975): 6–18, https://doi.org/10.1093/screen/16 .3.6. Mulvey's theory of the "male gaze" is revisited in Silverman, *The Acoustic Mirror*, 28–29; Elizabeth Hoffman, "'I'-Tunes: Multiple Subjectivities and Narrative Method in Computer Music," *Computer Music Journal* 36, no. 4 (December 2012): 40–58, https://doi.org/10.1162 /COMJ_a_00152.

18. Bosma, "Bodies of Evidence, Singing Cyborgs and Other Gender Issues," 12.

19. Eve Kosofsky Sedgwick, *Epistemology of the Closet* (Berkeley: University of California Press, 1990), 8.

20. Bosma, "The Electronic Cry," 54.

21. Butler, "Performative Acts and Gender Constitution," 520.

22. By using the term "transition," I intentionally draw a connection between gender reassignment and sound production.

23. Emmerson, *Living Electronic Music*, 14–16.

24. Connor, "Panophonia"; see also Connor, *Dumbstruck*, 353.

25. Truax, "The Inner and Outer Complexity of Music," 177.

26. Truax, "Homoeroticism and Electroacoustic Music," 119.

27. Smalley, "Defining Transformations," 279.

28. David M. Carr, "Gender and the Shaping of Desire in the Song of Songs and Its Interpretation," *Journal of Biblical Literature* 119, no. 2 (2000): 233–248.

29. Michael V. Fox, *The Song of Songs and the Ancient Egyptian Love Songs* (Madison, WI: University of Wisconsin Press, 1985), 296.

30. Fox, *The Song of Songs and the Ancient Egyptian Love Songs*, 299.

31. Fox, *The Song of Songs and the Ancient Egyptian Love Songs*, 323.

32. Fox, *The Song of Songs and the Ancient Egyptian Love Songs*, 255.

33. Source recordings for the environmental sounds, including the monk, were captured at the Santissima Annunziata monastery, near Amelia, Italy, in 1988. The warm summer weather reminded the Monk of Christmas in his native Argentinia, which is why he sings a popular holiday song, although it was the dead of summer. In this regard, Traux mentioned to me the erotic qualities of the Mediterranean climate, personal communication, Berlin, Germany, June 17, 2014.

34. Barry Truax, "The PODX System: Interactive Compositional Software for the DMX-1000," *Computer Music Journal* 9, no. 1 (1985): 29–38; Barry Truax, "Real-Time Granular Synthesis with a Digital Signal Processor," *Computer Music Journal* 12, no. 2 (1988): 14–26.

35. Barry Truax, "Granulation of Sampled Sound," February 2016, http://www.sfu.ca/.

36. Barry Truax, *Documentation DVD #3* (Cambridge Street Recordings, 2010).

37. V. I. Wolfe et al., "Intonation and Fundamental Frequency in Male-Female Transsexuals," *Journal of Speech and Hearing Disorders* 55 (1990): 43–50.

38. Findings based on Wolfe et al., "Intonation and Fundamental Frequency in Male-Female Transsexuals, 43–50.

39. P. S. J. Weston, "Discrimination of Voice Gender in the Human Audio Cortex," *NeuroImage* 105 (2015): 208–214.

40. For a primer on granular synthesis, including visual examples, see the composer's website Barry Truax, "Granular Synthesis," accessed April 27, 2021, https://www.sfu.ca/~truax/gran.html.

41. Weston, "Discrimination of Voice Gender in the Human Audio Cortex," 210.

42. Times indicate overall running time of the piece according to www.soundmakers.ca /soundstreams-commissions/song-of-songs-barry-truax. The time cue indicated in the score is included in parenthesis.

43. Alexa Lauren Woloshyn, "The Recorded Voice and the Mediated Body in Contemporary Canadian Electroacoustic Music" (Ottawa, Library and Archives Canada/Bibliothèque et Archives Canada, 2012), 159–160.

44. André LaCocque, "I Am Black and Beautiful," in *Scrolls of Love: Ruth and the Song of Songs*, ed. Peter S. Hawkins and Lesleigh Cushing Stahlberg (Fordham Univ Press, 2006), 170.

45. In lieu of bar numbers, system numbers are enumerated according to the score available from the composer, beginning anew with each movement.

46. In anonymized listening studies, upward inflection has been observed as a feminized trait, and is encouraged in "communication feminisation therapy" for individuals transitioning to the female gender, Adrienne Hancock, Lindsey Colton, and Fiacre Douglas, "Intonation and Gender Perception: Applications for Transgender Speakers," *Journal for Voice* 28, no. 2 (1997): 203–209.

47. Woloshyn, "The Recorded Voice and the Mediated Body," 152.

48. Sedgwick, *Epistemology of the Closet*, 1–2.

49. Truax, *Documentation DVD #3*.

50. The composer enriches the sound by overlaying many more grains than in the original, thus adding to the harmonic spectrum, which he harmonizes at a ratio of 4:2 (an octave below), 4:5 (a major third above), 4:6 (a perfect fifth above), and 4:12 (an octave plus a perfect fifth above). So, for example, the first note E flat sounds both at the fundamental (4) and one octave below, at harmonic 2, though I have not notated octave doubling.

51. Truax, *Documentation DVD #3*.

52. Truax, "Homoeroticism and Electroacoustic Music," 120.

53. Truax, "Homoeroticism and Electroacoustic Music," 119.

54. Sedgwick, *Epistemology of the Closet*, 89.

55. Holly Ingleton, "Recalibrating Fundamentals of Discipline and Desire through the Automatic Music Tent," *Contemporary Music Review* 35, no. 1 (2016): 80.

56. LaCocque, "I Am Black and Beautiful," 168.

57. Truax, "Homoeroticism and Electroacoustic Music," 119–120.

58. Truax, *Documentation DVD #3*.

59. McClary, "This Is Not the Story My People Tell," 142.

60. McClary, "This Is Not the Story My People Tell," 132.

61. McClary, "This Is Not the Story My People Tell," 136.

62. Similar to Truax's *Song of Songs*, in "O Superman" Anderson uses harmonization to populate the background "chorus" while speaking through a vocoder. Laurie Anderson, "How We Made Laurie Anderson's O Superman," *Guardian*, April 19, 2016, http://www.theguardian.com/culture/2016/apr/19/how-we-made-laurie-anderson-o-superman.

63. McClary, "This Is Not the Story My People Tell," 138.

64. McClary, "This Is Not the Story My People Tell," 139.

65. McClary, "Terminal Prestige," 74.

66. Truax, "The Aesthetics of Computer Music," 122.

67. Truax, "The Inner and Outer Complexity of Music," 185 emphasis added.

68. Truax, "The Inner and Outer Complexity of Music," 186.

69. Lewis, "Improvised Music after 1950."

CHAPTER 10

1. Lindsay Zoladz, "Refresh: The Lonely Futurism of TLC's FanMail," *Pitchfork*, May 4, 2012, https://pitchfork.com/features/article/8827-tlc/.

2. Lopes founded an orphanage in Honduras and devoted time there to promote holistic living and nourishment. This initiative toward healthy and balanced living continues today in Georgia, US, with The Lisa Lopes Foundation, supported in part by proceeds from her posthumously released album, *Eye Legacy* (2009). "The Lisa Lopes Foundation," accessed January 3, 2019, https://www.lisalopesfoundation.net.

3. "TLC—We're Resurrecting Left Eye . . . For a Reunion Tour," April 25, 2012, https://www.tmz.com/2012/04/25/tlc-reunion-tour-left-eye-lisa-lopes/.

4. Mitchell Morris, *The Persistence of Sentiment: Display and Feeling in Popular Music of the 1970s* (Berkeley, CA: University of California Press, 2013), 12.

5. Rose, *Black Noise*, 148.

6. Billboard Staff, "100 Greatest Girl Group Songs of All Time: Critics' Picks," *Billboard*, July 10, 2017, http://www.billboard.com/articles/columns/pop/7857816/100-greatest-girl-group-songs.

7. Nataki H. Goodall, "Depend on Myself: T.L.C. and the Evolution of Black Female Rap," *Journal of Negro History* 79, no. 1 (January 1, 1994): 85n, https://doi.org/10.2307/2717669.

8. Alexandria Lust, "'Understanding the Depth of the '90s Women': TLC's Political, Musical, and Artistic Complexities of Hip-Hop Feminist Thought, 1991–2002"

(master's thesis, Sarah Lawrence College, 2011), https://search-proquest-com.jproxy.nuim
.ie/docview/873565469?pq-origsite=summon.

9. Rose, *Black Noise*, 147.

10. Stephen Holden, "Donna Summer's Sexy Cinderella," *Rolling Stone*, January 12, 1978,
    54–56.

11. Rose, *Black Noise*, 147.

12. Rose, *Black Noise*, 148.

13. Justin Williams applies Serge Lacasse's term "allosonic quotation" to hip-hop sampling. Serge
    Lacasse, "Intertextuality and Hypertextuality in Recorded Popular Music," in *The Musical
    Work: Reality or Invention?*, ed. Michael Talbot (Liverpool: Liverpool University Press, 2000),
    35–58, https://doi.org/10.5949/liverpool/9780853238256.003.0003; quoted in Williams,
    Rhymin' and Stealin', 3.

14. Timing based on the first release of *FanMail* (5:26 duration version) https://open.spotify
    .com/track/76A1Fp65fO7oB0OQtyqz1O?si=1XTRnElsSiSSfIG5-jdJpA.

15. Timing based on the first release of *FanMail* (5:26 duration version) https://open.spotify
    .com/track/76A1Fp65fO7oB0OQtyqz1O?si=1XTRnElsSiSSfIG5-jdJpA.

16. Leibetseder, *Queer Tracks*, 36.

17. Leibetseder, *Queer Tracks*, 37.

18. Darren Levin, "How TLC's 'CrazySexyCool' Changed Everything," *FasterLouder*, April
    24, 2014, https://web.archive.org/web/20170219033559/http://fasterlouder.junkee.com
    /how-tlcscrazysexycool-changed-everything/837289.

19. Bradby, "Sampling Sexuality," 155.

20. Bradby, "Sampling Sexuality," 157.

21. Tricia Rose, *Longing to Tell: Black Women Talk about Sexuality and Intimacy* (New York: Far-
    rar, Straus and Giroux, 2003), 390–391.

22. Bradby, "Sampling Sexuality," 169.

23. Stratton, "Coming to the Fore," 117; 124.

24. Bradby, "Sampling Sexuality," 171.

25. Sofer, "Specters of Sex."

26. Corbett and Kapsalis, "Aural Sex," 103.

27. Collins, *Black Feminist Thought*, 81–82; see 85–86 in a musical context.

28. Brian R. Sevier, "Ways of Seeing Resistance: Educational History and the Conceptualiza-
    tion of Oppositional Action," *Taboo* 7, no. 1 (2003): 44.

29. Mark Anthony Neal, *What the Music Said: Black Popular Music and Black Public Culture*
    (New York: Routledge, 1999), 122.

30. Summer and Eliot, *Ordinary Girl*, 111.

31. Summer and Eliot, *Ordinary Girl*, 130.

32. For the legal ramifications of sampling, see Kembrew McLeod and Peter DiCola, *Creative License: The Law and Culture of Digital Sampling* (Durham, NC: Duke University Press, 2011); Paul D Miller, *Sound Unbound: Sampling Digital Music and Culture* (Cambridge, MA: MIT Press, 2008); and for an account specific to hip hop, see, Thomas G. Schumacher, "'This Is a Sampling Sport': Digital Sampling, Rap Music, and the Law in Cultural Production," in *That's the Joint! The Hip-Hop Studies Reader*, ed. Mark Anthony Neal and Murray Forman (New York: Routledge, 2004), 443–458.

33. Susana Loza, "Sampling (Hetero)Sexuality: Diva-Ness and Discipline in Electronic Dance Music," *Gender and Sexuality* 20, no. 3 (2001): 352.

34. Susana Loza, "Sampling (Hetero)Sexuality," 350.

35. Susana Loza, "Sampling (Hetero)Sexuality," 353.

36. Tom Briehan, "Rihanna: Good Girl Gone Bad," *Pitchfork*, June 15, 2007, https://pitchfork.com/reviews/albums/10320-good-girl-gone-bad/.

37. Leibetseder, *Queer Tracks*, 37.

38. Mel Stanfill, "Spinning Yarn with Borrowed Cotton: Lessons for Fandom from Sampling," *Cinema Journal* 54, no. 3 (2015): 131.

39. Hawkins, *Queerness in Pop Music*, 22.

40. Leibetseder, *Queer Tracks*, 41.

41. Dennis Denisoff, *Aestheticism and Sexual Parody 1840–1940* (Cambridge: Cambridge University Press, 2006), 3–4; quoted in Leibetseder, *Queer Tracks*, 41.

42. Fred E. Maus, "Glamour and Evasion: The Fabulous Ambivalence of the Pet Shop Boys," *Popular Music* 20, no. 3 (2001): 388.

43. Maus, "Glamour and Evasion," 389.

44. Holden, "Donna Summer's Sexy Cinderella."

45. Fink, "Do It ('til You're Satisfied)," 43.

46. Fink, "Do It ('til You're Satisfied)," 44.

47. Lust, "'Understanding the Depth of the '90s Women': TLC's Political, Musical, and Artistic Complexities of Hip-Hop Feminist Thought, 1991–2002," 74; Joan Morgan, "Revisit TLC's November 1994 Cover Story: 'The Fire This Time,'" *Vibe*, June 9, 2020, 66, https://www.vibe.com/featured/tlc-november-1994-cover-story-the-fire-this-time.

48. Morgan, "Revisit TLC's November 1994 Cover Story," 66.

49. Nosheen Iqbal, "TLC: 'I Will Never Forget the Day We Were Millionaires for Five Minutes,'" *Guardian*, June 24, 2017, http://www.theguardian.com/music/2017/jun/24/tlc-will-never-forget-day-we-were-millionaires-for-five-minutes.

50. Lust, "'Understanding the Depth of the '90s Women': TLC's Political, Musical, and Artistic Complexities of Hip-Hop Feminist Thought, 1991–2002," 74. In 1994, hip-hip scholar Joan Morgan, writing for *Vibe Magazine*, described the performer as "now celibate," Morgan, "Revisit TLC's November 1994 Cover Story," 66.

51. Rana A. Emerson, "'Where My Girls At?': Negotiating Black Womanhood in Music Videos," *Gender & Society* 16, no. 1 (February 2002): 126, https://doi.org/10.1177/0891243202 016001007.

52. Rose, *Black Noise*, 157.

53. Rose, *Black Noise*, 157–158.

54. Billboard Staff, "100 Greatest Girl Group Songs of All Time: Critics' Picks," *Billboard*, August 2, 2019, https://www.billboard.com/articles/columns/pop/7857816/100-greatest -girl-group-songs.

55. Deron Dalton, "Most Iconic '90s Girl Group: TLC or Spice Girls?," *The Tylt*, accessed August 2, 2019, https://thetylt.com/entertainment/tlc-spice-girls-90s-girl-groups.

56. Polly E. McLean, "Age Ain't Nothing But a Number: A Cross-Cultural Reading of Popular Music in the Construction of Sexual Expression among At-Risk Adolescents," *Popular Music and Society* 21, no. 2 (1997): 13.

57. Lauren Berlant, "Live Sex Acts (Parental Advisory: Explicit Material)," *Feminist Studies* 21, no. 2 (1995): 379–404.

58. Goodall, "Depend on Myself," 85.

59. Adam Krims, *Rap Music and the Poetics of Identity* (Cambridge: Cambridge University Press, 2000), 69.

60. Berlant, "Live Sex Acts (Parental Advisory: Explicit Material)," 398.

61. Berlant, "Live Sex Acts (Parental Advisory: Explicit Material)," 401.

62. Berlant, "Live Sex Acts (Parental Advisory: Explicit Material)," 400.

63. David M. Halperin, *How to Be Gay* (Cambridge, MA: Harvard University Press, 2012), 189–190.

64. Hawkins, *Queerness in Pop Music*, 22.

65. Hawkins, *Queerness in Pop Music*, 111.

66. Davis, *Blues Legacies and Black Feminism*, 129.

67. Davis, *Blues Legacies and Black Feminism*, 130.

68. Davis, *Blues Legacies and Black Feminism*, 131.

69. Davis, *Blues Legacies and Black Feminism*, 132.

70. Berlant, "Live Sex Acts (Parental Advisory: Explicit Material)," 382.

71. Berlant, "Live Sex Acts (Parental Advisory: Explicit Material)," 386.

72. Kevin Lynch, "Abstinence the Musical," accessed July 22, 2019, http://www.kevinlynchnj
.com/store/p18/%22Abstinence_the_Musical%22_-_Sheet_Music.html.

73. Collins, *Black Sexual Politics*, 196n30.

74. Breanne Fahs, "Radical Refusals: On the Anarchist Politics of Women Choosing Asexuality,"
Sexualities 13, no. 4 (August 1, 2010): 446, https://doi.org/10.1177/1363460710370650.

75. Eimert, "What Is Electronic Music?"

76. Myra T. Johnson, "Asexual and Autoerotic Women: Two Invisible Groups," in *The Sexually
Oppressed*, ed. Harvey L Gochros and Jean S Gochros (New York: Association Press, 1977),
96–107.

77. Sciatrix, "Asexual Journal Club: Johnson 1977," *The Asexual Agenda* (blog), May 17, 2013,
https://asexualagenda.wordpress.com/2013/05/16/asexual-journal-club-johnson-1977/.

78. Emily McCarty, "What It Means to Be on the Asexuality Spectrum," *Allure*, June 3, 2019,
https://www.allure.com/story/asexuality-spectrum-asexual-people-explain-what-it-means.

79. Samantha Bassler, "'But You Don't Look Sick': A Survey of Scholars with Chronic, Invisible
Illnesses and Their Advice on How to Live and Work in Academia," *Music Theory Online*
15, no. 3 and 4 (August 1, 2009), http://www.mtosmt.org/issues/mto.09.15.3/mto.09.15
.3.bassler.html.

80. Darlene Clark Hine, "For Pleasure, Profit, and Power: The Sexual Exploitation of Black
Women," in *African American Women Speak Out on Anita Hill—Clarence Thomas* (Detroit,
MI: Wayne State University Press, 1995), 382; quoted in Collins, *Black Feminist Thought*, 125.

81. Rupert Cornwell, "The New Voice of America," *Independent*, June 20, 1995, https://www
.independent.co.uk/news/the-new-voice-of-america-1587470.html

82. Rose, *Black Noise*, 176.

83. C. Riley Snorton, "Trapped in the Epistemological Closet," in *Nobody Is Supposed to Know:
Black Sexuality on the Down Low* (Minneapolis: University of Minnesota Press, 2014), 72,
https://www.upress.umn.edu/book-division/books/nobody-is-supposed-to-know.

84. Collins, *Black Feminist Thought*, 124.

85. Collins, *Black Sexual Politics*, 28.

86. Collins, *Black Sexual Politics*, 28.

87. Collins, *Black Sexual Politics*, 99.

88. E. Patrick Johnson, *Appropriating Blackness: Performance and the Politics of Authenticity*
(Duke University Press, 2003), 2.

89. For a discussion of pitch in the confluence of hip hop and R&B, see Alexander G. Weheliye, "'Feenin': Posthuman Voices in Contemporary Black Popular Music," *Social Text* 20,
no. 2 (June 1, 2002): 31–32; and specifically in rapping, see Mitchell Ohriner, "Analysing
the Pitch Content of the Rapping Voice," *Journal of New Music Research* 48, no. 5 (October
20, 2019): 2, https://doi.org/10.1080/09298215.2019.1609525.

CHAPTER 11

1. Roxane Gay, "Janelle Monáe's Afrofuture," *The Cut*, February 3, 2020, https://www.thecut.com/2020/02/janelle-monae-afrofuture.html.

2. Matt Diehl, "Janelle Monae: 'I Won't Be a Slave to My Own Belief System,'" *Rolling Stone* (blog), September 10, 2013, https://www.rollingstone.com/music/music-news/janelle-monae-i-wont-be-a-slave-to-my-own-belief-system-236096/.

3. Gary Graff and Janelle Monáe, "Janelle Monae Talks 'Electric Lady' Origins & Prince Collaboration," *Billboard*, September 9, 2013, https://www.billboard.com/articles/columns/the-juice/5687265/janelle-monae-talks-electric-lady-origins-prince-collaboration.

4. Janelle Monáe, *It's Code*, 2014, https://www.youtube.com/watch?v=SRZlDeSqcoM.

5. Gayle Murchison, "Let's Flip It! Quare Emancipations: Black Queer Traditions, Afrofuturisms, Janelle Monáe to Labelle," *Women and Music: A Journal of Gender and Culture* 22 (2018): 81â.

6. Sofer, "Specters of Sex."

7. In the United States, Title VII (the Civil Rights Act of 1964) "outlaws discrimination based on race, color, religion, sex, or national origin," but these rights did not extend to the LGBTQ+ community until very recently. The "Equality Act" bill passed only in February 2021.

8. García, "An Alternate History of Sexuality in Club Culture"; Luis-Manuel García, "Whose Refuge, This House?: The Estrangement of Queers of Color in Electronic Dance Music," in *The Oxford Handbook of Music and Queerness*, ed. Fred Everett Maus and Sheila Whiteley (Oxford: Oxford University Press, 2018), https://doi.org/10.1093/oxfordhb/9780199793525.013.49; Pwaangulongii Dauod, "Africa's Future Has No Space for Stupid Black Men," *Granta* (blog), July 13, 2016, https://granta.com/africas-future-has-no-space/; Shzr Ee Tan, "Performing the Closet: Gay Anti-Identities in Singapore's *a Cappella* Groups" (LBGTQ Study Group Business Meeting, Royal Musical Association, University of Liverpool, September 9, 2017).

9. See chapter 6. Currid, "'We Are Family': House Music and Queer Performativity." See also Jodie Taylor, "Making a Scene—Locality, Stylistic Distinction and Utopian Imaginations," in *Playing It Queer: Popular Music, Identity and Queer World-Making* (Peter Lang, 2012), 175–214.

10. "The Queerness of Hip Hop," special issue of *Palimpsest: A Journal on Women, Gender, and the Black International* 2, no. 2 (2013).

11. I thank my student David Buschmann for acquainting me with pornorap in our Music and Sexuality seminar. May his memory be a blessing.

12. C. Riley Snorton, *Nobody Is Supposed to Know: Black Sexuality on the Down Low* (Minneapolis: University of Minnesota Press, 2014), 77.

13. In 2019, these anti-Black currents running through music-theoretical discourses were exposed publicly in Philip Ewell's Plenary Presentation at the annual meeting of the

Society for Music Theory (Philip A. Ewell, "Music Theory and the White Racial Frame," https://vimeo.com/372726003). The subsequent racist backlash against Ewell's lecture, published in the Journal of Schenkerian Studies in 2020, became of public interest and broadcast on national new networks in the United States, for example NPR and CNN, and only reinforced that these issues are ever-present in the beliefs held by many music theorists today. Megan L. Lavengood, "Journal of Schenkerian Studies: Proving the Point," *Megan L. Lavengood* (blog), July 27, 2020, https://meganlavengood.com/2020/07/27/journal-of-schenkerian-studies-proving-the-point/. Ewell's findings have since been published, Philip A. Ewell, "Music Theory and the White Racial Frame," *Music Theory Online* 26, no. 2 (September 1, 2020), https://mtosmt.org/issues/mto.20.26.2/mto.20.26.2.ewell.html.

14. Francesca T. Royster, *Sounding Like a No-No? Queer Sounds and Eccentric Acts in the Post-Soul Era* (Ann Arbor: University of Michigan Press, 2012).

15. Rose, *Black Noise*, 21; quoted in Royster, *Sounding Like a No-No?*, 82.

16. Royster, *Sounding Like a No-No?*, 83.

17. Royster, *Sounding Like a No-No?*, 83.

18. Gwendolyn D. Pough, *Check It While I Wreck It: Black Womanhood, Hip-Hop Culture, and the Public Sphere* (Boston: Northeastern University, 2004), 78–83.

19. Royster, *Sounding Like a No-No?*, 154.

20. Elizabeth Wood, "Lesbian Fugue: Ethel Smyth's Contrapuntal Arts," in *Musicology and Difference: Music, Gender, and Sexuality in Music Scholarship*, ed. Ruth Solie (Berkeley: University of California Press, 1993), 173.

21. Murchison, "Let's Flip It! Quare Emancipations: Black Queer Traditions, Afrofuturisms, Janelle Monáe to Labelle," 86; Mark Dery, "Black to the Future: Interviews with Samuel R. Delany," in *Flame Wars: The Discourse of Cyberculture*, ed. Mark Dery (Duke University Press, 1994), 180.

22. George E. Lewis, "Foreword: After Afrofuturism," *Journal of the Society for American Music* 2, no. 2 (May 2008), https://doi.org/10.1017/S1752196308080048; James, "'Robo-Diva R&B.'"

23. Murchison, "Let's Flip It! Quare Emancipations," 84.

24. Murchison, "Let's Flip It! Quare Emancipations," 80.

25. Dery, "Black to the Future," 180.

26. Murchison, "Let's Flip It! Quare Emancipations," 89.

27. Murchison, "Let's Flip It! Quare Emancipations," 87, emphasis added.

28. Royster, *Sounding Like a No-No?*, 190.

29. E. Patrick Johnson, *Black. Queer. Southern. Women: An Oral History* (Durham: UNC Press Books, 2018), 2.

30. Jenna Wortham, "How Janelle Monáe Found Her Voice," *New York Times*, April 19, 2018, https://www.nytimes.com/2018/04/19/magazine/how-janelle-monae-found-her-voice.html.

CHAPTER 12

1. Topic Theory was conceived as a way of understanding references across musical types and styles in eighteenth-century music, but has since been deployed toward theorizing many kinds of musical phenomena. Danuta Mirka, "Introduction," in *The Oxford Handbook of Topic Theory* (Oxford: Oxford University Press, 2014), 1. In terms of expanded uses of Topic Theory, Thomas Johnson makes particularly salient points about the use of the C-major triad in twentieth-century music, Thomas Johnson, "Tonality as Topic: Opening A World of Analysis for Early Twentieth-Century Modernist Music," *Music Theory Online* 23, no. 4 (December 2017), https://doi.org/10.30535/mto.23.4.7.

2. Carrie Battan, "THEESatisfaction: AwE NaturalE," *Pitchfork*, March 29, 2012, https://pitchfork.com/reviews/albums/16444-awe-naturale/.

3. Kodwo Eshun, *More Brilliant Than the Sun: Adventures in Sonic Fiction* (London: Interlink Pub Group, 1999), 00[-006]-00[-005].

4. Rose, *Black Noise*, 21.

5. Paul von Hippel, "Questioning a Melodic Archetype: Do Listeners Use Gap-Fill to Classify Melodies?," *Music Perception: An Interdisciplinary Journal* 18, no. 2 (December 1, 2000): 139–153, https://doi.org/10.2307/40285906.

6. Robert S. Hatten, *Interpreting Musical Gestures, Topics, and Tropes: Mozart, Beethoven, Schubert* (Bloomington: Indiana University Press, 2004), 68.

7. Both E and B share four sharps C#-D#-F#-G#, where the song introduces A# only harmonically in the rap as punctuation.

8. Kyle Adams, "'People's Instinctive Assumptions and the Paths of Narrative': A Response to Justin Williams," *Music Theory Online* 15, no. 2 (June 1, 2009): paras. 12–13, http://www.mtosmt.org/issues/mto.09.15.2/mto.09.15.2.adams.html; Kyle Adams, "Further Discussion of Compositional Process," *Music Theory Online* 15, no. 2 (June 1, 2009), http://www.mtosmt.org/issues/mto.09.15.2/mto.09.15.2.adams_furtherdiscussion.php.

9. Justin A. Williams, "Beats and Flows: A Response to Kyle Adams, 'Aspects of the Music/Text Relationship in Rap,'" *Music Theory Online* 15, no. 2 (June 1, 2009), http://www.mtosmt.org/issues/mto.09.15.2/mto.09.15.2.williams.html.

10. Adams, "'People's Instinctive Assumptions and the Paths of Narrative,'" para. 6.

11. Adam Krims, "A Genre System for Rap Music," in *Rap Music and the Poetics of Identity* (Cambridge: Cambridge University Press, 2000), 65–66.

12. In the words of rapper Ice Cube, "Back then we was calling it 'reality rap'; 'gangster rap' is the name that the media coined." Kory Grow, "Ice Cube on Reliving N.W.A for

'Straight Outta Compton,'" *Rolling Stone*, April 15, 2015, https://www.rollingstone.com/music/music-features/ice-cube-on-n-w-as-reality-rap-and-straight-outta-compton-movie-106622/.

13. Pough, *Check It While I Wreck It*, 5.

14. Sound Check, "New Music: Electronic Psychedelic Soul Singer/Songwriter SassyBlack Thinks About Love on 'No More Weak Dates,'" *AFROPUNK*, June 2, 2016, https://afropunk.com/2016/06/new-music-electronic-psychedelic-soul-singersongwriter-sassyblack-thinks-about-love-on-no-more-weak-dates/; Hanif Abdurraqib et al., "10 Essential Psychedelic Soul Albums," *Treble* (blog), April 16, 2015, https://www.treblezine.com/22712-10-essential-psychedelic-soul-albums/.

15. Andrew Matson, "Album Review: 'Seattlecalifragilisticextrahelladopeness' by State of the Artist UPDATED," *Seattle Times*, May 24, 2010, https://www.seattletimes.com/entertainment/music/album-review-seattlecalifragilisticextrahelladopeness-by-state-of-the-artist-updated/.

16. Battan, "THEESatisfaction."

17. One of few academic exchanges engaging with musical-theoretical aspects of rap music (rather than sociocultural concerns or lyrics) occurred in *Music Theory Online* in 2009 between Kyle Adams and Justin Williams. In his rebuttal to Williams, Adams explains that including in "rap" Erykah Badu and Bone-Thugs-n-Harmony, artists who sing as well as rap, "speaks to the increasingly blurry boundary between rap, R&B, neo-soul, and other genres." Adams, "'People's Instinctive Assumptions and the Paths of Narrative.'"

18. Krims, "A Genre System for Rap Music," 65.

19. Krims, "A Genre System for Rap Music," 65.

20. Pough, *Check It While I Wreck It*, 85.

21. Tricia Rose, "Bad Sistas: Black Women Rappers and Sexual Politics in Rap Music," in *Black Noise: Rap Music and Black Culture in Contemporary America* (Hanover, NH: Wesleyan University Press, 1994), 146–182; Cheryl L. Keyes, "Empowering Self, Making Choices, Creating Spaces: Black Female Identity via Rap Music Performance," *Journal of American Folklore* 113, no. 449 (2000): 255–269; Pough, *Check It While I Wreck It*.

22. Keightley, "'Turn It Down!' She Shrieked: Gender, Domestic Space, and High Fidelity, 1948–59"; Nancy Guevara, "Women Writin' Rappin' Breakin'," in *Droppin' Science: Critical Essays on Rap Music and Hip Hop Culture*, ed. William Eric Perkins (Philadelphia, P.A: Temple University Press, 1996), 49–62; Christina Verán, "First Ladies," in *Hip Hop Divas*, ed. Vibe Magazine (New York: Three Rivers Press, 2001); Pough, *Check It While I Wreck It*, 84–85.

23. Philip A. Ewell, "Introduction to the Symposium on Kendrick Lamar's *To Pimp a Butterfly*," *Music Theory Online* 25, no. 1 (March 1, 2019): para. 9, http://mtosmt.org/issues/mto.19.25.1/mto.19.25.1.ewell.html.

24. Michael Veal, *Dub: Soundscapes and Shattered Songs in Jamaican Reggae* (Middletown, CT: Wesleyan University Press, 2013), 40.

25. Ramsey, *Race Music*, 3.

26. On the racialized division of American experimental music, see, for example, Lewis, "Improvised Music after 1950"; George E. Lewis, *A Power Stronger Than Itself: The AACM and American Experimental Music* (Chicago: University of Chicago Press, 2008); Sofer, "Categorising Electronic Music."

27. Alisha Lola Jones, "'You Are My Dwelling Place': Experiencing Black Male Vocalists' Worship as Aural Eroticism and Autoeroticism in Gospel Performance," *Women and Music: A Journal of Gender and Culture* 22 (2018): 5, https://doi.org/10.1353/wam.2018.0001.

28. Amy Coddington, "'Check Out the Hook While My DJ Revolves It,'" in *The Oxford Handbook of Hip Hop Music*, 2018, https://doi.org/10.1093/oxfordhb/9780190281090.013.35.

29. Robin James, "Dancing with Myself: Billie Eilish Is the Perfect Music for a Silent Disco," *Real Life*, September 26, 2019, https://reallifemag.com/dancing-with-myself/.

30. Robin James, "Dancing with Myself."

31. Robin James, "Dancing with Myself."

32. Robin James, "Dancing with Myself."

33. Allison McCracken, *Real Men Don't Sing: Crooning in American Culture* (Durham, NC: Duke University Press, 2015), https://www.dukeupress.edu/real-men-dont-sing.

34. James, "Dancing with Myself."

35. Zach Schonfeld, "How Billie Eilish Became an ASMR Icon," *Pitchfork*, April 16, 2019, https://pitchfork.com/thepitch/billie-eilish-asmr/.

36. Robin James, *The Sonic Episteme: Acoustic Resonance, Neoliberalism, and Biopolitics* (Durham NC: Duke University Press, 2019), 141.

37. James, *The Sonic Episteme* 141.

38. James, *The Sonic Episteme*, 136.

39. Alyssa Barna, "Perspective: These Are the Musicological Reasons Taylor Swift's New Album Sounds Dull," *Washington Post*, December 16, 2020, https://www.washingtonpost.com/outlook/2020/12/16/taylor-swift-musicology-evermore-boring/.

40. James, *The Sonic Episteme*, 144.

41. For an analysis of this work, see Toshie Kakinuma, "Sneezing Toward the Sun: The Human Voice in the Musique Concrète of Toru Takemitsu," *Contemporary Music Review* 37, nos. 1–2 (February 2018): 20–35, https://doi.org/10.1080/07494467.2018.1453334.

42. Ben Jakob, "The Best Israeli Hip Hop Artists You Need to Know," *Culture Trip*, May 10, 2018, https://theculturetrip.com/middle-east/israel/articles/the-best-israeli-hip-hop-artists-you-need-to-know/.

43. J. Griffith Rollefson, *Flip the Script: European Hip Hop and the Politics of Postcoloniality* (Chicago: University of Chicago Press, 2017), especially chapter 5 on "M.I.A.'s 'Terrorist Chic.'"

CONCLUSION

1. Story by Katharine Smyth, "The Tyranny of the Female-Orgasm Industrial Complex," *Atlantic*, April 26, 2021, https://www.theatlantic.com/health/archive/2021/04/weaponization-female-orgasm/618680/.

2. Suzanne Cusick, "On a Lesbian Relationship with Music: A Serious Effort Not to Think Straight," in *Queering the Pitch: The New Gay and Lesbian Musicology*, ed. Philip Brett, Elizabeth Wood, and and Gary C. Thomas (New York: Routledge, 2006), 69, emphasis in the original.

3. Cusick, "On a Lesbian Relationship with Music," 69.

4. Bosma, "Bodies of Evidence, Singing Cyborgs and Other Gender Issues in Electrovocal Music."

5. Andra McCartney, "Inventing Images: Constructing and Contesting Gender in Thinking about Electroacoustic Music," *Leonardo Music Journal* 5 (1995): 57–66, https://doi.org/10.2307/1513162; Hinkle-Turner, *Women Composers and Music Technology in the United States*; Rodgers, *Pink Noises*; Gluck, "Electroacoustic, Creative, and Jazz."

6. Sofer, "Categorising Electronic Music."

7. See Martha Mockus, "Lesbian Skin and Musical Fascination," in *Audible Traces: Gender, Identity, and Music*, edited by Elaine Barkin, Lydia Hamessley, and Benjamin Boretz, 51–69 (Zürich: Carciofoli Verlagshaus, 1999).

8. http://miyamasaoka.com/work/2017/vaginated-chairs-2/.

9. https://jenkutler.bandcamp.com/album/disembodied. I thank Amy Williamson for sharing this work with me.

10. Rachel Mundy, "Listening for Objectivity," in *Animal Musicalities: Birds, Beasts, and Evolutionary Listening*, Music/Culture (Middletown, CT: Wesleyan University Press, 2018), 125–145; Kate Crawford, *Atlas of AI: Power, Politics, and the Planetary Costs of Artificial Intelligence* (New Haven: Yale University Press, 2021), 4–5.

Bibliography

Abbate, Carolyn. "Music—Drastic or Gnostic?" *Critical Inquiry* 30, no. 3 (2004): 505–536.

Abdurraqib, Hanif, A.T. Bossenger, Paul Pearson, and Jeff Terich. "10 Essential Psychedelic Soul Albums." *Treble* (blog), April 16, 2015. https://www.treblezine.com/22712-10-essential -psychedelic-soul-albums/.

Adams, Kyle. "Further Discussion of Compositional Process." *Music Theory Online* 15, no. 2 (June 1, 2009). http://www.mtosmt.org/issues/mto.09.15.2/mto.09.15.2.adams_furtherdiscussion.php.

Adams, Kyle. "'People's Instinctive Assumptions and the Paths of Narrative': A Response to Justin Williams." *Music Theory Online* 15, no. 2 (June 1, 2009). http://www.mtosmt.org/issues/mto .09.15.2/mto.09.15.2.adams.html.

Adams, Norman. "Visualization of Musical Signals." In *Analytical Methods of Electroacoustic Music*, edited by Mary Simoni, 13–28. New York: Routledge, 2005.

Adorno, Theodor. "Sexual Taboos and Law Today." In *Critical Models: Interventions and Catchwords, Trans*, edited by Henry W. Pickford. New York: Columbia University Press, 1963.

Adorno, Theodor W. "The Form of the Phonograph Record." Translated by Thomas Y. Levin. *October* 55 (Winter 1990): 56–61.

Agamben, Giorgio. "Difference and Repetition: On Guy Debord's Films." In *Guy Debord and the Situationist International: Texts and Documents*, edited by Tom McDonough, translated by Brian Holmes, 313–319. Cambridge, MA: MIT Press, 1995.

Ahmed, Sara. *Living a Feminist Life*. Durham NC: Duke University Press, 2016.

Alderfer, Hannah, Beth Jaker, and Marybeth Nelson. *Diary of a Conference on Sexuality*. New York: Faculty Press, 1982.

Amin, Samir. "Unequal Development: An Essay on the Social Formations of Peripheral Capitalism." Translated by Brian Pierce. *ASA Review of Books* 4 (1978): 87–88. https://doi.org/10.2307/532256.

Andean, James. "Electroacoustic Mythmaking: National Grand Narratives in Electroacoustic Music." In *Confronting the National in the Musical Past*, edited by Elaine Kelly, Markus Mantere, and Derek Scott, 138–150. New York: Routledge, 2018.

Anderson, Laura. "Musique Concrète, French New Wave Cinema, and Jean Cocteau's Le Testament d'Orphée (1960)." *Twentieth-Century Music* 12, no. 2 (September 2015): 197–224. https://doi.org/10.1017/S1478572215000031.

Anderson, Laurie. "How We Made Laurie Anderson's O Superman." *Guardian*, April 19, 2016. www.theguardian.com/culture/2016/apr/19/how-we-made-laurie-anderson-o-superman.

Andrew Matson. "Album Review: 'Seattlecalifragilisticextrahelladopeness' by State of the Artist UPDATED." *Seattle Times*, May 24, 2010. https://www.seattletimes.com/entertainment/music /album-review-seattlecalifragilisticextrahelladopeness-by-state-of-the-artist-updated/.

Arena, James. *First Ladies of Disco: 29 Stars Discuss the Era and Their Singing Careers.* Jefferson, NC: McFarland , 2013.

Argott, Don. *Pornosonic.* Mini Mace Pro, 1999, compact disc. https://web.archive.org/web /19991013021409/http://pornosonic.com/.

Aristotle. *On Rhetoric: A Theory of Civic Discourse.* 2nd ed. Translated by George Alexander Kennedy. New York: Oxford University Press, 2007.

Artaud, Antonin. *The Theater and Its Double.* Translated by Mary Caroline Richards. New York: Grove Weidenfeld, 1958.

Attinello, Paul. "Performance and/or Shame." *Repercussion* 4, no. 2 (1996): 97–130.

Auner, Joseph. "'Sing It for Me': Posthuman Ventriloquism in Recent Popular Music." *Journal of the Royal Musical Association* 128, no. 1 (2003): 98–122.

Austern, Linda Phyllis. "'Forreine Conceites and Wandring Devises': The Exotic, the Erotic, and the Feminine." In *The Exotic in Western Music*, edited by Jonathan Bellman, 26–42. Boston: Northeastern University Press, 1998.

Babbitt, Milton. "The Composer as Specialist [First published as 'Who Cares If You Listen?']." *High Fidelity* 8 (1958): 38–40, 126–127.

Babbitt, Milton. "Twelve-Tone Rhythmic Structure and the Electronic Medium." *Perspectives of New Music* 1, no. 1 (1962): 49–79.

Barna, Alyssa. "Perspective: These Are the Musicological Reasons Taylor Swift's New Album Sounds Dull." *Washington Post*, December 16, 2020. https://www.washingtonpost.com/outlook /2020/12/16/taylor-swift-musicology-evermore-boring/.

Barthes, Roland. *Le Plaisir Du Texte.* Paris: Éditions du Seuil, 1973.

Barthes, Roland. "The Grain of the Voice." In *Image—Music—Text*, translated by Stephen Heath, 179–89. London: Fontana Press, 1977.

Bassler, Samantha. "'But You Don't Look Sick': A Survey of Scholars with Chronic, Invisible Illnesses and Their Advice on How to Live and Work in Academia." *Music Theory Online* 15, nos 3 and 4 (August 1, 2009). http://www.mtosmt.org/issues/mto.09.15.3/mto.09.15.3.bassler.html.

Bataille, Georges. "Kinsey, the Underworld and Work." In *Death and Sensuality: A Study of Eroticism and the Taboo,* translated by Mary Dalwood, 149–163. New York: Walker & Co., 1962.

Bataille, Georges. *L'Erotisme*. Paris: Les Editions de Minuit, 1957.

Battan, Carrie. "THEESatisfaction: AwE NaturalE." *Pitchfork*, March 29, 2012. https://pitchfork
.com/reviews/albums/16444-awe-naturale/.

Battier, Marc. "What the GRM Brought to Music: From Musique Concrète to Acousmatic
Music." *Organised Sound* 12, no. 3 (2007): 189–202.

Beauvoir, Simone de. *The Second Sex*. New York: Vintage Books, 1973.

Berlant, Lauren. "Live Sex Acts (Parental Advisory: Explicit Material)." *Feminist Studies* 21, no.
2 (1995): 379–404.

Billboard Staff. "100 Greatest Girl Group Songs of All Time: Critics' Picks." *Billboard*. Accessed
August 2, 2019. https://www.billboard.com/articles/columns/pop/7857816/100-greatest-girl-group
-songs.

Born, Georgina. *Music, Sound and Space: Transformations of Public and Private Experience*. Cambridge: Cambridge University Press, 2013.

Bosma, Hannah. "Bodies of Evidence, Singing Cyborgs and Other Gender Issues in Electrovocal Music." *Organised Sound* 8, no. 1 (April 2003): 5–17. https://doi.org/10.1017/S135577
180300102X.

Bosma, Hannah. "The Electronic Cry: Voice and Gender in Electroacoustic Music." PhD diss.,
University of Amsterdam, 2013.

Botstein, Leon. "Alban Berg and the Memory of Modernism." In *Alban Berg and His World*,
edited by Christopher Hailey. Princeton: Princeton University Press, 2010.

Boulez, Pierre. "Alea." *Perspectives of New Music* 3, no. 1 (Autumn–Winter 1964): 42–53.

Bradby, Barbara. "Sampling Sexuality: Gender, Technology and the Body in Dance Music."
Popular Music 12, no. 2 (1993): 155–176.

Brennan, Deirbhile. "CupcakKe Tearfully Announces That She Is Quitting Music for Good."
GCN, September 24, 2019. https://gcn.ie/cupcakke-tearfully-announces-quitting-music/.

Briehan, Tom. "Rihanna: Good Girl Gone Bad." *Pitchfork*, June 15, 2007. https://pitchfork.com
/reviews/albums/10320-good-girl-gone-bad/.

Brown, Jeffrey A. "Class and Feminine Excess: The Strange Case of Anna Nicole Smith." *Feminist
Review* 81 (2005): 74–94.

Buskin, Richard. "Donna Summer 'I Feel Love': Classic Tracks." *Sound On Sound*, October
2009. http://www.soundonsound.com/sos/oct09/articles/classictracks_1009.htm.

Butler, Judith. "Performative Acts and Gender Constitution: An Essay in Phenomenology and
Feminist Theory." *Theatre Journal* 40, no. 4 (1988): 519–531.

Cage, John. *The Selected Letters of John Cage*. Wesleyan University Press, 2016.

Carby, Hazel V. "It Jus Be's Dat Way Sometime: The Sexual Politics of Women's Blues." In *The
Jazz Cadence of American Culture*, edited by Robert G. O'Meally, 471–483. New York: Columbia
University Press, 1998.

Carmon, Iris. "Steven Tyler's Teenage Girlfriend Tells Her Side of the Abortion Story." *Jezebel*, May 24, 2011. https://jezebel.com/steven-tylers-teenage-girlfriend-tells-her-side-of-the-5805190.

Carr, David M. "Gender and the Shaping of Desire in the Song of Songs and Its Interpretation." *Journal of Biblical Literature* 119, no. 2 (2000): 233–248.

Caux, Daniel. "Les Danses Organiques du Luc Ferrari." *L'Art Vivant 43* (July 1973): 30–32.

Caux, Jacqueline. *Almost Nothing with Luc Ferrari*. Translated by Jérôme Hansen. Berlin/Los Angeles: Errant Bodies Press, 2012.

Cavarero, Adriana. *For More Than One Voice: Toward a Philosophy of Vocal Expression*. Stanford: Stanford University Press, 2005.

Cavicchi, Daniel. *Listening and Longing: Music Lovers in the Age of Barnum*. Middletown, CT: Wesleyan University Press, 2011.

Chadabe, Joel. *Electric Sound: The Past and Promise of Electronic Music*. Upper Saddle River, NJ: Prentice Hall, 1997.

Chaperon, Sylvie. "L'histoire contemporaine des sexualités en France." *Vingtième siècle. Revue d'histoire* 3 (2002): 47–59.

Chatterjee, Partha. *The Nation and Its Fragments: Colonial and Postcolonial Histories*. Princeton: Princeton University Press, 1993.

Chion, Michel. *Guide to Sound Objects: Pierre Schaeffer and Musical Research* (originally published as *Guide Des Objets Sonores: Pierre Schaeffer et La Recherche Musicale*). Translated by John Dack and Christine North. Paris: Editions Buchet/Chastel, 1983. http://monoskop.org/images/0/01/Chion_Michel_Guide_To_Sound_Objects_Pierre_Schaeffer_and_Musical_Research.pdf.

Cixous, Hélène. "The Laugh of the Medusa." Translated by Keith Cohen and Paula Cohen. *Signs* 1, no. 4 (Summer 1976): 875–893.

Clément, Catherine, and Or Opera. *The Undoing of Women*. Translated by Betsy Wing. Minneapolis: University of Minnesota Press, 1988.

Coddington, Amy. "Check Out the Hook While My DJ Revolves It: How the Music Industry Made Rap into Pop in the Late 1980s." In *The Oxford Handbook of Hip Hop Music*, edited by Justin D. Burton and Jason Lee Oakes. Oxford: Oxford University Press, 2018. https://doi.org/10.1093/oxfordhb/9780190281090.013.35.

Colin, Gordon, and Michel Foucault. "Prison Talk: An Interview." In *Power-Knowledge: Selected Interviews & Other Writings—1972–1977*, 37–54. New York: Pantheon Books, 1980.

Collins, Patricia Hill. *Black Feminist Thought: Knowledge, Consciousness and the Politics of Empowerment*. 2nd ed. New York: Routledge, 2000.

Collins, Patricia Hill. *Black Sexual Politics: African Americans, Gender, and the New Racism*. New York: Routledge, 2004.

Cone, Edward T. "The Creative Artist in the University." *American Scholar* 16, no. 2 (1947): 192–200.

Connor, Steven. *Dumbstruck: A Cultural History of Ventriloquism*. Oxford: Oxford University Press, 2000.

Connor, Steven. "Panophonia." Pompidou Centre, February 22, 2012. http://www.stevenconnor.com/panophonia/panophonia.pdf.

Cook, Susan C. "'R-E-S-P-E-C-T (Find Out What It Means to Me)': Feminist Musicology & the Abject Popular." *Women and Music: A Journal of Gender and Culture* 5 (2001): 140–44.

Corbett, John, and Terri Kapsalis. "Aural Sex: The Female Orgasm in Popular Sound." *TDR (1988–)* 40, no. 3 (1996): 102–111. https://doi.org/10.2307/1146553.

Cornwell, Rupert. "The New Voice of America." *Independent*, June 20, 1995. https://www.independent.co.uk/news/the-new-voice-of-america-1587470.html.

Cox, Geoffrey. "'There Must Be a Poetry of Sound That None of Us Knows . . .': Early British Documentary Film and the Prefiguring of Musique Concrète." *Organised Sound* 22, no. 2 (August 2017): 172–86. https://doi.org/10.1017/S1355771817000085.

Crane, Barbara B., and Jennifer Dusenberry. "Power and Politics in International Funding for Reproductive Health: The US Global Gag Rule." *Reproductive Health Matters* 12, no. 24 (2004): 128–37.

Crane, Larry. "Reinhold Mack: ELO, Queen, Black Sabbath & T. Rex | Tape Op Magazine | Longform Candid Interviews with Music Producers and Audio Engineers Covering Mixing, Mastering, Recording and Music Production." *Tape Op*, February 2011. https://tapeop.com/interviews/81/reinhold-mack/.

Crawford, Kate. *Atlas of AI: Power, Politics, and the Planetary Costs of Artificial Intelligence*. New Haven: Yale University Press, 2021.

Currid, Brian. "'We Are Family': House Music and Queer Performativity." In *Cruising the Performative: Interventions into the Representation of Ethnicity*, edited by Sue-Ellen Case, Philip Brett, and Susan Leigh Foster, 165–196. Bloomington: Indiana University Press, 1995.

Cusick, Suzanne. "On a Lesbian Relationship with Music: A Serious Effort Not to Think Straight." In *Queering the Pitch: The New Gay and Lesbian*, edited by Philip Brett, Elizabeth Wood, and Gary C. Thomas, 67–84. New York: Routledge, 2006.

Dalton, Deron. "Most Iconic '90s Girl Group: TLC or Spice Girls?" *The Tylt*. Accessed August 2, 2019. https://thetylt.com/entertainment/tlc-spice-girls-90s-girl-groups.

Dauod, Pwaangulongii. "Africa's Future Has No Space for Stupid Black Men." *Granta* (blog), July 13, 2016. https://granta.com/africas-future-has-no-space/.

Davies, Stephen. *Musical Meaning and Expression*. Ithaca, NY: Cornell University Press, 1994.

Davis, Angela Y. *Blues Legacies and Black Feminism: Gertrude "Ma" Rainey, Bessie Smith, and Billie Holiday*. New York: Vintage Books, 1995.

Defoe, Daniel. *The Adventures of Robinson Crusoe*. London, 1862.

DeJarnett, Megan. "The ICD Internal Review Part 1: There's No Policy Like No Policy." *Megan DeJarnett* (blog), March 20, 2021. https://megandejarnett.com/2021/03/20/icd-internal-review-part-1/.

DeLancey, Mark Dike, Rebecca Neh Mbuh, and Mark W. DeLancey. *Historical Dictionary of the Republic of Cameroon*. Lenham, MD: The Scarecrow Press, 2010.

Deleuze, Gilles. *Difference and Repetition*. Translated by Paul Patton. New York: Columbia University Press, 1994.

Deleuze, Gilles, and Félix Guattari. *A Thousand Plateaus: Capitalism and Schizophrenia*. Minneapolis, MN: University of Minnesota Press, 2007.

Deleuze, Gilles, and Félix Guattari. *Anti-Oedipus: Capitalism and Schizophrenia*. Translated by Robert Hurley, Mark Seem, and Helen R. Lare. New York: Viking, 1977.

Deleuze, Gilles, and Félix Guattari. *What Is Philosophy?* Translated by Hugh Tomlinson and Graham Burchell. New York: Columbia University Press, 1991.

Deleuze, Gilles, and Richard Howard. *Proust and Signs*. London: Continuum, 2008.

Denisoff, Dennis. *Aestheticism and Sexual Parody 1840–1940*. Cambridge: Cambridge University Press, 2006.

DeNora, Tia. "Music and Erotic Agency." *Body and Society* 3, no. 2 (1997): 43–65.

Derrida, Jacques. *Margins of Philosophy*. Translated by Alan Bass. Reprint edition. Chicago: University of Chicago Press, 1984.

Dery, Mark. "Black to the Future: Interviews with Samuel R. Delany." In *Flame Wars: The Discourse of Cyberculture*, 179–222. Durham, NC: Duke University Press, 1994.

Diehl, Matt. "Janelle Monae: 'I Won't Be a Slave to My Own Belief System.'" *Rolling Stone* (blog), September 10, 2013. https://www.rollingstone.com/music/music-news/janelle-monae-i-wont-be-a-slave-to-my-own-belief-system-236096/.

DiscoSaturdayNightTV. *Donna Summer—Love to Love You Baby (Original Extended Version) Oasis Records 1975*, 2013. https://www.youtube.com/watch?v=VRI1yzn2ugA.

Dolar, Mladen. *A Voice and Nothing More*. Cambridge, MA: MIT Press, 2006.

Douglass, Frederick. "The Color Line." *North American Review* 132 (June 1, 1881): 567–577.

Drott, Eric. "The Politics of Presque Rien." In *Sound Commitments: Avant-Garde Music and the Sixties*, edited by Robert Adlington, 145–166. Oxford: Oxford University Press, 2009.

Drucker, Johanna. "Review of Maria Beatty & Annie Sprinkle: *Sluts & Goddesses* Video Workshop." Edited by Kathy High and Marshall Reese. *Felix* 1, no. 3 (1993). http://www.e-felix.org/issue3/Sprinkle.html.

Dunn, Leslie C., and Nancy A. Jones. *Embodied Voices: Representing Female Vocality in Western Culture*. Cambridge: Cambridge University Press, 1994.

Dworkin, Andrea, Catherine MacKinnon, and Women Against Pornography. *The Reasons Why: Essays on the New Civil Rights Laws Recognizing Pornography as Sex Discrimination*. New York: Women Against Pornography, 1985.

Dyer, Richard. "In Defense of Disco." *Gay Left* 8 (Summer 1979): 20–23.

Dyer, Richard. *Only Entertainment*. London: Routledge, 1992.

Editors, The. "72 Songs to Add to Your Sex Playlist Right Now." *Cosmopolitan*, August 6, 2020. https://www.cosmopolitan.com/celebrity/news/songs-to-have-sex-to.

Eimert, Herbert. "What is Electronic Music?" *Die Reihe* 1 (1965): 1–10.

Emerson, Rana A. "'Where My Girls At?': Negotiating Black Womanhood in Music Videos." *Gender & Society* 16, no. 1 (February 2002): 115–35. https://doi.org/10.1177/0891243202016001007.

Emmerson, Simon. "EMAS and Sonic Arts Network (1979–2004): Gender, Governance, Policies, Practice." *Contemporary Music Review* 35, no. 1 (2016): 21–31.

Emmerson, Simon. *Living Electronic Music*. New York: Routledge, 2007.

Eshun, Kodwo. *More Brilliant Than the Sun: Adventures in Sonic Fiction*. London: Interlink Pub Group, 1999.

Ewell, Philip A. "Introduction to the Symposium on Kendrick Lamar's *To Pimp a Butterfly*." *Music Theory Online* 25, no. 1 (March 1, 2019). http://mtosmt.org/issues/mto.19.25.1/mto.19.25.1.ewell.html.

Ewell, Philip A. "Music Theory and the White Racial Frame." November 12, 2019. https://vimeo.com/372726003.

Ewell, Philip A. "Music Theory and the White Racial Frame." *Music Theory Online* 26, no. 2 (September 1, 2020). https://mtosmt.org/issues/mto.20.26.2/mto.20.26.2.ewell.html.

Fahs, Breanne. "Radical Refusals: On the Anarchist Politics of Women Choosing Asexuality." *Sexualities* 13, no. 4 (August 1, 2010): 445–461. https://doi.org/10.1177/1363460710370650.

Fink, Robert. "Do It ('til You're Satisfied): Repetitive Musics and Recombinant Desires." In *Repeating Ourselves: American Minimal Music as Cultural Practice*, 25–60. Berkeley: University of California Press, 2005.

Foucault, Michel. "Sex, Power, and the Politics of Identity." In *Ethics, Subjectivity, and Truth*, edited by Paul Rabinow, 163–175. New York: New Press, 1994.

Fox, Michael V. *The Song of Songs and the Ancient Egyptian Love Songs*. Madison: University of Wisconsin Press, 1985.

Fraley, Malaika. "Berkeley: Renowned Philosopher John Searle Accused of Sexual Assault and Harassment at UC Berkeley." *East Bay Times*, March 23, 2017. https://www.eastbaytimes.com/2017/03/23/berkeley-renowned-philosopher-john-searle-accused-of-sexual-assault-and-harassment-by-former-cal-aide/.

Frith, Simon, and Angela McRobbie. "Rock and Sexuality." In *On Record: Rock, Pop and the Written*, edited by Simon Frith and Andrew Goodwin, 371–389. New York: Routledge, 1978.

Galliano, Luciana. *Yogaku: Japanese Music in the 20th Century*. Scarecrow Press, 2002.

García, David F. *Listening for Africa: Freedom, Modernity, and the Logic of Black Music's African Origins*. Durham: Duke University Press, 2017.

García, Luis-Manuel. "An Alternate History of Sexuality in Club Culture." *Resident Advisor* (blog), January 28, 2014. https://www.residentadvisor.net/features/1927.

García, Luis-Manuel. "Whose Refuge, This House? The Estrangement of Queers of Color in Electronic Dance Music." In *The Oxford Handbook of Music and Queerness*, edited by Fred Everett Maus and Sheila Whiteley. Oxford: Oxford University Press, 2018. https://doi.org/10.1093/oxfordhb/9780199793525.013.49.

Gates, Henry Louis. *The Signifying Monkey: A Theory of African-American Literary Criticism*. Reprint edition. New York: Oxford University Press, 1989.

Gay, Roxane. "Janelle Monáe's Afrofuture." *The Cut*, February 3, 2020. https://www.thecut.com/2020/02/janelle-monae-afrofuture.html.

Gilman, Mikal. "Donna Summer: Is There Life After Disco?" *Rolling Stone*, March 23, 1978.

Gilman, Sander L. "Black Bodies, White Bodies: Toward an Iconography of Female Sexuality in Late Nineteenth Century Art, Medicine, and Literature." In *Race, Writing, and Difference*, edited by Henry Louis Gates, Jr., 223–261. Chicago: University of Chicago Press, 1986.

Gluck, Robert J. "Electroacoustic, Creative, and Jazz: Musicians Negotiating Boundaries." In *ICMC*, 2009.

Goh, Annie. "Sounding Situated Knowledges: Echo in Archaeoacoustics." *Parallax* 23, no. 3 (July 3, 2017): 283–304. https://doi.org/10.1080/13534645.2017.1339968.

Goodall, Nataki H. "Depend on Myself: T. L. C. and the Evolution of Black Female Rap." *Journal of Negro History* 79, no. 1 (January 1, 1994): 85–93. https://doi.org/10.2307/2717669.

Graff, Gary, and Janelle Monáe. "Janelle Monae Talks 'Electric Lady' Origins & Prince Collaboration." *Billboard*, September 9, 2013. https://www.billboard.com/articles/columns/the-juice/5687265/janelle-monae-talks-electric-lady-origins-prince-collaboration.

Graff, Gerald. *Professing Literature: An Institutional History*. Chicago: University of Chicago Press, 1989.

Grosz, Elizabeth. *Volatile Bodies: Toward a Corporeal Feminism*. Bloomington: Indiana University Press, 1994.

Grow, Kory. "Ice Cube on Reliving N.W.A. for 'Straight Outta Compton.'" *Rolling Stone*, April 15, 2015. https://www.rollingstone.com/music/music-features/ice-cube-on-n-w-as-reality-rap-and-straight-outta-compton-movie-106622/.

Guck, Marion A. "Edward T. Cone's 'The Composer's Voice': Beethoven as Dramatist." *College Music Symposium* 29 (1989): 8–18.

Guevara, Nancy. "Women Writin' Rappin' Breakin'." In *Droppin' Science: Critical Essays on Rap Music and Hip Hop Culture*, edited by William Eric Perkins, 49–62. Philadelphia: Temple University Press, 1996.

Halperin, David M. *How to Be Gay*. Cambridge, MA: Harvard University Press, 2012.

Hancock, Adrienne, Lindsey Colton, and Fiacre Douglas. "Intonation and Gender Perception: Applications for Transgender Speakers." *Journal for Voice* 28, no. 2 (1997): 203–209.

Hanna, Judith Lynne. *Dance, Sex, and Gender: Signs of Identity, Dominance, Defiance, and Desire.* Chicago: University of Chicago Press, 1988.

Hanninen, Dora A. "Associative Sets, Categories, and Music Analysis." *Journal of Music Theory* 48, no. 2 (2004): 147–218.

Harris, Sophie. "The 50 Best Sexy Songs Ever Made." *Time Out New York*, January 18, 2021. https://www.timeout.com/newyork/music/best-sexy-songs.

Hassard, John. "Representing Reality: *Cinéma Vérité.*" In *Organization-Representation: Work and Organizations in Popular Culture*, edited by John Hassard and Ruth Holliday, 41–66. London: Sage Publications, 1998.

Hasty, Christopher. "Segmentation and Process in Post-Tonal Music." *Music Theory Spectrum* 3, no. 1 (1981): 54–73.

Hatten, Robert S. *Interpreting Musical Gestures, Topics, and Tropes: Mozart, Beethoven, Schubert.* Bloomington: Indiana University Press, 2004.

Hawkins, Stan. *Queerness in Pop Music: Aesthetics, Gender Norms, and Temporality.* New York: Routledge, 2015.

Hayes, Eileen M. "Black Women and 'Women's Music.'" In *Black Women and Music: More than the Blues*, edited by Linda F. Williams and Eileen M. Hayes, 153–176. Urbana: University of Illinois Press, 2007.

Hayles, Katherine. *How We Became Posthuman: Virtual Bodies in Cybernetics, Literature, and Informatics.* Chicago: University of Chicago Press, 1999.

Hess, Mickey. "Was Foucault a Plagiarist? Hip-Hop Sampling and Academic Citation." *Computers and Composition* 23 (2006): 280–295.

Hickman, Roger. "Wavering Sonorities and the Nascent Film Noir Musical Style." *Journal of Film Music* 2, nos. 2–4 (2009): 165–174. https://doi.org/10.1558/jfm.v2i2-4.165.

Hine, Darlene Clark. "For Pleasure, Profit, and Power: The Sexual Exploitation of Black Women." In *African American Women Speak Out on Anita Hill—Clarence Thomas*, 168–77. Detroit: Wayne State University Press, 1995.

Hinkle-Turner, Elizabeth. *Women Composers and Music Technology in the United States.* Aldershot: Ashgate, 2006.

Hippel, Paul von. "Questioning a Melodic Archetype: Do Listeners Use Gap-Fill to Classify Melodies?" *Music Perception: An Interdisciplinary Journal* 18, no. 2 (December 1, 2000): 139–153. https://doi.org/10.2307/40285906.

Hisama, Ellie M. "Getting to Count." *Music Theory Spectrum* 43, no. 2 (2021): 349–363. https://doi.org/10.1093/mts/mtaa033.

Hoby, Hermoine. "Ronnie Spector Interview: 'The More Phil Tried to Destroy Me, the Stronger I Got.'" *Telegraph*, March 6, 2014. https://www.telegraph.co.uk/culture/music/rockandpopfeatures/10676805/Ronnie-Spector-interview-The-more-Phil-tried-to-destroy-me-the-stronger-I-got.html.

Hodgkinson, Tim. "An Interview with Pierre Schaeffer—Pioneer of Musique Concrète." *Recommended Records Quarterly* 2, no.1 (1987).

Hodkinson, Juliana. "Sonic Writing: A Vibrational Practice." *Seismograf/DMT* 31 (March 2014). http://seismograf.org/node/5502.

Hodkinson, Juliana, and Niels Rønsholdt. *Fish & Fowl*. Da Capo, 2011, compact disc.

Hoffman, Elizabeth. "'I'-Tunes: Multiple Subjectivities and Narrative Method in Computer Music." *Computer Music Journal* 36, no. 4 (December 2012): 40–58. https://doi.org/10.1162/COMJ_a_00152.

Holden, Stephen. "Donna Summer's Sexy Cinderella." *Rolling Stone*, January 12, 1978, 54–56.

Hubbs, Nadine. "'Jolene,' Genre, and the Everyday Homoerotics of Country Music: Dolly Parton's Loving Address of the Other Woman." *Women and Music: A Journal of Gender and Culture* 19, no. 1 (2015): 71–76. https://doi.org/10.1353/wam.2015.0017.

Hugill, Andrew. "The Origins of Electronic Music." In *The Cambridge Companion to Electronic Music*, 2nd ed., edited by Nick Collins and Julio d'Escrivan, 7–24. Cambridge: Cambridge University Press, 2017.

Ihde, Don. *Listening and Voice: Phenomenologies of Sound*. Albany: State University of New York Press, 2007.

Ingleton, Holly. "Recalibrating Fundamentals of Discipline and Desire Through the Automatic Music Tent." *Contemporary Music Review* 35, no. 1 (2016): 71–84.

Iqbal, Nosheen. "TLC: 'I Will Never Forget the Day We Were Millionaires for Five Minutes.'" *Guardian*, June 24, 2017. http://www.theguardian.com/music/2017/jun/24/tlc-will-never-forget-day-we-were-millionaires-for-five-minutes.

Irigaray, Luce. *This Sex Which Is Not One*. Translated by Catherine Porter. Ithaca: Cornell University Press, 1985.

Irigaray, Luce. "Veiled Lips." Translated by Sara Speidel. *Mississippi Review* 11, no. 3 (1983): 93–131.

Iverson, Jennifer. *Electronic Inspirations: Technologies of the Cold War Musical Avant-Garde*. New Cultural History of Music. Oxford: Oxford University Press, 2019.

Jackson, Zakiyyah Iman. *Becoming Human: Matter and Meaning in an Antiblack World*. New York: NYU Press, 2020.

Jakob, Ben. "The Best Israeli Hip Hop Artists You Need to Know." *Culture Trip*. Last modified May 10, 2018. https://theculturetrip.com/middle-east/israel/articles/the-best-israeli-hip-hop-artists-you-need-to-know/.

James, Robin. "Dancing with Myself: Billie Eilish Is the Perfect Music for a Silent Disco." *Real Life*, September 26, 2019. https://reallifemag.com/dancing-with-myself/.

James, Robin. "'Robo-Diva R&B': Aesthetics, Politics, and Black Female Robots in Contemporary Popular Music." *Journal of Popular Music Studies* 20, no. 4 (December 2008): 402–423. https://doi.org/10.1111/j.1533-1598.2008.00171.x.

James, Robin. *The Sonic Episteme: Acoustic Resonance, Neoliberalism, and Biopolitics*. Durham, NC: Duke University Press, 2019.

Jameson, Frederic. "Of Islands and Trenches: Neutralization and the Production of Utopian Discourse." In *The Ideologies of Theory*, 386–414. London: Verso, 2008.

Jameson, Frederic. "World Reduction in Le Guin: The Emergence of Utopian Narrative." *Science Fiction Studies* 2, no. 3 (1975): 221–230.

Janelle Monáe. "It's Code." YouTube video, 4:05, November 5, 2014, https://www.youtube.com/watch?v=SRZlDeSqcoM.

Jardine, Alice. *Gynesis: Configurations of Women and Modernity*. Ithaca: Cornell University Press, 1985.

Jenny, Herbert J. "Perotin's 'Viderunt Omnes.'" *Bulletin of the American Musicological Society* 6 (August 1, 1942): 20. https://doi.org/10.2307/829204.

Johnson, E. Patrick. *Appropriating Blackness: Performance and the Politics of Authenticity*. Durham: Duke University Press, 2003.

Johnson, E. Patrick. *Black. Queer. Southern. Women. An Oral History*. Chapel Hill: The University of North Carolina Press, 2018.

Johnson, Myra T. "Asexual and Autoerotic Women: Two Invisible Groups." In *The Sexually Oppressed*, edited by Harvey L Gochros and Jean S Gochros, 96–107. New York: Association Press, 1977.

Johnson, Thomas. "Tonality as Topic: Opening a World of Analysis for Early Twentieth-Century Modernist Music." *Music Theory Online* 23, no. 4 (December 2017). https://doi.org/10.30535/mto.23.4.7.

Johnston, Jill. *Lesbian Nation: The Feminist Solution*. New York: Simon and Schuster, 1973.

Jones, Alisha Lola. "'You Are My Dwelling Place': Experiencing Black Male Vocalists' Worship as Aural Eroticism and Autoeroticism in Gospel Performance." *Women and Music: A Journal of Gender and Culture* 22 (2018): 3–21. https://doi.org/10.1353/wam.2018.0001.

Jordan, Stephanie. "British Modern Dance: Early Radicalism." *Dance Research: The Journal of the Society for Dance Research* 7, no. 2 (1989): 3–15. https://doi.org/10.2307/1290769.

Kakinuma, Toshie. "Sneezing Toward the Sun: The Human Voice in the Musique Concrète of Toru Takemitsu." *Contemporary Music Review* 37, nos. 1–2 (February 2018): 20–35. https://doi.org/10.1080/07494467.2018.1453334.

Kalinak, Kathryn. *Settling the Score: Music and the Classical Hollywood Film*. Madison: University of Wisconsin Press, 1992.

Kane, Angela. "Richard Alston: Twenty-One Years of Choreography." *Dance Research: The Journal of the Society for Dance Research* 7, no. 2 (1989): 16–54. https://doi.org/10.2307/1290770.

Kane, Brian. "L'Objet Sonore Maintenant: Pierre Schaeffer, Sound Objects and the Phenomenological Reduction." *Organised Sound* 12, no. 1 (April 2007): 15–24. https://doi.org/10.1017/S135577180700163X.

Kane, Brian. *Sound Unseen: Acousmatic Sound in Theory and Practice*. Oxford: Oxford University Press, 2014.

Katz, Mark. "Sampling Before Sampling: The Link between DJ and Producer." *Samples: Online-Publikationen Des Arbeitskreis Studium Populärer Musik* 9 (2010): 1–11.

Keightley, Keir. "'Turn It Down!' She Shrieked: Gender, Domestic Space, and High Fidelity, 1948–59." *Popular Music* 15, no. 2 (1996): 149–177.

Kerton, Sarah. "Too Much, Tatu Young: Queering Politics in the World of Tatu." In *Queering the Popular Pitch*, edited by Sheila Whiteley and Jennifer Rycenga, 155–168. New York: Routledge, 2013.

Keyes, Cheryl L. "Empowering Self, Making Choices, Creating Spaces: Black Female Identity via Rap Music Performance." *Journal of American Folklore* 113, no. 449 (2000): 255–269.

Kierkegaard, Soren. "The Immediate Erotic Stages or the Musical Erotic." In *Either/Or: A Fragment of Life*, translated by Alastair Hannay, 113–284. London: Penguin UK, 2004.

Kinsey, Alfred Charles, Wardell B. Pomeroy, and Clyde E. Martin. *Sexual Behavior in the Human Male*. Bloomington: Indiana University Press, 1948.

Kinsey, Alfred Charles, Wardell B. Pomeroy, Clyde E. Martin, and Paul H. Gebhard. *Sexual Behavior in the Human Female*. Bloomington: Indiana University Press, 1953.

Kofman, Sarah, and Mara Dukats. "Rousseau's Phallocratic Ends." *Hypatia* 3, no. 3 (1988): 123–136.

Krims, Adam. "A Genre System for Rap Music." In *Rap Music and the Poetics of Identity*, 46–92. Cambridge: Cambridge University Press, 2000.

Krims, Adam. *Rap Music and the Poetics of Identity*. Cambridge: Cambridge University Press, 2000.

Kröpfl, Francisco. "Electronic Music: From Analog Control to Computers." *Computer Music Journal* 21, no. 1 (1997): 26–28. https://doi.org/10.2307/3681212.

Kumar, Anita. "What's the Matter?: Shakti's (Re)Collection of Race, Nationhood, and Gender." *TDR: The Drama Review* 50, no. 1 (2006): 72–95.

Lacan, Jacques. *The Four Fundamental Concepts of Psychoanalysis: The Seminar of Jacques Lacan*. Edited by Jacques-Alain Miller. Translated by Alan Sheridan. Book 11. New York: W. W. Norton, 1981.

Lacasse, Serge. "Intertextuality and Hypertextuality in Recorded Popular Music." In *The Musical Work: Reality or Invention?*, edited by Michael Talbot, 35–58. Liverpool: Liverpool University Press, 2000. https://doi.org/10.5949/liverpool/9780853238256.003.0003.

LaCocque, André. "I Am Black and Beautiful." In *Scrolls of Love: Ruth and the Song of Songs*, edited by Peter S. Hawkins and Lesleigh Cushing Stahlberg, 162–171. New York: Fordham University Press, 2006.

Lang, Cady. "President Trump Has Attacked Critical Race Theory. Here's What to Know About the Intellectual Movement." *Time*, September 29, 2020. https://time.com/5891138/critical-race -theory-explained/.

Langer, Susanne K. *Philosophy in a New Key: A Study in the Symbolism of Reason, Rite, and Art.* Cambridge, MA: Harvard University Press, 1949.

Latham, Clara Hunter. "Rethinking the Intimacy of Voice and Ear." *Women and Music: A Journal of Gender and Culture* 19 (2015): 125–132.

Lavengood, Megan L. "Journal of Schenkerian Studies: Proving the Point." *Megan L. Lavengood* (blog), July 27, 2020. https://meganlavengood.com/2020/07/27/journal-of-schenkerian-studies -proving-the-point/.

Lavitt, Pamela Brown. "First of the Red Hot Mamas: 'Coon Shouting' and the Jewish Ziegfeld Girl." *American Jewish History* 87, no. 4 (1999): 253–290.

Leibetseder, Doris. *Queer Tracks: Subversive Strategies in Rock and Pop Music.* Farnham: Ashgate, 2013.

Lemoine-Luccioni, Eugénie. *Partage Des Femmes.* Paris: Editions du Seuil, 1976.

Lentjes, Rebecca. "Surreal Conjunctions." *VAN Magazine*, July 26, 2017. https://van-us.atavist .com/surreal-conjunctions.

Levin, Darren. "How TLC's 'CrazySexyCool' Changed Everything." *FasterLouder*, April 24, 2014.

Levinson, Jerrold. "Erotic Art and Pornographic Pictures." *Philosophy and Literature* 29, no. 1 (2005): 228–240. https://doi.org/10.1353/phl.2005.0009.

Lewis, George E. *A Power Stronger Than Itself: The AACM and American Experimental Music.* Chicago: University of Chicago Press, 2008.

Lewis, George E. "Foreword: After Afrofuturism." *Journal of the Society for American Music* 2, no. 2 (May 2008). https://doi.org/10.1017/S1752196308080048.

Lewis, George E. "Improvised Music after 1950: Afrological and Eurological Perspectives." *Black Music Research Journal* 16, no. 1 (1996): 91. https://doi.org/10.2307/779379.

Lewis, Waylon. "Beautiful Agony," September 19, 2013. http://www.elephantjournal.com/2013 /04/beautiful-agony-8-songs-that-parallel-the-rhythmic-path-to-orgasm/.

Locey, Elizabeth. *The Pleasure of the Text: Violette Leduc and Reader Seduction.* Lanham, MD: Rowman, 2002.

Lochhead, Judy. "Hearing 'Lulu.'" In *Audible Traces: Gender, Identity, and Music*, edited by Elaine Barkin, Lydia Hamessley, and Benjamin Boretz, 233–252. Zürich: Carciofoli Verlagshaus, 1999.

Lockwood, Annea. *Early Works 1967–82.* EM1046CD, 2007, compact disc.

Lockwood, Annea. "Sound Mapping the Danube River from the Black Forest to the Black Sea: Progress Report, 2001-03." *Soundscape: Journal of Acoustic Ecology* 5, no. 1 (2004): 32–34.

Lockwood, Annea. "What Is a River? Down the Danube by Ear." *Soundscape: Journal of Acoustic Ecology* 7, no. 1 (2007): 43–44.

Loza, Susana. "Sampling (Hetero)Sexuality: Diva-Ness and Discipline in Electronic Dance Music." *Gender and Sexuality* 20, no. 3 (2001): 349–357.

Luening, Otto. "An Unfinished History of Electronic Music." *Music Educators Journal* 55, no. 3 (1968): 42–49.

Lust, Alexandria. "'Understanding the Depth of the '90s Women': TLC's Political, Musical, and Artistic Complexities of Hip-Hop Feminist Thought, 1991–2002." MA diss., Sarah Lawrence College, 2011. https://search-proquest-com.jproxy.nuim.ie/docview/873565469?pq-origsite =summon.

Lynch, Kevin. "Abstinence the Musical." Accessed July 22, 2019. http://www.kevinlynchnj.com /store/p18/%22Abstinence_the_Musical%22_-_Sheet_Music.html.

MacKinnon, Catherine. "Feminism, Marxism, Method, and the State: An Agenda for Theory." *Signs* 7, no. 3 (1982): 515–544.

Maes, Hans. "Why Can't Pornography Be Art?" In *Art and Pornography: Philosophical Essays*, edited by Jerrold Levinson and Hans Maes, 17–47. Oxford: Oxford University Press, 2012.

Maher, Natalie, and Nicole DeMarco. "40 Songs That Probably Aren't on Your Sex Playlist (But Should Be)." *Harper's Bazaar,* May 16, 2020. https://www.harpersbazaar.com/culture/art-books -music/a23538483/best-sexy-songs/.

Manning, Peter. *Electronic and Computer Music*. Oxford: Oxford University Press, 2013.

Maus, Fred E. "Glamour and Evasion: The Fabulous Ambivalence of the Pet Shop Boys." *Popular Music* 20, no. 3 (2001): 379–393.

McAdams, Stephen, and Albert Bregman. "Hearing Musical Streams." *Computer Music Journal* 3, no. 4 (1979): 26–42.

McCann, James January. "Laoidh Chab an Dosáin." Master's thesis, Aberystwyth University, 2011.

McCartney, Andra. "'Creating Worlds for My Music to Exist': How Women Composers of Electroacoustic Music Make Place for Their Voices." Master's thesis, York University, 1997.

McCartney, Andra. "Inventing Images: Constructing and Contesting Gender in Thinking about Electroacoustic Music." *Leonardo Music Journal* 5 (1995): 57–66. https://doi.org/10.2307/1513162.

McCarty, Emily. "What It Means to Be on the Asexuality Spectrum." *Allure*, June 3, 2019. https://www.allure.com/story/asexuality-spectrum-asexual-people-explain-what-it-means.

McCaskill, Barbara, and Caroline Gebhard. *Post-Bellum, Pre-Harlem: African American Literature and Culture, 1877–1919*. New York: NYU Press, 2006.

McClary, Susan. *Feminine Endings: Music, Gender, and Sexuality*. Minneapolis: University of Minnesota Press, 1991.

McClary, Susan. "Terminal Prestige: The Case of Avant-Garde Music Composition." *Cultural Critique* (1989): 57–81.

McClary, Susan. "This Is Not the Story My People Tell: Musical Time and Space According to Laurie Anderson." In *Feminine Endings: Music, Gender, and Sexuality*, 132–147. Minneapolis: University of Minnesota Press, 1991.

McCracken, Allison. *Real Men Don't Sing: Crooning in American Culture*. Durham, NC: Duke University Press, 2015. https://www.dukeupress.edu/real-men-dont-sing.

McLaughlin, Mark. *Disco: Spinning the Story*, 2005.

McLean, Polly E. "Age Ain't Nothing But a Number: A Cross-Cultural Reading of Popular Music in the Construction of Sexual Expression among At-Risk Adolescents." *Popular Music and Society* 21, no. 2 (1997): 1–16.

McLeod, Kembrew, and Peter DiCola. *Creative License: The Law and Culture of Digital Sampling*. Durham, NC: Duke University Press, 2011.

Meduri, Avanthi. "Bharata Natyam—What Are You?" *Asian Theatre Journal* 5, no. 1 (1988): 1–22.

Merrich, Johann. *Le Pioniere della Musica Elettronica*. Milan: Auditorium Edizioni, 2012.

Meyer, Leonard. *Style and Music: Theory, History, and Ideology*. Chicago: University of Chicago Press, 1989.

Miller, Paul D. *Sound Unbound: Sampling Digital Music and Culture*. Cambridge, MA: MIT Press, 2008.

Mini, Darshana Sreedhar. "'Un-Sound' Sounds: Pornosonics of Female Desire in Indian Mediascapes." *Music, Sound, and the Moving Image* 13, no. 1 (September 11, 2019): 3–30.

Mirka, Danuta. "Introduction." In *The Oxford Handbook of Topic Theory*, 1–57. Oxford: Oxford University Press, 2014.

Mockus, Martha. *Sounding Out: Pauline Oliver's and Lesbian Musicality*. New York: Routledge, 2008.

Monson, Ingrid. "Music and the Anthropology of Gender and Cultural Identity." *Women and Music: A Journal of Gender and Culture* 1, no. 1 (1997): 58–67.

Montano, Linda, Annie Sprinkle, and Veronica Vera. "Summer Saint Camp 1987: With Annie Sprinkle and Veronica Vera." *TDR* 33, no. 1 (Spring 1989): 94–103. https://doi.org/10.2307/1145947.

Morgan, Frances. "Pioneer Spirits: New Media Representations of Women in Electronic Music History." *Organised Sound* 22, no. 02 (August 2017): 238–249. https://doi.org/10.1017/S1355771817000140.

Morgan, Joan. "Revisit TLC's November 1994 Cover Story: 'The Fire This Time.'" *Vibe*, June 9, 2020. https://www.vibe.com/featured/tlc-november-1994-cover-story-the-fire-this-time.

Morris, Mitchell. *The Persistence of Sentiment: Display and Feeling in Popular Music of the 1970s*. Berkeley: University of California Press, 2013.

Morrison, Matthew D. "Sound in the Construction of Race: From Blackface to Blacksound in Nineteenth-Century America." PhD diss., Columbia University, 2014.

Mulvey, Laura. "Visual Pleasure and Narrative Cinema." *Screen* 16, no. 3 (October 1, 1975): 6–18. https://doi.org/10.1093/screen/16.3.6.

Mundy, Rachel. "Listening for Objectivity." In *Animal Musicalities: Birds, Beasts, and Evolutionary Listening*, 125–145. Middletown, CT: Wesleyan University Press, 2018.

Murchison, Gayle. "Let's Flip It! Quare Emancipations: Black Queer Traditions, Afrofuturisms, Janelle Monáe to Labelle." *Women and Music: A Journal of Gender and Culture* 22 (2018): 79–90.

Nakamura, Lisa. "Indigenous Circuits: Navajo Women and the Racialization of Early Electronic Manufacture." *American Quarterly* 66, no. 4 (2014): 919–941. https://doi.org/10.1353/aq.2014.0070.

Nancy, Jean-Luc. *The Muses*. Translated by Peggy Kamuf. Stanford: Stanford University Press, 1996.

Neal, Mark Anthony. *What the Music Said: Black Popular Music and Black Public Culture*. New York: Routledge, 1999.

Nielsen Music, and Erin Crawford. "2017 Year-End Music Report: U.S.," January 3, 2018. https://www.nielsen.com/us/en/insights/report/2018/2017-music-us-year-end-report/.

Nietzsche, Friedrich, and Josefine Naukoff. *The Gay Science*. Cambridge: Cambridge University Press, 2001.

Normandeau, Robert. "Jeu." electroCD. Accessed April 21, 2021. https://electrocd.com/en/oeuvre/14380/Robert_Normandeau/Jeu.

Normandeau, Robert. "Jeu de Langues." electroCD, 2012. https://electrocd.com/en/piste/imed_12116-1.3.

Normandeau, Robert. *Lieux inouïs*. Experientes DIGITALes, 1998. IMED-CD 9802, compact disc.

Nyman, Michael. *Experimental Music: Cage and Beyond*. Cambridge: Cambridge University Press, 1974.

Ogborn, David. "Interview with Robert Normandeau." *EContact!* 11, no. 2 (March 26, 2009). http://cec.sonus.ca/econtact/11_2/normandeauro_ogborn.html.

O'Reilly, Patrick. *Dancing Tahiti*. Paris: Nouvelles Editions Latines, 1977.

O'Shea, Janet. "At Home in the World? The Bharatanatyam Dancer as Transnational Interpreter." *TDR* 47, no. 1 (2003): 182.

Oterion, Frank J. "Annea Lockwood beside the Hudson River." *NewMusicBox*, January 1, 2004. https://nmbx.newmusicusa.org/annea-lockwood-beside-the-hudson-river/.

Ouzounian, Gascia. "Sound Installation Art: From Spatial Poetics to Politics, Aesthetics to Ethics." In *Music, Sound and Space: Transformations of Public and Private Experience*, edited by Georgina Born, 73–89. Cambridge: Cambridge University Press, 2013.

Ouzounian, Gascia. *Stereophonica: Sound and Space in Science, Technology, and the Arts*. Cambridge, MA: MIT Press, 2021.

Paglia, Camille and Art. *Sex and American Culture: Essays*. New York: Vintage, 1992.

Parker, Holt N. "Love's Body Anatomized: The Ancient Erotic Handbooks and the Rhetoric of Sexuality." In *Pornography and Representation in Greece and Rome*. New York: Oxford University Press, 1992.

Penley, Constance. "Crackers and Whackers: The White Trashing of Porn." In *White Trash: Race and Class in America*, edited by Annalee Newitz and Matt Wray, 89–112. New York: Routledge, 1996.

Peraino, Judith. *Listening to the Sirens: Musical Technologies of Queer Identity from Homer to Hedwig*. Berkeley: University of California Press, 2006.

Peters, Deniz, Gerhard Eckel, and Andreas Dorschel, eds. *Bodily Expression in Electronic Music: Perspectives on Reclaiming Performativity*. New York: Routledge, 2012.

"Pornography, n." In *Oxford English Dictionary*. Oxford: Oxford University Press, June 2020. http://www.oed.com/viewdictionaryentry/Entry/148012.

Pough, Gwendolyn D. *Check It While I Wreck It: Black Womanhood, Hip-Hop Culture, and the Public Sphere*. Boston: Northeastern University, 2004.

Poulenc, Francis. "In Praise of Banality." In *Francis Poulenc: Articles and Interviews: Notes from the Heart*, edited by Nicolas Southon, 27–30. Farnham: Ashgate, 2014.

Proust, Marcel. *A la Recherche du temps perdu*. Paris: Bibliothèque de la Pleide, 1988.

Ramsey, Guthrie P. *Race Music: Black Cultures from Bebop to Hip-Hop*. Berkeley: University of California Press, 2004.

Ramsey, Guthrie P. "Who Hears Here? Black Music, Critical Bias, and the Musicological Skin Trade." *Musical Quarterly* 85, no. 1 (2001): 1–52.

Rangacharya, Adya. *Introduction to Bharata's Nātya-Śāstra*. Bombay: Popular Prakashan, 1966.

Rea, Michael C. "What Is Pornography?" *Noûs* 35, no. 1 (2001): 118–145.

Robert Dixon and John Godrich. *Recording the Blues*. New York: Stein and Day Books, 1970. http://archive.org/details/RecordingTheBlues.

Robindoré, Brigitte. "Interview with an Intimate Iconoclast." *Computer Music Journal* 22, no. 3 (1998): 8–16.

Rodgers, Tara. *Pink Noises: Women on Electronic Music and Sound*. Durham, NC: Duke University Press, 2010.

Rodgers, Tara. "Tinkering with Cultural Memory: Gender and the Politics of Synthesizer Historiography." *Feminist Media Histories* 1, no. 4 (October 1, 2015): 5–30. https://doi.org/10.1525/fmh.2015.1.4.5.

Rogin, Michael. *Blackface, White Noise: Jewish Immigrants in the Hollywood Melting Pot*. Berkeley: University of California Press, 1996.

Rollefson, J. Griffith. *Flip the Script: European Hip Hop and the Politics of Postcoloniality*. Chicago: University of Chicago Press, 2017.

Rose, Tricia. "Bad Sistas: Black Women Rappers and Sexual Politics in Rap Music." In *Black Noise: Rap Music and Black Culture in Contemporary America*, 146–182. Hanover, NH: Wesleyan University Press, 1994.

Rose, Tricia. *Black Noise: Rap Music and Black Culture in Contemporary America*. Hanover, NH: Wesleyan University Press, 1994.

Rose, Tricia. *Longing to Tell: Black Women Talk about Sexuality and Intimacy*. New York: Farrar, Straus & Giroux, 2003.

Rose, Tricia. *The Hip Hop Wars: What We Talk about When We Talk about Hip Hop—and Why It Matters*. New York: Perseus Book Group, 2008.

Rothenberg, David. "Three Ways Toward Deep Listening in the Natural World." *Sound American*, Artist Essays on Deep Listening, 7. Accessed November 24, 2020. http://archive.soundamerican .org/sa_archive/sa7/sa7-artist-essays-on-deep-listening.html.

Rousseau, Jean-Jacques. *Seconde partie des confessions de J. J. Rousseau, citoyen de Geneve. Edition enrichie d'un nouveau recueil de ses lettres . . .* Vol. 4. A Neuchatel: l'imprimerie de L. Fauche-Borel, imprimeur du Roi, 1790.

Rousseau, Jean-Jacques. *The Confessions of J. J. Rousseau, . . . Part the Second. To Which Is Added, a New Collection of Letters from the Author. Translated from the French. In Three Volumes . . .* Translated by Anonymous. Vol. 1. London: Printed for G. G. J. and J. Robinson, and J. Bew, 1790.

Royster, Francesca T. *Sounding Like a No-No? Queer Sounds and Eccentric Acts in the Post-Soul Era*. Ann Arbor: University of Michigan Press, 2012.

Rubin, Gayle. "Blood under the Bridge: Reflections on 'Thinking Sex.'" In *Deviations: A Gayle Rubin Reader*, 194–223. Durham, NC: Duke University Press, 2011.

Rubin, Gayle. "Thinking Sex: Notes for a Radical Theory of the Politics of Sexuality." In *Deviations: A Gayle Rubin Reader*, 137–182. Durham, NC: Duke University Press, 2011.

Sanders, Charles L. "Donna Summer." *Ebony* (October 1977): 33–42.

Sauvagnargues, Anne. "Deleuze and Guattari as VJay: Digital Art Machines." Presented at the The Dark Precursor: International Conference on Deleuze and Artistic Research, Ghent, Belgium, November 10, 2015.

Schaeffer, Pierre. *In Search of Concrete Music*. Translated by Christine North and John Dack. Berkeley: University of California Press, 2012.

Schaeffer, Pierre. *La musique concrète*. Paris: Presses Universitaires de France, 1973.

Schaeffer, Pierre. *L'Œuvre musicale*. Vol. 2. INA-GRM ina c 1006/7/8, 1990.

Schaeffer, Pierre. *Traité Des Objets Musicaux: Essai Interdisciplines*. Paris: Editions du Seuil, 1966.

Schafer, R. Murray. *The Soundscape: Our Sonic Environment and the Tuning of the World*. Rochester, VT: Destiny Books, 1977.

Schafer, R. Murray. "Schizophonia." In *The New Soundscape: A Handbook for the Modern Music Teacher*, 43. Ontario: Don Mills, 1969.

Schedel, Margaret. "Electronic Music and the Studio." In *The Cambridge Companion to Electronic Music*, 2nd ed., edited by Nick Collins and Julio d'Escrivan, 25–39. Cambridge: Cambridge University Press, 2017. https://doi.org/10.1017/9781316459874.004.

Schloss, Joseph Glenn. *Making Beats: The Art of Sample-Based Hip-Hop*. Middletown, CT: Wesleyan University Press, 2014.

Schonfeld, Zach. "How Billie Eilish Became an ASMR Icon." *Pitchfork*, April 16, 2019. https://pitchfork.com/thepitch/billie-eilish-asmr/.

Schumacher, Thomas G. "'This Is a Sampling Sport': Digital Sampling, Rap Music, and the Law in Cultural Production." In *That's the Joint!: The Hip-Hop Studies Reader*, edited by Mark Anthony Neal and Murray Forman, 443–458. New York: Routledge, 2004.

Scott, Derek B. *From the Erotic to the Demonic: On Critical Musicology*. New York: Oxford University Press, 2003.

Scruton, Roger. *Beauty: A Very Short Introduction*. Oxford: Oxford University Press, 2011.

Scruton, Roger. "Representation in Music." *Philosophy* 51, no. 197 (1976): 273–287.

Searle, John R. *Intentionality: An Essay in the Philosophy of Mind*. Cambridge: Cambridge University Press, 1983.

Searle, John R. *Making the Social World: The Structure of Human Civilization*. Oxford: Oxford University Press, 2010.

Sedgwick, Eve Kosofsky. *Epistemology of the Closet*. Berkeley: University of California Press, 1990.

Sevier, Brian R. "'Ways of Seeing Resistance: Educational History and the Conceptualization of Oppositional Action.'" *Taboo* 7, no. 1 (2003): 87–107.

Shields, Alice. "'Patient and Psychotherapist: The Music.'" In *The Psychoaesthetic Experience: An Approach to Depth-Oriented Treatment*, edited by Arthur Robbins, 57–77. New York: Human Sciences Press, 1989.

Shields, James Mark. "Eros and Transgression in an Age of Immanence: George Bataille's (Religious) Critique of Kinsey." *Journal of Religion & Culture* 13 (2000): 175–186.

Siclait, Aryelle, and Madeline Howard. "Guess How Many Rihanna Songs Are on This Sex Playlist." *Women's Health*, August 4, 2020. https://www.womenshealthmag.com/sex-and-love/a19935020/sex-playlist/.

Silverman, Kaja. *The Acoustic Mirror: The Female Voice in Psychoanalysis and Cinema*. Bloomington: Indiana University Press, 1988.

Singerling, L. "Donna Summer, Queen of Disco." July 5, 2012. http://donnasummequeenofdisco.blogspot.com/2012_07_05_archive.html.

Smalley, Denis. "Defining Transformations." *Interface* 22, no. 4 (November 1993): 279–300. https://doi.org/10.1080/09298219308570638.

Smalls, Shanté Paradigm. "Queer Hip Hop: A Brief Historiography." In *The Oxford Handbook of Music and Queerness*, edited by Fred Everett Maus and Sheila Whiteley. Oxford: Oxford University Press, 2018. https://doi.org/10.1093/oxfordhb/9780199793525.013.103.

Smith, Jacob. *Vocal Tracks: Performance and Sound Media*. Berkeley: University of California Press, 2008.

Smith, Patricia Juliana. *The Queer Sixties*. Hoboken: Taylor and Francis, 2013.

Smyth, Katherine. "The Tyranny of the Female-Orgasm Industrial Complex." *Atlantic*, April 26, 2021. https://www.theatlantic.com/health/archive/2021/04/weaponization-female-orgasm /618680/.

Snorton, C. Riley. *Nobody Is Supposed to Know: Black Sexuality on the Down Low*. Minneapolis: University of Minnesota Press, 2014.

Sofer, Danielle. "Categorising Electronic Music." *Contemporary Music Review* 39, no. 2 (2020): 231–251.

Sofer, Danielle. "Confined Spaces/Erupted Boundaries: Crowd Behavior in Prokofiev's *The Gambler*." Master's thesis, Stony Brook University, 2012.

Sofer, Danielle. "Eroticism and Time in Computer Music: Juliana Hodkinson and Niels Rønsholdt's *Fish & Fowl*." In *Music Technology Meets Philosophy: From Digital Echoes to Virtual Ethos*, 1 (2014): 148–153.

Sofer, Danielle. *Preface to Study Score "Erwin Schulhoff, Symphony No. 5."* Munich: Musikproduktion Jürgen Höflich, 2014.

Sofer, Danielle. "Specters of Sex: Tracing the Tools and Techniques of Contemporary Music Analysis." *Zeitschrift Der Gesellschaft Für Musiktheorie [Journal of the German-Speaking Society of Music Theory]* 17, no. 1 (2020): 31–63. https://doi.org/10.31751/1029.

Sofer, Danielle. "Strukturelles Hören? Neue Perspektiven Auf Den 'idealen' Hörer." In *Geschichte und Gegenwart ses Musikalischen Hörens. Diskurse—Geschichente(n)—Poetiken*, edited by Klaus Aringer, Franz Karl Praßl, Peter Revers, and Christian Utz, 107–132. Freiburg: Rombach Verlag, 2017.

Solie, Ruth, ed. *Musicology and Difference: Music, Gender, and Sexuality in Music Scholarship*. Berkeley: University of California Press, 1993.

Sommerstein, Alan H. *Aeschylean Tragedy*. London: Bloomsbury, 2010.

Soneji, Davesh. *Unfinished Gestures: Devadasis, Memory, and Modernity in South India*. Chicago: University of Chicago Press, 2012.

Sound Check. "NEW MUSIC: Electronic Psychedelic Soul Singer/Songwriter SassyBlack Thinks About Love on 'No More Weak Dates.'" *AFROPUNK*, June 2, 2016. https://afropunk.com/2016 /06/new-music-electronic-psychedelic-soul-singersongwriter-sassyblack-thinks-about-love-on-no -more-weak-dates/.

Southern, Eileen. *Biographical Dictionary of Afro-American and African Musicians*. Westport, CT: Greenwood Press, 1982. http://archive.org/details/biographicaldict00sout.

Spivak, Gayatri Chakravorty. "Theory in the Margin: Coetzee's Foe Reading Defoe's Crusoe/ Roxana." In *Consequences of Theory*, edited by Jonathan Arac and Barbara Johnson. Baltimore, MD: Johns Hopkins University Press, 1991.

Stalarow, Alexander John. "Listening to a Liberated Paris: Pierre Schaeffer Experiments with Radio." PhD diss., University of California, Davis, 2017.

Stanfill, Mel. "Spinning Yarn with Borrowed Cotton: Lessons for Fandom from Sampling." *Cinema Journal* 54, no. 3 (2015): 131–137.

Stiehl, Pamyla A. "Bharatanatyam: A Dialogical Interrogation of Feminist Voices in Search of the Divine Dance." *Journal of Religion and Theatre* 3, no. 2 (2004): 275–302.

Stoever, Jennifer Lynn. *The Sonic Color Line: Race and the Cultural Politics of Listening*. New York: NYU Press, 2016.

Stratton, Jon. "Coming to the Fore: The Audibility of Women's Sexual Pleasure in Popular Music and the Sexual Revolution." *Popular Music* 33, no. 1 (January 2014): 109–128.

Summer, Donna. The Fresh Air Interview: "Queen of Disco" Donna Summer. Interview by Terry Gross, November 4, 2003. https://www.npr.org/2003/11/04/1491690/queen-of-disco-donna-summer.

Summer, Donna, and Marc Eliot. *Ordinary Girl: The Journey*. New York: Villard, 2003.

Tan, Shzr Ee. "Performing the Closet: Gay Anti-Identities in Singapore's *a Cappella* Groups." Presented at the LBGTQ Study Group Business Meeting, Royal Musical Association, University of Liverpool, September 9, 2017.

Taylor, Jodie. "Making a Scene—Locality, Stylistic Distinction and Utopian Imaginations." In *Playing It Queer: Popular Music, Identity and Queer World-Making*, 175–214. Bern: Peter Lang, 2012.

Teruggi, Daniel. "Technology and Musique Concrète: The Technical Developments of the Groupe de Recherches Musicales and Their Implication in Musical Composition." *Organised Sound* 12, no. 3 (2007): 213–231.

Thapalyal, Ranjana. "Sringara Rasa: The Feminine Call of the Spiritual/Erotic Impulse in Indian Art." In *The Sacred and the Feminine: Imagination and Sexual Difference*, edited by Griselda Pollock and Victoria Urvey-Sauron, 135–149. London: I. B. Tauris, 2007.

"TLC—We're Resurrecting Left Eye . . . For a Reunion Tour." April 25, 2012. https://www.tmz.com/2012/04/25/tlc-reunion-tour-left-eye-lisa-lopes/.

Tormey, Alan. *The Concept of Expression: A Study in Philosophical Psychology and Aesthetics*. Princeton: Princeton University Press, 1971.

Truax, Barry. "The Aesthetics of Computer Music: A Questionable Concept Reconsidered." *Organised Sound* 5, no. 3 (2000): 119–126.

Truax, Barry. "Composing with Real-Time Granular Synthesis." *Perspectives of New Music* 28, no. 2 (1990): 120–134.

Truax, Barry. "Editorial: Context-Based Composition." *Organised Sound* 23, no. 1 (April 2018): 1–2. https://doi.org/10.1017/S1355771817000218.

Truax, Barry. "Electroacoustic Music and the Soundscape: The Inner and the Outer World." In *Companion to Contemporary Musical Thought*, edited by John Paynter, 374–398. New York: Routledge, 1992.

Truax, Barry. "Granular Synthesis." Accessed April 27, 2021. https://www.sfu.ca/~truax/gran.html.

Truax, Barry. "Granulation of Sampled Sound," February 2016. http://www.sfu.ca/.

Truax, Barry. "Homoeroticism and Electroacoustic Music: Absence and Personal Voice." *Organised Sound* 8, no. 1 (2003): 117–124.

Truax, Barry. "The Inner and Outer Complexity of Music." *Perspectives of New Music* 32, no. 1 (1994): 176–193.

Truax, Barry. "The PODX System: Interactive Compositional Software for the DMX-1000." *Computer Music Journal* 9, no. 1 (1985): 29–38.

Truax, Barry. "Real-Time Granular Synthesis with a Digital Signal Processor." *Computer Music Journal* 12, no. 2 (1988): 14–26.

Truax, Barry. "Sound, Listening and Place: The Aesthetic Dilemma." *Organised Sound* 17, no. 3 (2012): 193–201.

Tucker, Sherrie. "When Subjects Don't Come Out." In *Queer Episodes in Music and Modern Identity*, edited by Lloyd Whitesell and Sophie Fuller, 293–310. Champaign: University Press of Illinois, 2002.

Ussachevsky, Vladimir. "Musical Timbre Mutation by Means of the—Klangumwandler, a Frequency Transposition Device." A Paper Delivered at the Audio Engineering Society Convention, White Plains, NY, 1958.

Veal, Michael. *Dub: Soundscapes and Shattered Songs in Jamaican Reggae*. Middletown, CT: Wesleyan University Press, 2013.

Verán, Christina. "First Ladies." In *Hip Hop Divas*, edited by *Vibe Magazine*. New York: Three Rivers Press, 2001.

Vérin, Nicolas, and Pierre Henry. "Entretien Avec Pierre Henry." *Ars Sonora* 9 (1999). http://www.ars-sonora.org/html/numeros/numero09/09c.htm.

Voigt-Lund, Jens. "The Demon of Intimacy: About the Music," December 15, 2005. http://www.nielsroensholdt.dk/uploads/2/3/2/1/23214662/demonofintimacy.rtf.

Ward, Brian. *Just My Soul Responding: Rhythm and Blues, Black Consciousness, and Race Relations*. Berkeley: University of California Press, 1998.

Watson, Seosamh. "Laoi Chab an Dosáin: Background to a Late Ossianic Ballad." *Eighteenth-Century Ireland/Iris an Dá Chultúr* 5 (1990): 37–44.

Weheliye, Alexander G. "'Feenin': Posthuman Voices in Contemporary Black Popular Music." *Social Text* 20, no. 2 (June 1, 2002): 21–47.

Weheliye, Alexander G. *Phonographies: Grooves In Sonic Afro-Modernity*. Durham: Duke University Press, 2005.

Westlund, Andrea. "Love and the Sharing of Ends." In *Twenty-First Century Papers: On-Line Working Papers from the Center for 21st Century Studies*. University of Wisconsin-Milwaukee, WI: Center for 21st Century Studies, 2005. http://www4.uwm.edu/c21/pdfs/workingpapers/westlund.pdf.

Weston, P.S.J. "Discrimination of Voice Gender in the Human Audio Cortex." *NeuroImage* 105 (2015): 208–214.

Whitburn, Joel. *Hot Dance Disco 1974–2003.* Milwaukee, WI: Record Research Incorporated, 2004. https://www.amazon.co.uk/Joel-Whitburns-Dance-Disco-1974-2003/dp/089820156X.

White, Deborah Gray. *Ar'n't I a Woman? Female Slaves in the Plantation South.* New York: W. W. Norton, 1985.

Whiteley, Sheila, and Jennifer Rycenga. *Queering the Popular Pitch.* New York: Routledge, 2013.

Williams, Bernard, ed. *Obscenity and Film Censorship: An Abridgement of the Williams Report.* Cambridge: Cambridge University Press, 1982.

Williams, Justin A. "Beats and Flows: A Response to Kyle Adams, 'Aspects of the Music/Text Relationship in Rap.'" *Music Theory Online* 15, no. 2 (June 1, 2009). http://www.mtosmt.org /issues/mto.09.15.2/mto.09.15.2.williams.html.

Williams, Justin A. *Rhymin' and Stealin': Musical Borrowing in Hip-Hop.* Ann Arbor: University of Michigan Press, 2013.

Williams, Linda. "A Provoking Agent: The Pornography and Performance Art of Annie Sprinkle." *Social Text*, no. 37 (1993): 117–133. https://doi.org/10.2307/466263.

Williams, Linda. *Hard Core: Power, Pleasure, and the "Frenzy of the Visible."* Berkeley, CA: University of California Press, 1989.

Wolfe, V. I., D. L. Ratusnik, F. H. Smith, and G. Northrop. "Intonation and Fundamental Frequency in Male-Female Transsexuals." *Journal of Speech and Hearing Disorders* 55 (1990): 43–50.

Woloshyn, Alexa. "Interview with Robert Normandeau." *EContact!* 13, no. 3 (2010). http:// econtact.ca/13_3/woloshyn_normandeau_2011.html.

Woloshyn, Alexa. "Wallace Berry's Structural Processes and Electroacoustic Music." *EContact!* 13, no. 3 (2010). https://econtact.ca/13_3/woloshyn_onomatopoeias.html.

Woloshyn, Alexa Lauren. "The Recorded Voice and the Mediated Body in Contemporary Canadian Electroacoustic Music." PhD diss., University of Toronto, 2012.

Wood, Elizabeth. "Lesbian Fugue: Ethel Smyth's Contrapuntal Arts." In *Musicology and Difference: Music, Gender, and Sexuality in Music Scholarship*, edited by Ruth Solie, 164–183. Berkeley: University of California Press, 1993.

Wood, Liz, and Mitchell Morris. "Flirting with 'Theory,' Flirting with Music: A Discussion in Advance of ForePlay." In *Gay and Lesbian Study Group (of the American Musicological Society) Newsletter* (1994): 3–7.

Wortham, Jenna. "How Janelle Monáe Found Her Voice." *New York Times*, April 19, 2018, https://www.nytimes.com/2018/04/19/magazine/how-janelle-monae-found-her-voice.html.

Young, John. "Source Recognition of Environmental Sounds in the Composition of Sonic Art with Field-Recordings: A New Zealand Viewpoint." PhD diss., University of Canterbury, 1989.

Ziplow, Steven. *The Film Maker's Guide to Pornography.* Portage, MI: Drake Publishers, 1977.

Zoladz, Lindsay. "Refresh: The Lonely Futurism of TLC's FanMail." *Pitchfork*, May 4, 2012. https://pitchfork.com/features/article/8827-tlc/.

Index

ping pong (in the work of Hodkinson/
Rønsholdt), 73, 75, 279
scorching iron (in the Vishnu Purana),
144
Banality, 78–87, 265. *See also* Moderation
vs. innovation, 80–81
Banner, Christopher, 31. *See also Tiger Balm*
(Lockwood)
Bardeleben, Anka, 84. *See also Fish & Fowl*
(Hodkinson/Rønsholdt)
Bardot, Brigitte, 10, 117, 209
Barraud, Francis, *His Master's Voice*, 52.
See also Medium, recorded/sounding
Barthes, Roland
on pleasure and loss, 66
on the voice and individuality, 38, 180
(*see also* Voice)
Bataille, Georges, xxiv, 5
Battan, Carrie, on the "alien" features of
THEESatisfaction's "Enchantruss," 248.
See also Alienation; THEESatisfaction
Beatty, Maria, 102, 164, 167, 172. *See also*
Sluts & Goddesses (Sprinkle)
Beauvoir, Simone de, on becoming a woman,
19. *See also* Becoming-woman (Deleuze/
Guattari)
Bebey, Francis, xiii
Becoming-woman (Deleuze/Guattari), 15,
18–19, 78. *See also* Deleuze, Gilles; Grosz,
Elizabeth; Guattari, Félix; Jardine, Alice;
Beethoven, Ludwig von
and gender, xxii–xxiii, 267
sampled by Nas, 59
Béjart, Maurice, and *Symphonie pour un
homme seul*, 10. See also Dance; *Sym-
phonie pour un homme seul* (Schaeffer/
Henry)
Belching, 156, 310n81. *See also* Apocalypse
(Shields); Matus, Jim
Bellotte, Pete, 101, 106–107, 115, 121.
See also "Love to Love You Baby" (Sum-
mer); Summer, Donna

Benjamin, Walter, on cliché, 81. *See also*
Banality
Berg, Alban, Lulu, 71, 78–80
Berio, Luciano, xxv, 86
Berlant, Lauren
on live sex acts vs. dead citizenship, 225
on Tipper Gore's Parental Advisory labels,
222 (*see also* Parental Advisory labels)
on sex law, 227 (*see also* Law)
Betty, 248, 251. *See also* "Wobble warp"
Beyoncé, 99–100, 103, 216
and Summer's "Love to Love You Baby,"
100, 123, 211
Bharata Muni, on *Navarasa* theory, 131–137,
307n28, 307n34
Bharatanatyam, 101, 126, 131–43, 147–150,
160. *See also* Apocalypse (Shields); Dance
Bhise, Swati, 137–138. *See also*
Bharatanatyam; Shield, Alice
Birkin, Jane, 209
Black Eyed Peas, 258
Black Sheep, 258
Blackface minstrelsy, 95–98, 241
Blanco, Mykki, "Wavvy," 239
Blood, Erik, 257. *See also* THEESatisfaction
Blue discs/records, xix, 5
Blues, the, li
and "race records," 93 (*see also* Race records)
and sexuality, 99, 104, 223–224
Bogart, Neil, 106, 109–10, 118, 121–22.
See also "Love to Love You Baby"
(Summer); Summer, Donna
Born, Georgina, 285
Bosma, Hannah, on gender disparity in
electronic music, xxix, xli–xlii, 180,
268–269
Botstein, Leon, on Berg's *Lulu*, 79–80.
See also Berg, Alban
Boulez, Pierre, xii, xiv
Boundaries
and Afro-modernism, 1
between background and foreground, 171

213, 216, 261, 263, 265, 267–268, 271, 273–274. *See also* Orgasms

Cock rock, 143–147, 158. *See also Apocalypse* (Shields)

Cocteau, Jean, and Schaeffer/Henry, 4

Coddington, Amy, on the desegregation of popular music charts, 260

Code. *See* "It's Code" (Monáe)

code-switching (as protection), 244

Coetzee, J. M., *Foe*, 14–17. *See also Robinson Crusoe* (Defoe)

Collective, 27, 29, 247

desire, 19 (*see also* Desire)

imagination, 19, 95

intentionality, 49–54, 65, 98 (*see also* Intentionality)

memory, 53, 201 (*see also* Memory)

nonconformity, 238

participation, 294n15

silence, xiii (*see also* Silence)

thought, 91

Collins, Patricia Hill

on (a)sexual pathology and race, 226, 228–230

on Black feminist thought/expression, 91, 99, 229, 242

Colonialism, 12–17. *See also* Exoticism

and dance, 101 (*see also* Dance)

and "found sound"/*musique concrète*, xliii, 5, 12–17

Color

"Black Seal" records, xlvi

"blue discs/records," xix, 5

"Red Seal" records, xlvi

Columbia-Princeton Electronic Music Center, xiii, 101, 127, 306n7, 306n9, 309n58

Cone, Edward T., on composer desire/voice, 30

Connor, Steven

on *Fish & Fowl*, 71 (*see also Fish & Fowl* [Hodkinson/Rønsholdt])

on the recorded voice, 41, 181

Consent, lack of, xxxviii–xxxix, 6, 37, 40, 46–47, 64, 67, 97, 201, 206, 271–273. *See also* Erasure; Uncredited performance

Context-dependency

of affect, 237

of content, 22

of history, 22

of identity, 17

of listener perceptions of gender and sexuality, 182

of shame, 161

of sound, xviii, 21, 28, 71, 163, 267

of Truax's compositions, 177–179, 195–197 (*see also* Truax, Barry)

Cook, Susan, on hierarchical categories in music, xlix–l

Corbett, John and Terri Kapsalis, on aural sex, xxxvi, 173, 210–211, 267

Corrie, Rachel, 128–129. *See also* Shields, Alice

Cox, Ida, 99

Crane, Barbara B. and Jennifer Dusenberry, on opposition to abortion in the '80s and '90s, 150. *See also* Abortion

Critical race studies/theory, xl–xli, 230–231

Cultural retrospectivism, 201, 233. *See also* TLC

CupcakKe, 89, 238

Currid, Brian, on dance clubs, 114, 120, 238. *See also* Disco

Cusick, Suzanne, on sex and music, 268

Dance. *See also* Disco

Alston choreography, 23–24, 30–32, 164 (*see also* Alston, Richard; *Tiger Balm* [Lockwood])

Balinese, 148

Bharatanatyam, 101, 126, 131–43, 147–150, 160. *See also Apocalypse* (Shields)

and spatiality, 142

and *Symphonie pour un homme seul*, 10

Tahitian 'upa'upa dance, 13

and therapy, 142

Dandridge, Dorothy, eyes of, 254. *See also* Monáe, Janelle; Spalding, Esperanza
Danses organiques, Les (Ferrari), xxxix, 37, 54–55, 64. *See also* Ferrari, Luc
Darwin, Charles
    on kin vs. resemblance, 78
    on sound-source recognition and evolution, 21
Davidovsky, Mario, 127. *See also* Columbia-Princeton Electronic Music Center
Davies, Stephen, on Langer's idea of music and feeling, xxxiv
Davis, Angela, on church and sex in Bessie Smith, 223–224. *See also* Smith, Bessie
De La Soul, 258
Deleuze, Gilles, 17–21, 66
    on art (with Guattari), 37, 50
    on becoming (with Guattari), 15, 18–21, 78
    on duration (with Guattari), 77–78
    on Proust's writing/eroticism, xxvii–xxviii, 17–18
    on theater, 138
DeNora, Tia: on gender in music, xxiii
Derrida, Jacques, 17, 40, 294n15
Dery, Mark, on Afrofuturism, 241–242. *See also* Afrofuturism; Monáe, Janelle
Desire, 17–21, 79–80, 92, 118–121, 283n38
    desiring machines, 18–20
    for the performer, 101, 114, 212 (*see also* Availability)
    and silence, 66–67
    stretched, 193
    and the voice, 52–53 (*see also* Voice)
    of women, 166, 175
de Sousa Dias, António, 61
Deuber, Franz. *See also* "Love to Love You Baby" (Summer)
Devi, Rukmini, 136. *See also Bharatanatyam*
Deviance, xlii–xliii, 114, 152, 212, 222–225, 238

Difference
    and identity, xxiii–xxiv
    and repetition, 77–78
    racial, 230
    sexual, 72, 180, 229
    vectors of, 204, 231
    and voice, 10
Digable Planets, 262
Digitality, and TLC, 199–200, 202, 213, 221, 231. *See also* TLC
Dilla, J. *See* J Dilla
Disco, 112–116. *See also* "Love to Love You Baby" (Summer)
    "disco sucks," 113
    as expensive to produce, 115
    and minimalism, 118–120
Distance, xxvi–xxviii. *See also* Spatiality
    and colonialism, 14 (*see also* Colonialism)
    and communication, 84
    and concert etiquette, 98 (*see also* Etiquette)
    and gender, 265–266
    and lovers, xxviii
    and moderation, 264
    and "safety," 101, 211–212, 224
    and sampling, 211 (*see also* Sampling)
    sonified, 272
    and sound, 8, 44
    and utopian vision, 158
    and women, xxvi–xxviii
Diversity work, 51. *See also* Institution
Dodge, Charles, 127. *See also* Columbia-Princeton Electronic Music Center
Dolar, Mladen, on the voice, 38, 82, 98. *See also* Voice
Domesticity
    and electronic music, xii
    and sexual repression, 152
Douglass, Frederick, 93–94
Drake, "I Get Lonely Too," 201. *See also* TLC
Drott, Eric, on Ferrari's "anecdotal music," 35–37. *See also* Ferrari, Luc

Ferreira, José Luís, 61
Ferreyra, Beatriz, and Schaeffer, 35, 61
Fidelity, 147, 292n21
Fink, Robert, on Summer's "Love to Love
    You Baby," 118–120, 206, 216, 231.
    See also "Love to Love You Baby"
    (Summer); Repetition
Fire
    and the hair of Shakti/Shiva, 133,
        138
    sound of, 24, 178, 184 (see also Lockwood,
        Annea; Truax, Barry)
    Wings of Fire (Truax), 196
Fish & Fowl (Hodkinson/Rønsholdt), xxxvii,
    4, 69–87, 170, 261–263, 268, 271, 279.
    See also Hodkinson, Juliana; Rønsholdt,
    Niels
For its own sake
    the extraordinary, 30
    sex as spontaneous event, 172 (see also
        Spontaneity)
    sound, xiv, 8, 11
Foucault, Michel
    on citation, xliv
    on contemporary art and sexual expression,
        129, 131 (see also Apocalypse [Shields])
Fox, Michael V., on the Song of Songs,
    183–184. See also Song of Songs (Truax)
Freud, Sigmund, 20, 38
Frith, Simon and Angela McRobbie, on rock
    and sexuality, 143–144. See also Cock
    rock
Fukushima, Kazuo, xiii

Game
    courtship, 219 (see also TLC: "Kick Your
        Game")
    electroacoustic, 26–27, 43, 46–54, 57–67,
        92, 101, 121, 165, 211, 271
    pornographic, 165–71 (see also Pornogra-
        phy; Sprinkle, Annie)
    racializing, 97

Gaze
    heterosexual, 271
    of the male producer, 180, 210, 257 (see
        also Moroder, Giorgio; Summer, Donna)
    of the video camera, 218
Genres
    boundaries of, xvi, 265, 325n17
    as institutions, 50 (see also Institution)
    of rap, 238–239, 257–260 (see also Rap)
Gesture, 13, 90, 101, 131–139, 143, 148,
    160–161. See Apocalypse (Shields); Dance
Getting off. See also Orgasms
    on sound, xi, xxxvi, 110, 268
    and temporality, 11
Gibb, Bill, 121. See also "Love to Love You
    Baby" (Summer)
Gluck, Robert, 269
Goodall, Nataki H., on TLC, 202–203, 221.
    See also TLC
Gore, Tipper, and Parental Advisory labels,
    221–222
Gospel, 117
    and music education, 260
Grosz, Elizabeth, on Deleuzian desire, 20
Guattari, Félix
    on art (with Deleuze), 37, 50
    on becoming (with Deleuze), 15, 17–21,
        78
    on duration (with Deleuze), 77–78
Guayaba, 239
Guck, Marion, on composer authority/per-
    sona, 30–32, 293n32
Guevara, Nancy, on women in the history of
    electronic music, 259

Halperin, David
    on camp/queer music, 223
    on refusal as an act, 225 (see also
        Abstinence)
Hanninen, Dora, on musical segmentation
    and associations, 26. See also Music theory
Hanson, 201. See also TLC

Harris-White, Catherine. *See* SassyBlack

Harrison, Martin, 115. *See also* "Love to Love You Baby" (Summer)

Hasty, Christopher, on segmentation, 26. *See also* Music theory

Hatten, Robert, on musical topics and tropes, 249

Hawkins, Stan, 103

  on the aspirations of camp, 223 (*see also* Camp)

  on parody, 214 (*see also* Parody)

Hayes, Eileen, on the neglect of Black women in music scholarship, 304n28

Hayles, N. Katherine, on the posthuman, xlvi–xlvii. *See also* Posthuman

Hegamin, Lucille, 99

Henderson, Rosa, 99

Henry, Pierre, xiv, xvi, xxv, xxxviii, li, 3–22, 25, 28–30, 35, 67, 78–79, 90, 172, 270–271, 273. *See also Symphonie pour un homme seul* (Schaeffer/Henry)

  on cinema and sound, 4

Hess, Mickey, on sampling, xliv. *See also* Sampling

Hickman, Roger, on suspense-generating "waver" drones, 170. *See also* Cinema

Hill, Anita, 228

Hine, Darlene Clark, on the sexual exploitation of Black women, 228

Hinkle-Turner, Elizabeth, 269

  on Anderson/Lockwood/Oliveros, 23, 164

  on Shields's lack of recognition, 127

Hip hop, xlii–xliv, 255–266

  and alienation, 248 (*see also* Alienation)

  and drum breaks, 115

  and sampling, xliii–xliv, 32, 59, 114, 256–257 (*see also* Sampling)

  and sexual content, xlii, 103

Hirschfeld, Magnus, 5. *See also* Erasure

Hisama, Ellie, on sexual misconduct by music theorists, xiv–xv

Historiography of electronic/electroacoustic music, xi–xix, xli, xliv–xlvii, 22, 89, 91, 152, 259

Hodkinson, Juliana, 90, 261–263

  Fish & Fowl, xxxvii, 4, 69–87, 170, 261–263, 268, 271, 279

  sagte er, dachte ich, 76, 279

  on "sonic writing," 71–72

Homoeroticism. *See* Truax, Barry

hooks, bell, 99, 242

Horror, 79, 170. *See also* Cinema

Hron, Terri, and Normandeau's *Jeu de langues*, 64. *See also Jeu de langues* (Normandeau)

Hughes, Langston, "Songs to the Dark Virgin," 185. *See also Song of Songs* (Truax)

Hugill, Andrew, on electronic music and the creative imagination, xvii–xviii

Humor

  in *Apocalypse* (Shields), 144, 150, 157 (*see also Apocalypse* [Shields])

  in *Sluts & Goddesses* (Sprinkle), 102, 167–169 (*see also Sluts & Goddesses* [Sprinkle])

  of TLC, 206–208, 221 (*see also* TLC)

Hunter, Alberta, 99

Hypersexuality/hypersexualization, 90, 100, 103, 109–112, 117–118, 121, 193, 203, 206, 208, 211–213, 217–218, 242, 264

Iandoli, Kathy, on gender role reversals in TLC, 219. *See also* TLC

Idealized

  Bharatanatyam dancers, 136–137, 148 (*see also Bharatanatyam*; Dance)

  Blackness, 95

  listener, 27, 32 (*see also* Listener)

  representation of female sexuality, xlviii, 71–72, 79–80, 170, 173, 209–212

Identity, 17–21. *See also* Uncredited performance

  and difference, xxiii–xxiv

Jones, Alisha Lola, on music education at HBCUs, 260

Jones, Grace, 233, 239–240, 248, 251

Jones, Nancy A., and Leslie C. Dunn, on the embodied voice, 180. *See also* Voice

Jubilee Singers, xviii

Jungle Brothers, 258

Kalinak, Kathryn, 87

Kamasutra Vatsayana, xxxiii, 101, 138

Kane, Angela, on Alston's choreography pre-Lockwood, 24. *See also* Alston, Richard

Kane, Brian
  on acousmatic sound, xviii–xxi, 21, 52–53, 80 (*see also* Acousmatic sound; Veiling)
  on Ferrari's anecdotal music, 37–46, 55, 80 (*see also* Ferrari, Luc)
  on the voice, 52–53, 294n16

Kane, Sarah, *Blasted*, 65. *See also Pluie Noire* (Normandeau)

Kapsalis, Terri, and John Corbett, on aural sex, xxxvi, 173, 210–211, 267

Karaoke, 213–214. *See also* Allosonic quotation

Katz, Mark, on the break, 118

Keightley, Keir, on women in the history of electronic music, 259

Kerton, Sarah, on boybands, 103. *See also* Asexuality

Keyes, Cheryl, on women's voices in hip hop, 259

King, Dave, 115, 122. *See also* "Love to Love You Baby" (Summer)

Kinsey, Alfred, xxiv, 5

Klein, Melanie, on repression, 142

Koenig, Gottfried Michael, 23

Komachi, 128. *See also* Shields, Alice

Krenek, Ernst, xiii

Krims, Adam, on rap genres, 257–258, 262 (*see also* Genres)

Kutler, Jen, *Disembodied*, 272–273

Lacan, Jacques
  on desire, 19–20, 52, 79–80, 120–21 (*see also* Desire)
  on the voice, 38 (*see also* Voice)

Lacasse, Serge, "allosonic quotation," 318n13. *See also* Allosonic quotation

Langer, Susanne
  on artwork and purpose, xxxvii
  on (erotic) representation in art/music, xxxiv–xxxv (*see also* Representation)

Lavitt, Pamela Brown, on blackface and "coon shouting," 95–96, 98

Law. *See also* Summer, Donna: lawsuits of
  and gender, 322n7
  obscenity, 5
  sex, 227

"Left Eye," 199–201, 202–204, 206–208, 215–216, 219, 231. *See also* TLC
  as composer, 211, 221
  death of, 200–201
  Eye Legacy, 317n2
  philanthropic work of, 317n2
  and safe-sex advocacy, 221 (*see also* Safe sex)

Le Guin, Ursula K., *The Left Hand of Darkness*, 158

Leibetseder, Dorus, 103
  on parody/pastiche and gender, 207, 214–215

Levin, Darren, on "Kick Your Game" 207–208. *See also* TLC

Levinson, Jerrold, on pornography vs. art, xxv. *See also* Pornography

Lewis, George E., xlv

Lewis, Waylon, on songs and the rhythmic path to orgasm, xxi. *See also* Orgasms

Lima, Cândido, 61

Listener
  "ideal," 27, 32
  imagination, 58
  participation, 26–27, 29, 31, 50, 67, 73, 178
  performer as, 91, 96

Rap (cont.)
 and gender, 209–211
 genres of, 238–239, 257–260 (*see also*
  Genres)
 and queerness, 239
Rea, Michael, on pornography. *See*
  Pornography
Reagan, Nancy, "Just Say No" campaign,
  149–150, 222
Reagan, Ronald, 126, 131, 140
 and the "global gag rule," 149–150, 152
Realism, electroacoustic, 39, 42–44
Recognizability
 as desirable, 7–8, 12
 of gender, 185–192 (*see also Song of Songs*
  [Truax])
 of the human voice, xviii, 45, 177–186,
  192
 of samples, 59–60, 73–77, 177–184, 196,
  256 (*see also* Sampling)
Reibel, Guy, 35
Remix, 261–265
 of one's own music, 69–83, 261–263,
  279 (*see also Fish & Fowl* [Hodkinson/
  Rønsholdt])
Repetition, xix, xxi, 77–82
 and banality, 80–84
 of breathing, 55, 69–70, 74–78 (*see also*
  Breath)
 and difference, 77–78
 in disco, 118–119
 as enticing, 12
 in *Fish & Fowl* (Hodkinson/Rønsholdt),
  73–87
 and listening, 92
 looping, 8–12, 25–26
 of purring, 25–26 (*see also Tiger Balm*
  [Lockwood])
 and sex, xix, 9, 65
 and technology, xix (*see also* Technology)
 and temporality, 9, 77–78 (*see also*
  Temporality)

 and threes, 142 (*see also* Three)
 and understanding, 73
Representation, xxix–xxxvii, 39–43, 46,
  102–103, 273
 erotic, xxix–xxxvii, 9, 117, 182
 gender, 130
 vs. presentation, 39, 42
 of women's sexuality, 71–72
Repression, 142–143, 148, 151–52
 and racial uplift, xliii, 104
Rhythm
 and breath, 85–86
 queering, 253
 and race, 98
 and sexuality, 127, 209, 211
Rihanna, 99, 103, 216
 "Umbrella," 121
Risk, 24, 261, 263–264
Rist, Simone, 35
Robindoré, Brigitte, on Ferrari and consent,
  47. *See also* Consent; Ferrari, Luc;
  Uncredited performance
Robinson, Sylvia, as producer, 259
Robinson Crusoe (Defoe), 6, 14–16, 107.
  See also Colonialism
Rodgers, Tara, 269
 on work excluded from cultural memory,
  xvii, 269
 on working with sound, 27–28, 30
Rollefson, J. Griffith, on hip hop and power
  imbalances, 32
Rønsholdt, Niels, 90, 196, 261–263
 Die Wanderin, 70
 Fish & Fowl, xxxvii, 4, 69–87, 170,
  261–263, 268, 271, 279
 Hammerfall, 70
 Inside Your Mouth, Sucking the Sun, 85
 Triumph, 82–83, 85–86, 279
Roots, The, 258
Rose, Tricia, 239–240
 on Bad Sistas, 201, 203–204, 213,
  228–229, 259

Scott, Derek, on erotic representation in music, xxix–xxx. *See also* Representation

Screaming, 117, 143–144, 154, 157. *See also* *Apocalypse* (Shields); Moaning

Scruton, Roger, on representation, 39. *See also* Representation

Searle, John
on "collective intentionality," 49–50, 54, 98 (*see also* Collective; Intentionality)
and sexual misconduct, 51

Sedgwick, Eve Kosofsky, 238
on gender and bodies, 180
on sexual definitions, 190
on transitivity, 192 (*see also* Transitivity)

Segmentation, 8, 26

Semegen, Daria, 125–127, 141, 306n9. *See also* Columbia-Princeton Electronic Music Center; Shields, Alice

Sender, Ramon, and the San Francisco Tape Music Center, 164

Sevier, Brian R., on disco divas, 211. *See also* Disco

Sex work, xxxiii, 112, 121, 136, 151, 166, 168

Shabazz Palaces, 257, 262

Shadow
and psychoanalysis, l, 142–143 (*see also* Psychoanalysis)
and temporality, 184–185 (*see also* Temporality)

Shame, and performance, 161–162

Sheeran, Ed, "Shape of You," 201. *See also* TLC

Shields, Alice, xxix, l, 101, 125–162, 197, 264, 269, 271–272, 306n4, 306n7, 309n58. *See also* *Apocalypse* (Shields); Columbia-Princeton Electronic Music Center
Ave Maris Stella, 128
Criseyde, 128
Komachi at Sekidera, 128
Kyrielle, 128

Mass for the Dead, 160–161, 307n17

Mioritza—Requiem for Rachel Corrie, 128–129

Sahityam, 128

Shivatanz, 128

Shock value, of naked men, 31. *See also* Banner, Christopher; *Tiger Balm* (Lockwood)

Silence
collective, xiii (*see also* Collective)
and desire, 66–67 (*see also* Desire)
of (happy) women, xxvii
and limiting sexual freedoms, 152
and listening rituals, 50 (*see also* Institution)
and pornography, 44 (*see also* Pornography)
voicing of, 20

Silverman, Kaja, on the "female voice" in cinema, 51–52

Simon, Pierre, 5

Sluts & Goddesses (Sprinkle), xxxiii, 102–103, 163–176, 271, 273. *See also* Sprinkle, Annie
and "numbers," 169–170

Smalley, Denis, on digital transformations, 182. *See also* Transformation

Smalls, Shanté Paradigm, on "Rapper's Delight," 22

Smallwood, Reverend Richard, and gospel in music education, 260

Smiley, Pril, 127, 306n9. *See also* Columbia-Princeton Electronic Music Center

Smith, Anna Nicole, 173

Smith, Bessie, 103. *See also* Blues, the
"Moan, You Moaners," 223
"You've Got to Give Me Some," 99

Smith, Clara, 99

Smith, Jacob
on the labeling of high and low culture in the phonograph industry, xlvi
on the legislation of "obscene" records, 5

Smith, Trixie, 99

Voice (cont.)
   and nothing more, 82, 107
   racialized, xlii
   and technology, 105, 107
   virtual, 98
   "voice-body," 29, 41, 48, 82, 131, 172
Voigt-Lund, Jens, on Rønsholdt's compositions, 86

Wallace, Sippie, 99
Walser, Robert, on hip hop categorization,
   256
Walther, Gitta, 115. See also "Love to Love
   You Baby" (Summer)
Ward, Brian, on disco records, 112, 304n29.
   See also Disco
Waters, Ethel, 99
Watkins, Tionne. See "T-Boz"
Watson, Seosamh, on the third episode of
   Laoidh Chab an Dosáin, 155
Wedel, Robby, 115. See also "Love to Love
   You Baby" (Summer)
Weezer, "No Scrubs," 201. See also TLC
Weheliye, Alexander G.
   on Afro-modernism in electronic music, l
   on Hayles's analysis of the posthuman, xlvii
   (see also Posthuman)
   on the virtual voice, 98
Westlund, Andrea, on "plural subject-hood,"
   65
Whiteley, Sheila, on asexuality, 103. See also
   Asexuality
Williams, Bernard, on pornography, xxxiii.
   See also Pornography
Williams, Justin, 318n13. See also Allosonic
   quotation
   on collaboration in rap, 256–257
Williams, Linda, 165–167, 172–174. See also
   Pornography
   on the art/pornography divide, xl, 165–167
   on the illusion of spontaneity in pornographic film, 173

on overdubbing in pornographic films, 172
   (see also Spatiality)
on Sprinkle, 165–167, 174
on the visual in men's sexual pleasure, xxxvi
Willis, Ellen, and the "Sexuality" conference,
   xl
Wilson, Edith, 99
Wirrick, James "Tip," on drum breaks, 115.
   See also Disco
"Wobble warp," 233–266
Woloshyn, Alexa
   on Normandeau's "onomatopoeias cycle,"
   62–63
   on Truax's Song of Songs, 188 (see also Song
   of Songs [Truax])
Women Against Pornography, xxxviii
Wonder, Stevie, xlv, 248
   Journey Through the Secret Life of Plants,
   239–140
Wood, Elizabeth, on hidden messages in
   music by lesbian composers, 241. See also
   "It's Code" (Monáe)
Woodland, Nick, 115. See also "Love to Love
   You Baby" (Summer)
Wuorinen, Charles, 127. See also Columbia-
   Princeton Electronic Music Center

Young, John, 25
   on Lockwood's Tiger Balm, 26, 29 (see also
   Tiger Balm [Lockwood])
Yuasa, Joji, xiii

Zemlinsky, Alexander von, 274
Zinovieff, Peter, 23
Ziplow, Steven, The Film Maker's Guide to
   Pornography, 169